Springer Proceedings in Mathematics & Statistics

Volume 46

For further volumes:
http://www.springer.com/series/10533

Springer Proceedings in Mathematics & Statistics

This book series features volumes composed of select contributions from workshops and conferences in all areas of current research in mathematics and statistics, including OR and optimization. In addition to an overall evaluation of the interest, scientific quality, and timeliness of each proposal at the hands of the publisher, individual contributions are all refereed to the high quality standards of leading journals in the field. Thus, this series provides the research community with well-edited, authoritative reports on developments in the most exciting areas of mathematical and statistical research today.

Ratan Dasgupta
Editor

Advances in Growth Curve Models

Topics from the Indian Statistical Institute

Springer

Editor
Ratan Dasgupta
Indian Statistical Institute
Professor, Theoretical Statistics
and Mathematics Unit
Kolkata, India

ISSN 2194-1009 ISSN 2194-1017 (electronic)
ISBN 978-1-4614-6861-5 ISBN 978-1-4614-6862-2 (eBook)
DOI 10.1007/978-1-4614-6862-2
Springer New York Heidelberg Dordrecht London

Library of Congress Control Number: 2013943266

Printed on acid-free paper

Springer is part of Springer Science+Business Media (www.springer.com)

Preface

A growth curve is an empirical model of the evolution of a quantity over time. Growth curve models (GCM) in longitudinal studies are widely used in many disciplines besides biology, particularly in statistics, population studies, economics, biological sciences, statistical quality control, environment, sociology, nano-biotechnology, fluid mechanics and for quantities such as population size, body height, biomass, and fungal growth. An important precursor of the GCM is the classical GCM considered by S N Roy and R. Potthoff in early 1960s and C R Rao about the same time. That leads to the development of repeated measurement designs, longitudinal models, and related evolutionary models in epidemiology and bioinformatics. Even the Chaos theory comes under such models. It has important applications in psychometry and psychiatry. The evolutionary models are also akin to GCM. Growth and nutrition of Indian children has not improved much in spite of India's economic prosperity.

This conference proceeding presents some ideas about the research works on GCM that is going on by the scientists of Indian Statistical Institute in different branches of science. The genesis of this work started several years back when the editor and his colleagues conducted growth experiments in the agricultural firm at Indian Statistical Institute, Giridih, Jharkhand; a tribal area. At that time Editor took academic & administrative responsibility of ISI Giridih as Coordinator on the third tier in a three-tier administrative system. Continued research for several years on plant growth posed some theoretical and applied problems that are recorded in this proceeding. We also thought it will be a nice idea if the researchers working in GCM had an opportunity to exchange ideas about their field of interest. To this end, a workshop was organized in the year 2011 that was followed by a national conference on GCM in the year 2012 at Giridih. Another workshop on GCM was conducted at ISI Giridih during 21-22 March, 2013. We invited some well-known researchers to contribute to this conference proceeding and further invited the participants of the conference to submit more than one paper, if possible for the proceedings. All the papers were peer reviewed. The result is the compilation of 15 papers in different branches of science in this proceeding. The endeavor will be considered successful if this can give some idea about solving theoretical and practical problems in this broad area of GCM to which many researchers are interested in.

December 2012

Ratan Dasgupta
Kolkata, India

Contents

Fig. 1 Yam corm sorting before plantation in Giridih farm

Fig. 2 Ecological observatory in ISI Giridih Farm

Fig. 3 A 10.642 kg Yam out of a seed weight 800 g shown by Ratan Dasgupta in conference 2012

Fig. 4 Seed corm orientation shown in experimental plots

Fig. 5 Yam corm plantation by a field worker

Fig. 6 Dry region in Giridih and Ushri River in winter

Fig. 7 Rock structure of almost dried Ushri falls in winter. This is fiercely turbulent in monsoon, see https://www.youtube.com/watch?v=WECZ8y9BXM4

Fig. 8 Grown Elephant-foot-yam plantations in Giridih farm

Contributors

Premananda Bharati Indian Statistical Institute, Kolkata, India

Susmita Bharati Indian Statistical Institute, Kolkata, India

Ratan Dasgupta Theoretical Statistics and Mathematics Unit, Indian Statistical Institute, Kolkata, India

Avinash Dharmadhikari Tata Motors, Pune, India

S. Krithika University of Toronto, Toronto, ON, Canada

H. Maity Indian Statistical Institute, Kolkata, India

B.S. Mazumder Indian Statistical Institute, Kolkata, India

Nicholas Mesue School of Health Sciences, University of Tampere, Tampere, Finland

S. Mirzaei Salehabadi Indian Statistical Institute, Kolkata, India

Tapio Nummi School of Health Sciences, University of Tampere, Tampere, Finland

Manoranjan Pal Indian Statistical Institute, Kolkata, India

Prasanta Pathak Indian Statistical Institute, Kolkata, India

P.S.S.N.V.P. Rao Indian Statistical Institute, Kolkata, India

Pranab K. Sen University of North Carolina, Chapel Hill, NC, USA

Debasis Sengupta Indian Statistical Institute, Kolkata, India

Bikas K. Sinha Indian Statistical Institute, Kolkata, India

T.S. Vasulu Indian Statistical Institute, Kolkata, India

Vivek Verma Indian Statistical Institute, Kolkata, India

Chapter 1
Yam Growth Experiment and Above-Ground Biomass as Possible Predictor

Ratan Dasgupta

Summary. Prediction problem of agricultural yield in terms of observable quantities in an efficient manner raises a number of theoretical issues that involve identification of important predictors and subsequent interpretation. For underground crops, one such possible predictor is above-ground biomass. Yam is a staple food for poor tribes. Growth experiments with Elephant foot yam are conducted at Indian Statistical Institute's Giridih, Jharkhand Farm to study the relationship between yield of Yam with initial seed weight and seed surface texture. Five different grading of skin texture levelled as 1–5 are considered, 1 being roughest (and smallest surface area) and 5 being smoothest (and largest) skin surface of cut seed-corms. For each of these five grading of seed skin, five different levels of seed weight are also considered viz., 200g, 350g, 500g, 650g, 800g. Yam yield is seen to have association with weight and surface texture of seed corm. The third level of surface i.e., moderately rough skin texture of the seed corm with medium-sized surface area, having weight of about 650 g is recommended for best production in lateritic gravel soil like that of Giridih, Jharkhand. A Yam-corm cut with weight of 200g is seen to be critical for sprouting & survival of the plant, irrespective of skin texture of the seed. Estimated ratios of final yield vs. seed weight are as follows. A little bit of organic manure like vermicompost and cow-dung when added in pit preparation of the Yam corm resulted in the ratio 4.79 of final yield to initial seed-weight. The ratio was significantly high at 5.84, when a little bit of coal ash was additionally administered as manure in the pits at initial stage additionally and sprouted Yam corms were planted, which resulted in a massive growth of plant biomass at later stages. As many as 14 sprouts from a single seed at a time in this situation were observed. "Squared residual" for error of prediction seemed to decrease with increase in the number of yam sprouts, indicating higher prediction accuracy of yield with large number of sprouts.

R. Dasgupta (✉)
Theoretical Statistics and Mathematics Unit, Indian Statistical Institute, Kolkata, India
e-mail: ratandasgupta@gmail.com

R. Dasgupta (ed.), *Advances in Growth Curve Models*, Springer Proceedings in Mathematics & Statistics 46, DOI 10.1007/978-1-4614-6862-2_1,

The ratio of Yam yield to initial seed weight was 3.03 when cut seed corms before sprouting were used and no manure was given. In a fertile land the expected return of Yam is to the tune of 5 times the initial weight. Thus Giridih farm soil is about 60% fertile compared to fertile land of Nadia in West Bengal for Yam cultivation. The production can be greatly enhanced to 5.84 times the initial weight when already sprouted (cut) Yam corm having moderately rough skin texture is used as seed and a little bit of organic manure along with coal ash are applied. Seed skin textures that are too smooth or too rough seem inferior to moderately rough skin for high yield.

The estimated growth curve takes a sharp upward turn towards end of the production period, even with slight increase in age of yam plant. Thus Yam deposition is highly dependent on the length of production season, although farmers sometimes prefer early harvesting from financial considerations. This special feature of yam growth curve indicates presence of a spike towards end.

Yam yield is observed to be approximately linear with maximum height of stem in logarithmic scale. Stem girth at base is highly correlated with height of stem. Thus above-ground biomass that can be approximated by nonlinear combination of observable predictors like product of stem-density and approximately cylindrical stem-volume, along with expansive leaf structures on the top that can be recorded continuously over time, may turn out to be a good predictor of underground Yam deposition. Available data indicates such possibilities in the light of theoretical considerations. The procedures developed may be adopted in general cases while analyzing hidden variables based on the observable variables. A lower bound of the expected yield based on the proposed log linear model is obtained.

Positive effect of initial weight (X) on final weight (Y) is further analysed by considering the logarithm of ratio, $k = \log K = \log(Y/X)$. Invoking a theorem on nonstationary Gaussian process, we investigate a model $Y = e^u X$, where $u \sim N(\mu, \sigma^2)$. Bootstrapped distribution of the mean of $k = \log(Y/X)$ seems to be normal; estimated 2σ confidence interval of percentage return of yam is $(440, 520)\%$ in that region.

Model sensitivity is quantified as change in distribution of random variables under slight change in the values of the parameters. Such change of distribution may be measured by generalised Mahalanobis distance (Dasgupta, 2008, Proc. of ISI Platinum Jubilee conference, World Scientific, pp 367–382) that induces a quadratic differential metric. The same is also induced by Fisher's information matrix. For correlated normal random variables Fisher information matrix has a simpler form, thus facilitating the calculation of model sensitivity.

MSC classification 2010: Primary 62P10; Secondary 62J02

1.1 Introduction

Amorphophallus paeoniifolius (Densst) Nicolson(Aracea), also known as Elephant foot yam, is a potential tropical tuber crop. The Yam tubers are rich in nutrients. The trace elements and key minerals present in Elephant foot yam are calcium, iron, copper, zinc, selenium, phosphorus, magnesium and potassium. Vegetable curry, pickles and many indigenous medicinal preparations are made using its tubers. Stems of yam are edible vegetable. Yam being a cash crop, area under its cultivation is increasing. It has origin in South-east Asia and grows wild in the Philippines, Malaysia, Indonesia and South-eastern Asian countries. The production of Elephant foot yam is a boon to farmers in barren land, see Dasgupta (2013a), Bretona and Musiela (1987), Venkatram et al. (2007). In fertile land of West Bengal the production can be five times the initial seed weight. It can grow even in lateritic soil texture full of gravels like that of Giridih, Jharkhand although the production may be comparatively less in such type of soil. For poor tribes Elephant foot yam is a staple food when boiled with tamarind leaves and/or a sour taste local fruit named red *kudrum* to reduce itching sensation due to calcium oxalate raphides present in yam. Besides wild variety of yam available in forest, tribals also cultivate yam along with *ragi*, maize, *kudrum*, pulses, niger, *gondli* a rice like crop, horse gram, etc.; apart from seasonal vegetables in their land in forest and plains. Government financial help is scare for land cultivation in forest zone; so is scarcity of irrigation in dry season of Jharkhand. Benefit of "Kisan Credit Card" is yet to reach that poor segment without financial sureties. In this socioeconomic background cultivation of Elephant foot yam as a staple food plays a major role for tribal areas like Jharkhand. *Kaun* paddy, a similar crop like *gondli*, is still cultivated in Bangladesh and nearby areas, as a rice substitute. Cheaply available yam has no serious side effect in regular consumption as food.

Cut yam, much like a vertical slice of apple, containing a part of "main eye" from whole yam is usually planted by farmers around February–April and one irrigation is given after plantation. Sprouting starts in the next month. Massive vegetative growth of Yam stem above surface is an indication of much deposition of carbohydrates i.e., underground yam due to higher level of photosynthesis. Yam-stem also has good market value as a tasty edible vegetable. Usually within 7–10 months the stems and leaves of the plant become yellowish and the crop is ready to harvest. If the corm is left undisturbed under earth, then this sprouts once again in the next season to produce a larger yam. Some farmers wait for several years after planting seed corms. These are initially about 20–25 g each and take about 3–4 seasons to develop into corms weighing 8–10 kg each, as sometimes seen in agricultural exhibitions. Mature yam may be stored in the field of cultivation and farmers dig it up at times of demand of market supply, so as to avoid low price. There is no loss of quality and nutrients for such storage till harvest even if it is delayed for several months after mature yam plants die at the end of season. At very low humidity, dehydration takes place and at high humidity rotting is a problem for storage of harvested yam.

In growth experiments of Elephant foot yam, the weight and surface texture of corm are recorded before plantation, and subsequently the growth of main-stem and the secondary sprouts may be frequently recorded to monitor the growth of Yam developed underground.

With the variety Bidhan-Kusum of Elephant foot yam, experiments are conducted with four 5×5 Graeco-Latin square design with 5 Yam seed weight levels viz., 200g, 350g, 500g, 650g, 800g and five levels of skin texture for Yam seed from roughest to smoothest, to see the effect of seed weight and skin texture on final Yam yield.

In Sect. 1.2 we analyse the yield data from farm and observe that (cut) seed corm of moderately rough skin texture and moderately large surface area of skin, having seed weight of about 650g may be recommended for lateritic soil full of gravels as in Giridih, Jharkhand. The study takes into account skin texture of Yam seed-corm in analysis. A strong correlation between the growth of main stem (which can be continuously recorded over time) and the yield of yam is observed in logarithmic scale. Height of plant and girth of plant at base is seen to be highly correlated. These are some of the indications that yield and above-ground biomass may be highly correlated, as we shall see later in Sect. 1.3 that the latter variable can be expressed in terms of former variables in an approximate nonlinear form that can be linearized in logarithmic scale.

In Sect. 1.2 our analysis suggests that early sprouting of planted seed-corm has positive effect on yield, the growth curve takes a sharp upward turn towards end, even with slight increase in age of yam plant.

A lot of vegetative growth is observed when sprouted seed-corms are planted, resulting in high number of stems per seed-corm and higher yield of yam. With large number of stems for a seed corm, one is sure about higher yield, but for less number of stems per seed-corm it can be either way; yield may be less or high. Magnitude of least square residuals show decreasing trend as number of sprouts increases indicating higher precision of yield prediction for large number of sprouts. Invoking a theorem (Theorem 1) of Dasgupta (2013b) on nonstationary Gaussian process we estimate the parameters of a proposed log-linear model of yam based on initial weight. Yam plant accumulates much carbohydrates towards end. Sensitive yam plant if critically endangered by accident, takes a drastic turnaround to store yam underground during the remaining short lifespan in a significantly faster rate compared to other healthy plants, possible explanation is that the injured plant could foresee the end while facing terminal illness, and adopt an intelligent decision to hurry up unfinished task. This is observed in a recent experiment.

Section 1.3 provides a motivation to consider above-ground biomass as a possible good predictor. Under certain assumptions above-ground biomass over time is computed based on some observed variables that can be collected frequently over growth period. Tenacity of the assumption that biomass is a potentially good predictor for underground yam is planned to be tested in future field experiments. A lower bound for expected yam yield is obtained with live data.

Interrelations between model sensitivity, Fisher information matrix and generalised Mahalanobis distance are investigated in Sect. 1.4.

1.2 Data Analysis and Search for a Nonlinear Predictor

Two Latin squares are said to be mutually orthogonal if the two squares when superimposed have the property that each pair of letters or symbols from two squares appears only once. The superimposed square is called a Graeco-Latin square. A 5×5 Graeco-Latin square is shown below.

$$\begin{pmatrix} A\alpha & B\beta & C\gamma & D\delta & E\epsilon \\ B\gamma & C\delta & D\epsilon & E\alpha & A\beta \\ C\epsilon & D\alpha & E\beta & A\gamma & B\delta \\ D\beta & E\gamma & A\delta & B\epsilon & C\alpha \\ E\delta & A\epsilon & B\alpha & C\beta & D\gamma \end{pmatrix}.$$

In the field experiment two characteristics of the corm i.e., seed yam tested are represented by Latin and Greek letters viz., weight and surface area, respectively. Here the Latin letters $A = 200, B = 350, C = 500, D = 650, E = 800$ are weights in grams of the seed-corm (levelled as 1–5, respectively), and the Greek letter α represents largest and smoothest surface area of the seed (level 5), β being the second largest and smoothest (level 4) and so on; ϵ represents the smallest and roughest surface area (level 1).

In the years reported below, the experiment is replicated four times in nearby plots with $25 \times 4 = 100$ combinations of corm size and skin texture. We provide a brief summary of data analysis associated with graphical presentation.

Analysis of Yam Production Data for the Year 2008. The growth experiment was conducted in an unfertile piece of land, where for the first time yam is grown. All the $25 \times 4 = 100$ combination germinated to fully grown plants except two $A\epsilon$ combinations out of four such and one $A\beta$ combination, thus indicating presence of main effect and possibly interaction of weight and surface area. The data, main-variable: yield, along with some auxiliary variables of multiple-stem growth like height of the plant (x_1), girth at the base (x_2), at middle (x_3), and at top of the stem (x_4), are analysed in the following.

An estimate of the ratio of final production to the total initial weight is 3.03.

In Fig. 1.1 we plot the three-dimensional picture of Final average yield for the $5 \times 5 = 25$ combinations of seed weight and skin texture, based on least squares estimates in linear model. The size and roughness of yam corm affect the final yield. We observe that γ i.e., area level 3, with D (650 g) i.e., weight level 4, may be recommended to farmers. In the same category of skin texture γ, next level of seed weight 800 g provides an additional yield weight of about 82 g on the average whereas initial additional investment in seed weight is 150 g. Thus the combination $D\gamma$ is recommended to farmers instead of the highest bar in the diagram corresponding to $E\gamma$, in Fig. 1.1.

In Fig. 1.2 we plot Principal component 1 (PC 1) with coefficients (0.468, 0.517, 0.511, 0.503) vs. Principal component 2 (PC 2) with coefficients (−0.853, 0.0, 0.347, 0.388) for the variables x_1, x_2, x_3, x_4, respectively. PC 1 seems to represent mean of all the variables, an average growth or size characteristic. PC 2 assigns zero coefficient for x_2. The four eigen values are 1.896688, 0.5379271, 0.2798609 and 0.1867836. The first two PC explain about 84% of the variation in data. No visible trend is found in the scatterplot of PC 1 vs. PC 2, in Fig. 1.2.

Figure 1.3 shows that height of plant and girth of plant at base are highly correlated ($r = 0.8476$). Lowess curve and least square regression line fit are also shown to be close. Similar features are revealed in Figs. 1.4 and 1.5 where height of the plant vs. girth in middle and at the top of stem are considered, respectively.

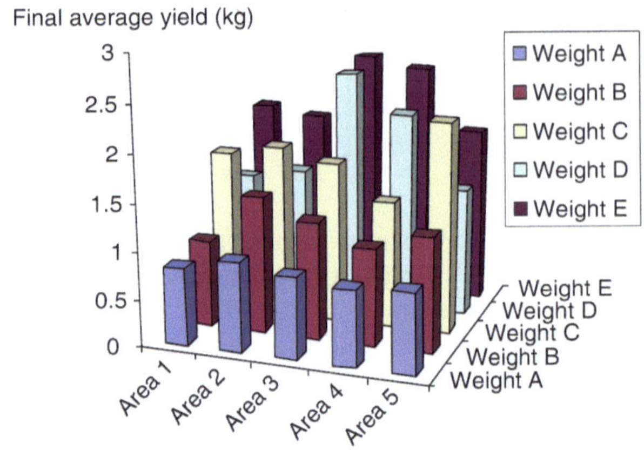

Fig. 1.1 Final average yield (kg)

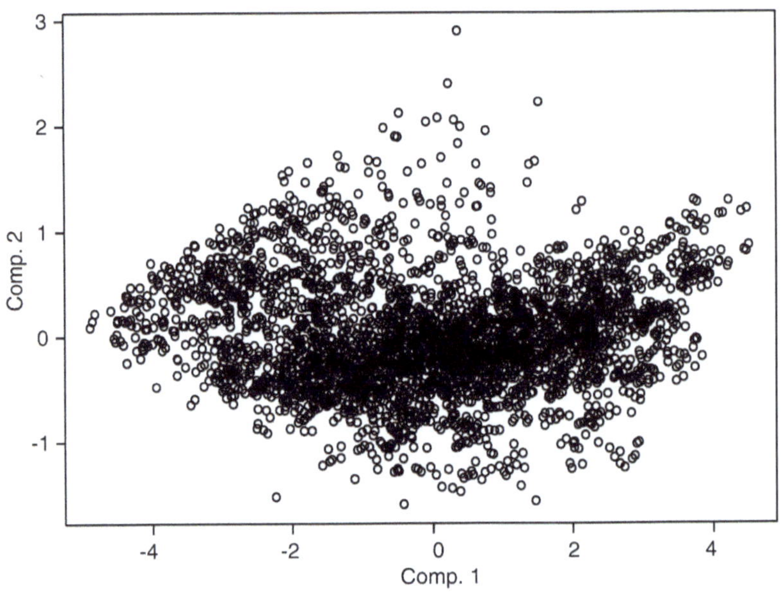

Fig. 1.2 Principal component 1 vs. principal component 2

We observed a strong correlation between the growth of main stem (which can be continuously recorded over time) and the yield of yam in Fig. 1.6. The relationship is more prominent in logarithmic scale in Fig. 1.7. The retransformed least square fit on log-data is shown in Fig. 1.8. The least square curve almost overlaps with nonparametric lowess curve except at the far end, indicating that log-scale is appropriate for the concerned variables, without any model assumptions.

We also observed a relationship between the "Age of the plant at yield" and the yield-amount; see Fig. 1.9. While fitting a nonparametric lowess regression,

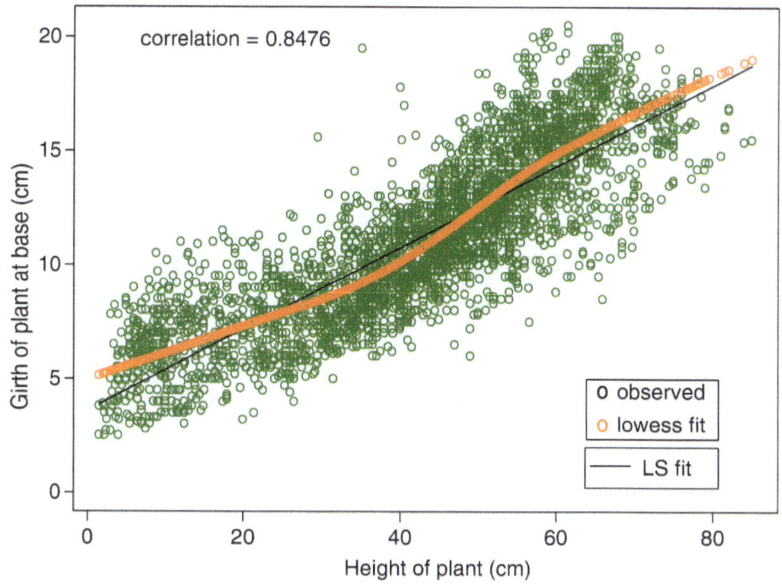

Fig. 1.3 Scatter plot of girth at base vs. plant height

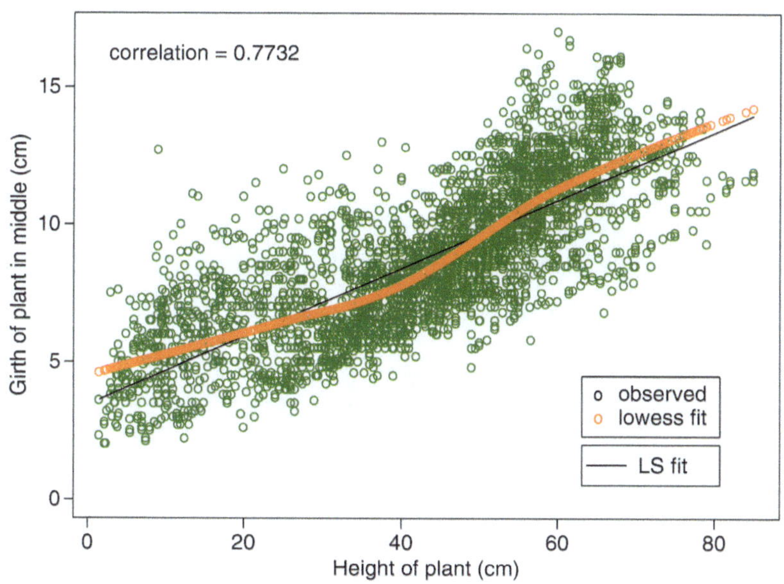

Fig. 1.4 Scatter plot of girth in the middle vs. plant height

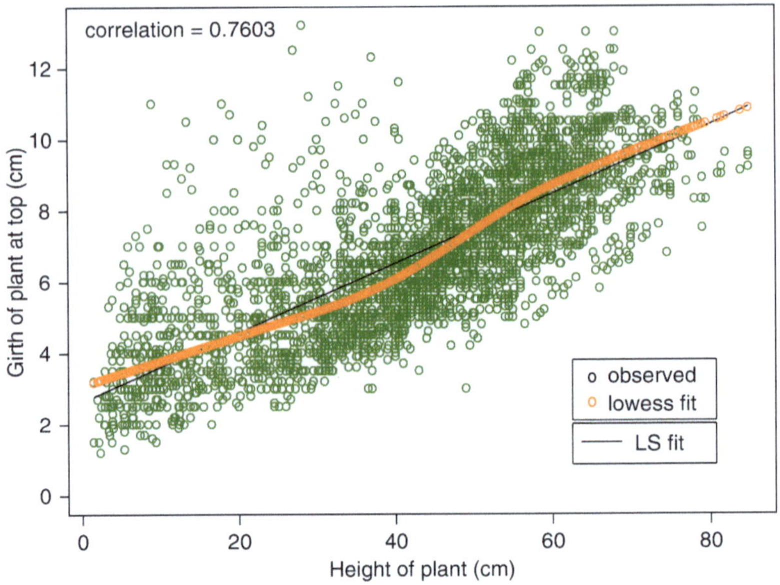

Fig. 1.5 Scatter plot of girth in the top vs. plant height

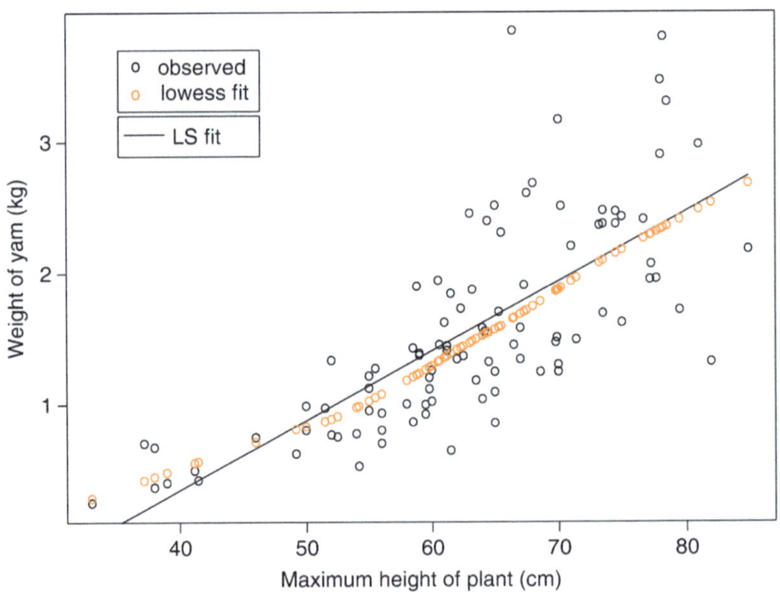

Fig. 1.6 Regression of yam yield on maximum plant height

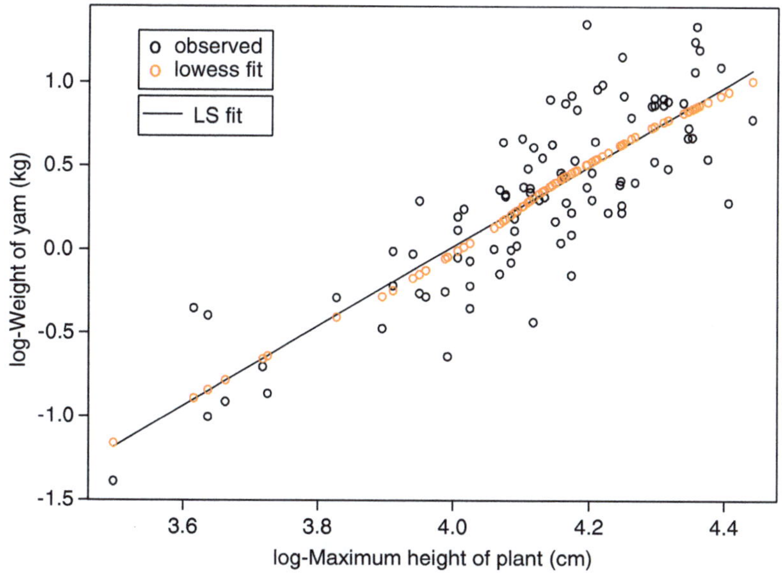

Fig. 1.7 Regression of log(yam yield) on log(maximum plant height)

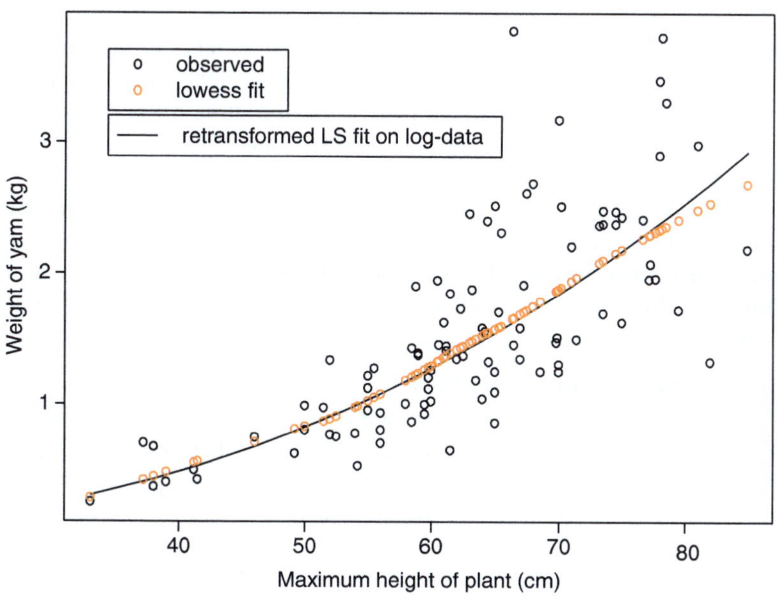

Fig. 1.8 Regression of yam yield on maximum plant height (retransformed)

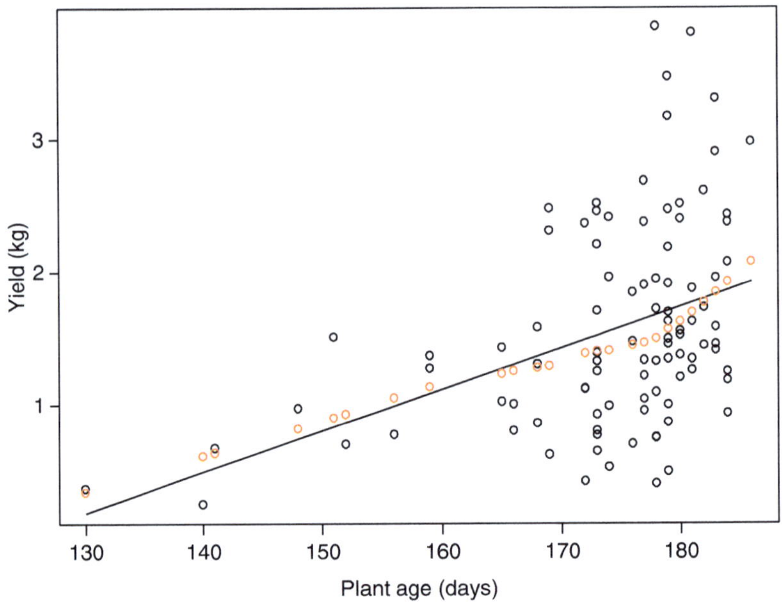

Fig. 1.9 Final yield vs. Age of plant

towards the right of the scatterplot on the top, the curve takes a sharp upward turn indicating that a small increase of "Age of the plant at yield" near the extreme has sharp increasing effect on yam yield. Thus, early sprouting has a sharp increasing effect on the yield-amount. The relationships become more prominent in the log scale of yield as seen in Fig. 1.10. The features of Figs. 1.9, 1.10 remain unchanged in the corresponding Figs. 1.11, 1.12, where initial seed weight is deducted from the final yield for regression.

Yam corm with surface level 3, i.e., moderate roughness is seen to cause early sprouting with high upward trend of lowess curve of yield vs. age of plant; see Fig. 1.13. The feature remains same even in log scale of yield as seen from Fig. 1.14.

Analysis of Data for the Year 2009. A significant difference here is that the experiment started late, and there was already a tendency of sprouting from the "surface eyes" *before* plantation. As a result the numbers of sprouts per corm were higher compared to earlier year. There were a lot of vegetative growth and the final Yam yield had a positive correlation with the number of sprouts for each level of surface area.

Positive associations of number of sprouts vs. yield are seen in Figs. 1.15–1.19 for seed area levels 1–5, respectively. Figure 1.20 shows the same for all data combined. Also see Figs. 1.21–1.25 for "Number of sprouts" vs. "Residual-squared" plots at seed area levels 1–5, respectively. Except for level 2 in Fig. 1.22, all other figures show downward trend in lowess regression (black curves) and cubic spline regression (slightly curly yellow curves going beyond data range that may be used

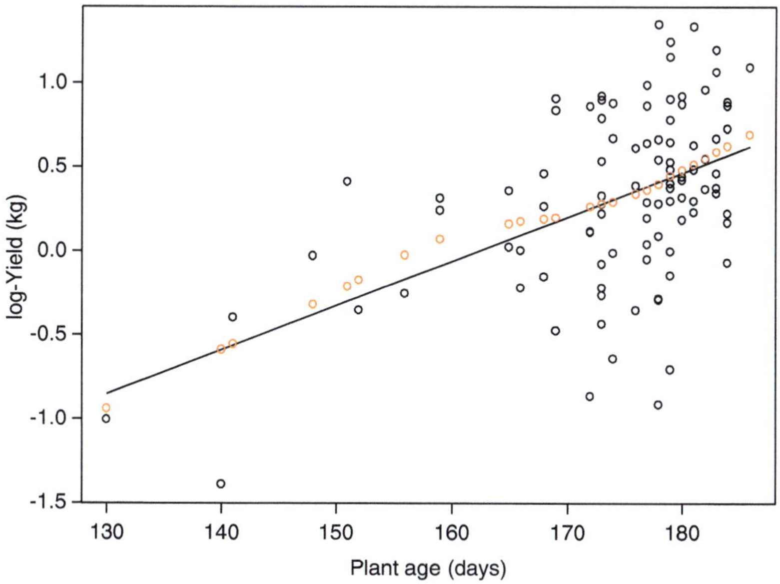

Fig. 1.10 log(Yield) vs. Age of plant

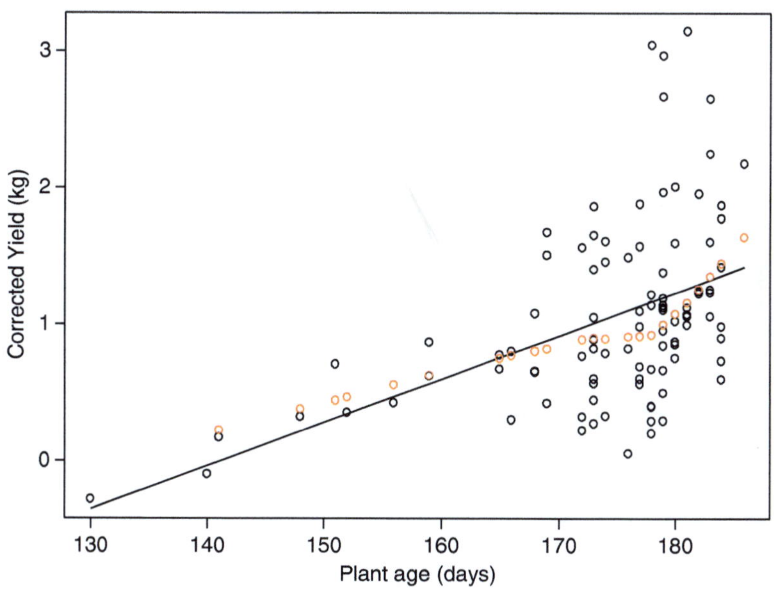

Fig. 1.11 (Yield–Seed weight) vs. Age of plant

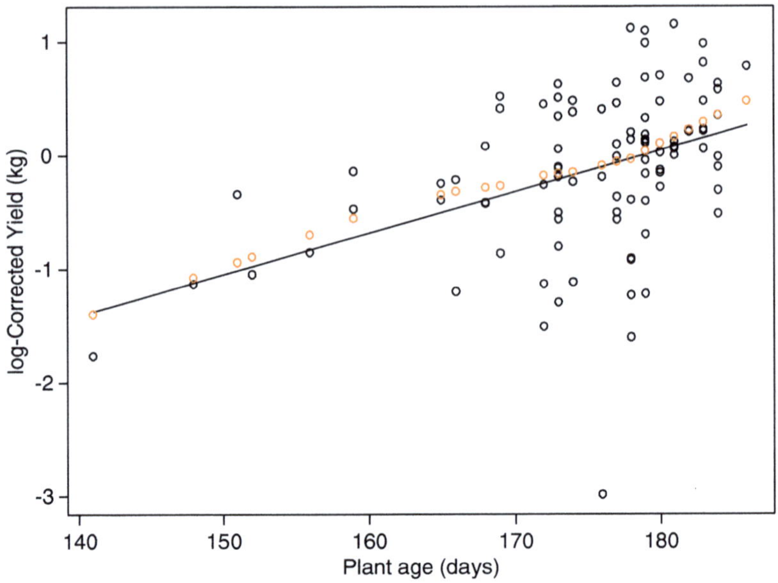

Fig. 1.12 log(Yield–Seed wt) vs. Age of plant

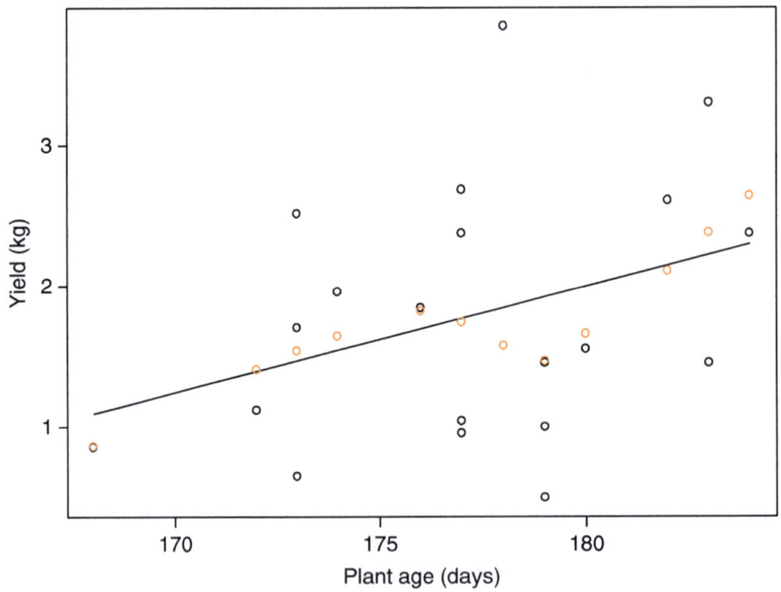

Fig. 1.13 Yield vs. Age of plant (level 3 seed area only)

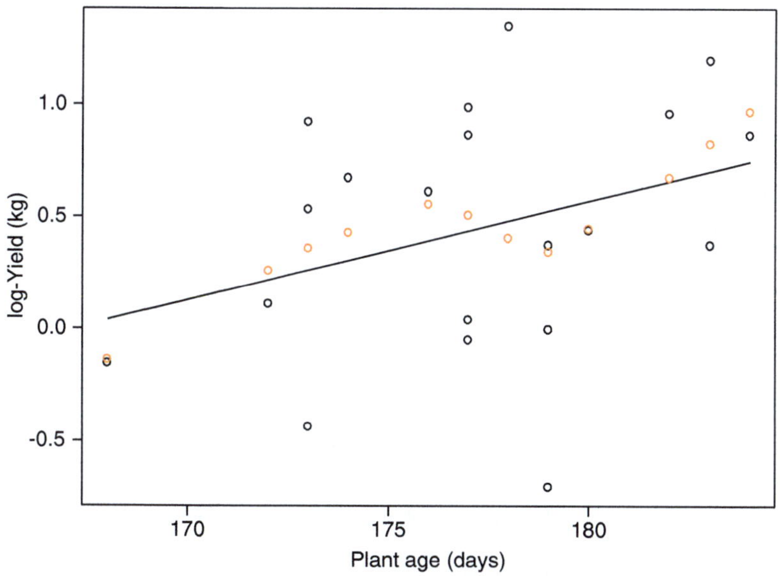

Fig. 1.14 log(Yield) vs. Age of plant (level 3 seed area only)

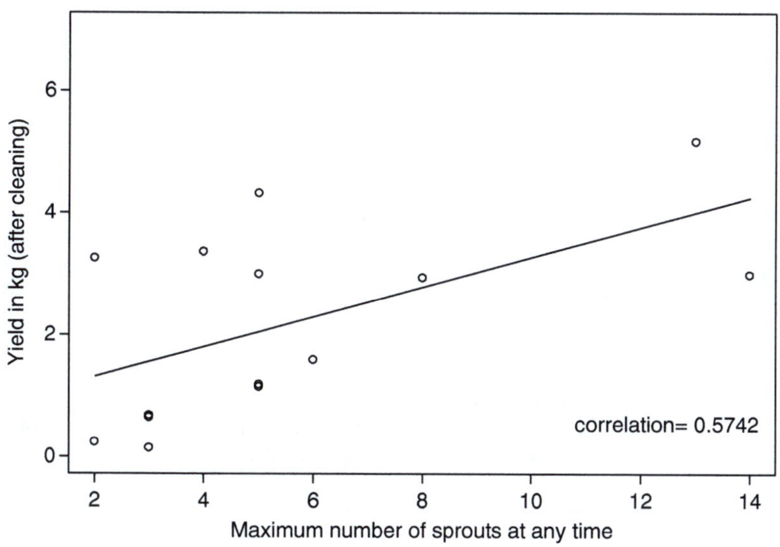

Fig. 1.15 Yield vs. maximum number of sprouts (Area 1)

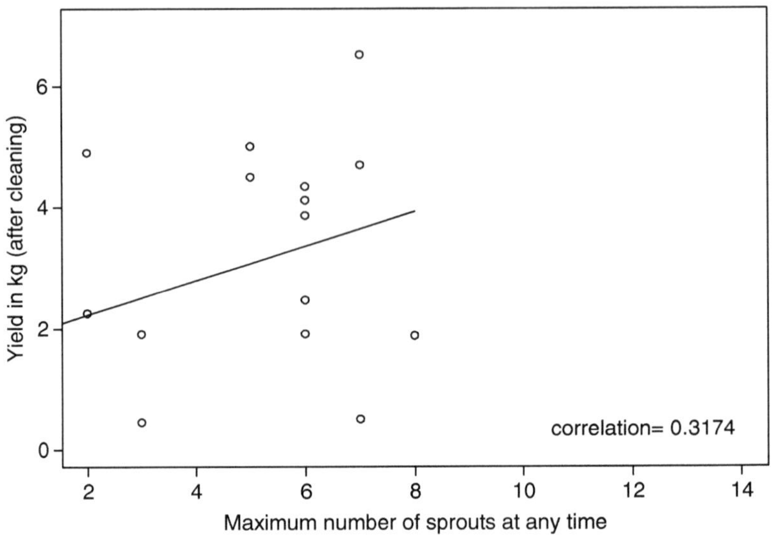

Fig. 1.16 Yield vs. maximum number of sprouts (Area 2)

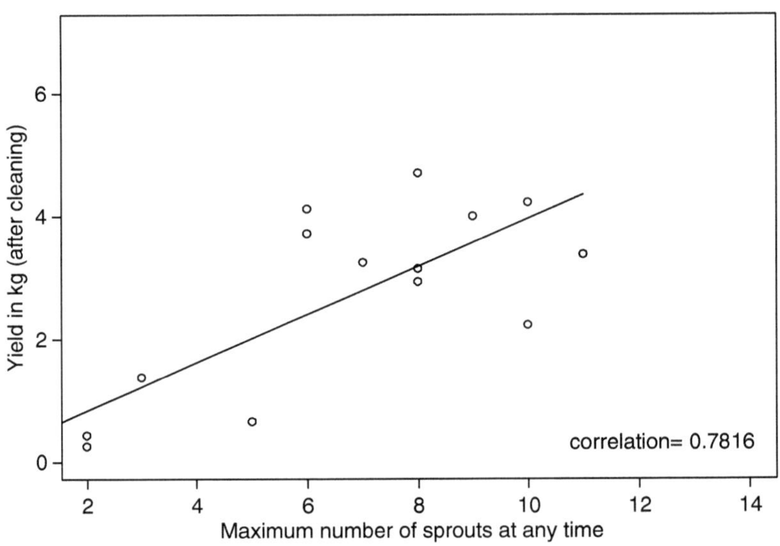

Fig. 1.17 Yield vs. maximum number of sprouts (Area 3)

for prediction purpose), indicating that magnitude of residual in prediction is less for large number of sprouts. With high number of sprouts one can be reasonably sure that the final yield is high, whereas for small number of sprouts, there is more variation in final yield, as evident from the plot of number of sprouts vs. squared

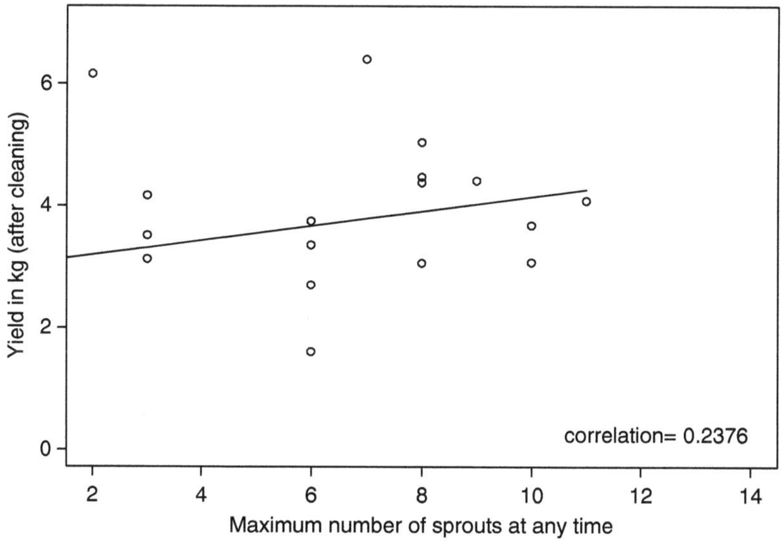

Fig. 1.18 Yield vs. maximum number of sprouts (Area 4)

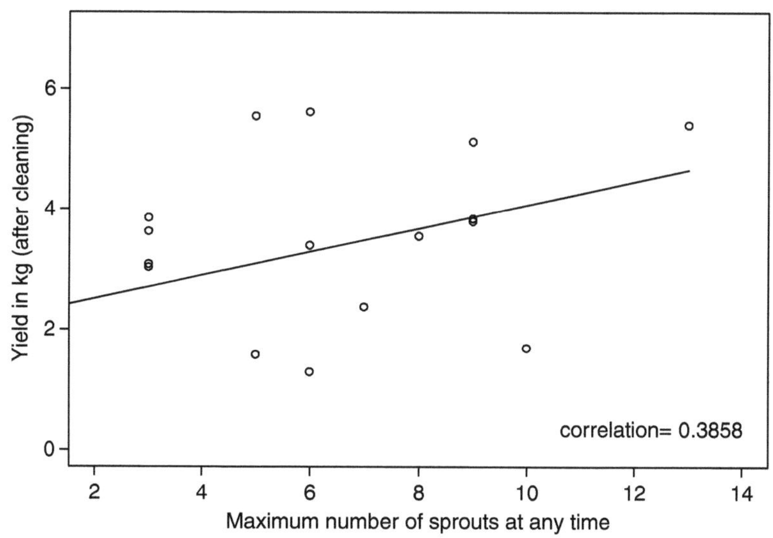

Fig. 1.19 Yield vs. maximum number of sprouts (Area 5)

residuals of prediction. Figure 1.26 shows the similar downward trend for all data points merged together.

Row, column, their interaction, four replication effect of Latin Squares, area, interaction of initial weight and surface area were all insignificant. So we are analysing an (unbalanced) design with 19 missing observations out of total 100

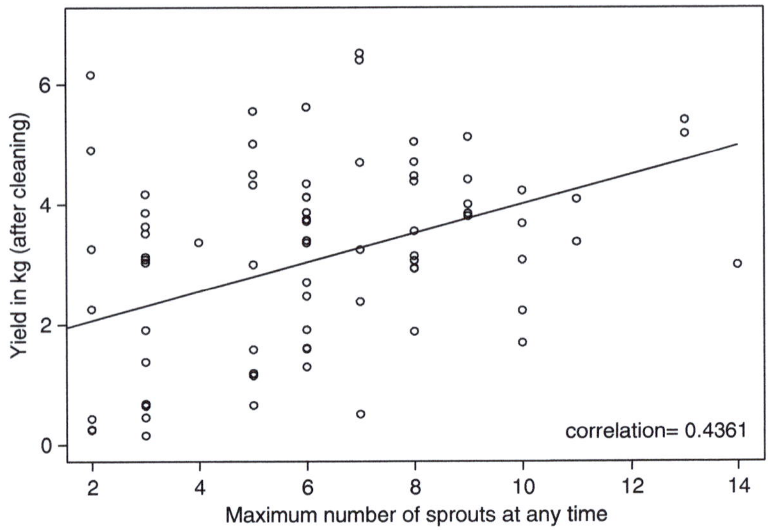

Fig. 1.20 Yield vs. maximum number of sprouts (All areas)

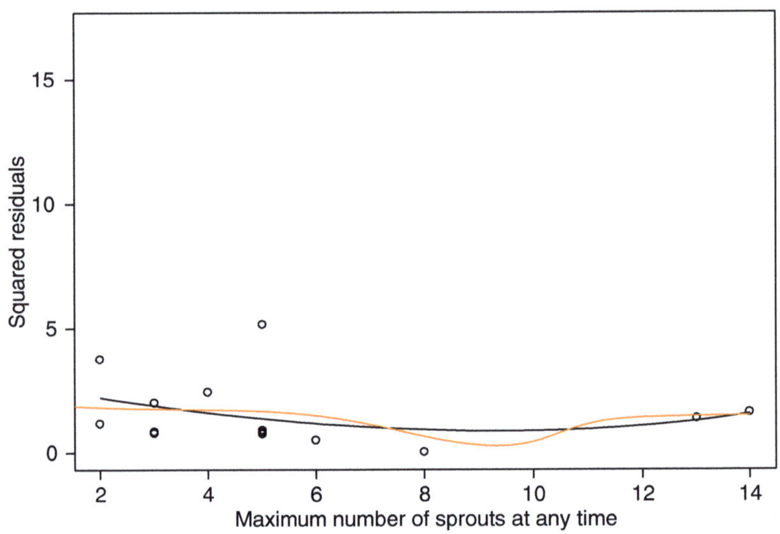

Fig. 1.21 Squared residuals vs. maximum number of sprouts (Area 1)

observations. Multiple R^2 of Yam yield on maximum number of sprouts at a time, and on all 25 weight-surface area combinations of seed corm is 0.5556. It is further seen that the effect of number of sprouts, eliminating the effect of initial Yam corm weight is insignificant (P value is .02415263). However initial weight remained

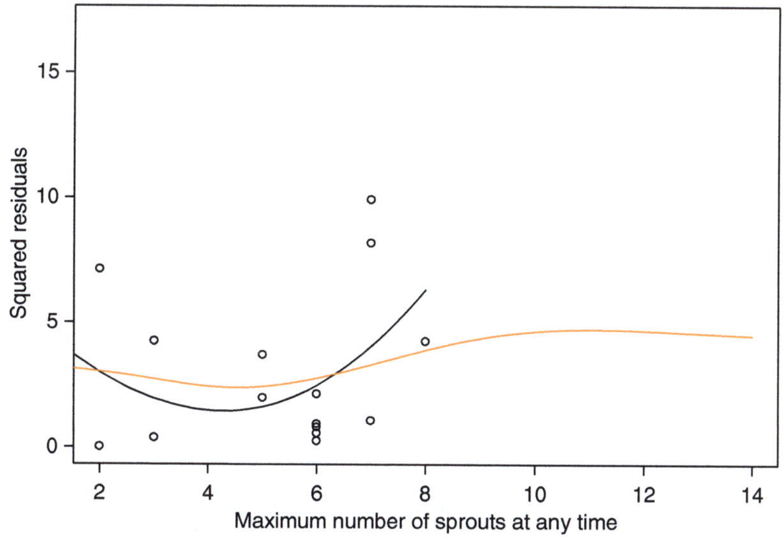

Fig. 1.22 Squared residuals vs. maximum number of sprouts (Area 2)

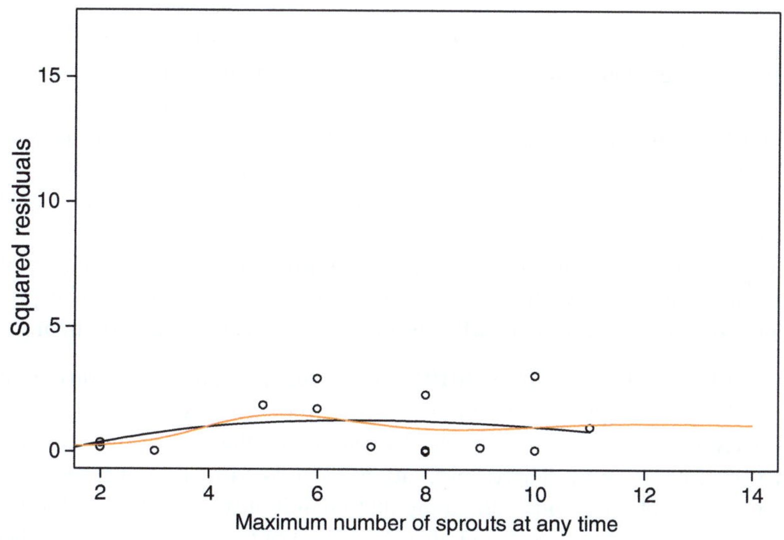

Fig. 1.23 Squared residuals vs. maximum number of sprouts (Area 3)

significant even after sequentially eliminating the effect of maximum number of sprouts and surface area of seed (*P* value is .00081304).

Thus, the initial weight has the dominant effect on yield (given that the sprouts are germinating from surface eyes before plantation).

An estimate of the ratio of final production to the total initial weight is 5.84, considering only surviving plants. The ratio is 5.01, when "no yield" is taken as

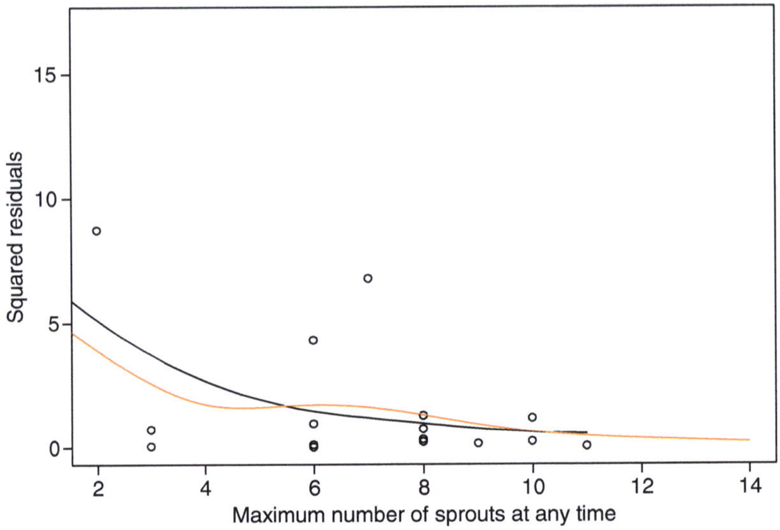

Fig. 1.24 Squared residuals vs. maximum number of sprouts (Area 4)

zero and all the 100 pit observations are considered. The ratio is higher than the previous year's estimate 3.03, when there were early plantations with no indication of sprouting from surface eyes before plantation. No significant difference is seen if weight-levels are taken separately in regression or a least square fit is made for initial weight vs. final weight of Yam, $R^2 = 0.313(0.2913)$; broken lines joining the five group means are quite close to the least square regression line of whole data set (Fig. 1.27).

Thus we conclude that the yield may be almost double if sprouting-seeds are planted, in comparison with non-sprouted seeds before plantation. The previous experiment also suggested that early sprouting has a positive effect on final yield.

Analysis of Data for the Year 2010. Some organic manure like vermicompost and cow dung were applied while preparing the plots for cultivation. This resulted in improved production to the ratio 4.79, even though the seed corms were planted before sprouting this time. Three observations with $A\epsilon$, $A\alpha$, $A\gamma$ combination had nil yield out of 100 observations, indicating that the initial weight of $A = 200$ gram is critical irrespective of the condition of surface level of the cut seed corm for lateritic soil with gravels in Giridih.

The following data relates to weights in kilogram of 100 yams from a growth experiment conducted in the year 2010 at Indian Statistical Institute, Giridih farm in the above mentioned serial order from 1 to 100.

4.50, 3.20, 2.60, 3.15, 2.05, 2.10, 2.65, 0.80, 1.70, 1.15, 2.90, 3.50, 4.35, 3.85, 3.60, 1.30, 2.20, 1.70, 3.70, 2.50, 3.40, 3.10, 4.45, 5.60, 4.15, 1.50, 1.90, 2.00, 3.10, 3.00, 3.10, 2.25, 2.65, 2.90, 3.60, 1.50, 1.20, 0.70, 2.80, 2.70, 3.75, 2.05, 1.60, 1.50, 3.60,

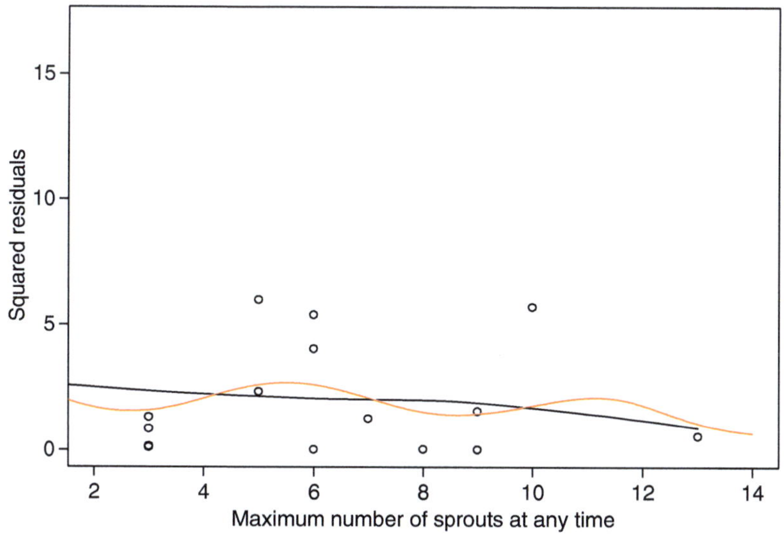

Fig. 1.25 Squared residuals vs. maximum number of sprouts (Area 5)

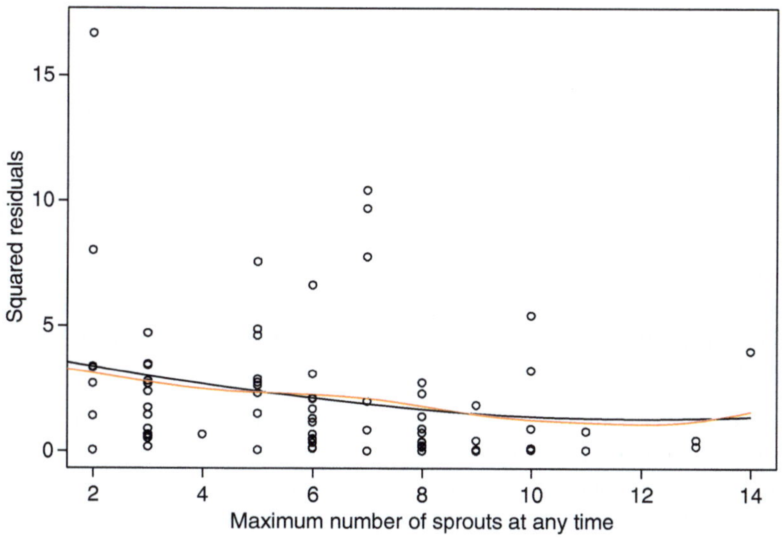

Fig. 1.26 Squared residuals vs. maximum number of sprouts (All areas)

2.20, 1.40, 1.20, 0.00, 2.40, 2.50, 1.45, 1.05, 0.70, 0.00, 2.25, 2.00, 2.45, 1.55, 0.90, 0.75, 2.65, 2.25, 1.20, 2.25, 2.00, 3.80, 3.00, 3.00, 2.35, 1.05, 0.80, 3.80, 2.30, 3.80, 1.60, 0.00, 3.60, 1.60, 4.00, 3.00, 1.95, 2.00, 3.65, 3.60, 1.40, 1.40, 1.30, 3.90, 3.60, 5.50, 2.90, 2.60, 1.70, 2.80, 1.90, 1.70, 1.80, 1.10, 2.80.

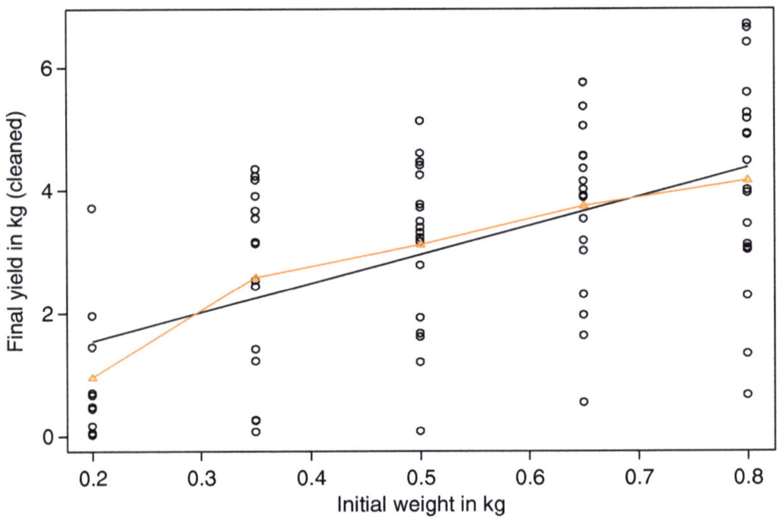

Fig. 1.27 Group means and least squares regression line

We may analyse this yam data in terms of proportionate return. Ratio K of the final weight (Y) to initial weight (X) is given below in same serial.

5.625, 4.923077, 5.2, 9.0, 10.25, 2.625, 4.076923, 1.6, 4.857143, 5.75, 14.5, 4.375, 6.692308, 7.7, 10.285714, 6.5, 2.75, 2.615385, 7.4, 7.142857, 9.714286, 15.5, 5.5625, 8.615385, 8.3, 4.285714, 9.5, 2.5, 4.769231, 6.0, 6.2, 6.428571, 13.25, 3.625, 5.538462, 3.0, 3.428571, 3.5, 3.5, 4.153846, 5.769231, 4.1, 4.571429, 7.5, 4.5, 3.384615, 2.8, 3.428571, 0.0, 3.0, 3.125, 2.230769, 2.1, 2.0, 0.0, 2.8125, 3.076923, 4.9, 4.428571, 4.5, 3.75, 3.3125, 3.461538, 2.4, 6.428571, 10.0, 4.75, 4.615385, 6.0, 6.714286, 3.0, 4.0, 4.75, 3.538462, 7.6, 4.571429, 0.0, 4.5, 2.461538, 8.0, 6.0, 5.571429, 10.0, 4.5625, 5.538462, 2.8, 4.0, 6.5, 4.875, 5.538462, 8.461538, 5.8, 7.428571, 8.5, 3.5, 2.923077, 3.4, 5.142857, 5.5, 3.5

One should be cautious in interpreting a very high value of ratio, especially those grater than 10. These refer to the lowest seed weight 200 g, which is also critical weight for survival of the plants at Giridih farm. If such a plant survives over time it may deposit a moderate-sized yam underground, due to post survival care of administering manure and nutrients. The ratio is then quite high as seed size appearing in the denominator is small, although the respective yam yield is moderate or small.

Histogram of the ratio $K = Y/X$ is seen to be positively skew as shown in Fig. 1.28. The median is at 4.68, mode is at 5.0. A model of the type $Y = KX + E$ with error component E to be normal may not be appropriate.

However, histogram of logarithm of the ratio, $k = \log(Y/X) = \log K$ as shown in Fig. 1.29, ignoring the nil yields, is seen to be symmetric. The quantile–quantile plot of Fig. 1.30 suggests k to be normally distributed, indicating a strong possibility

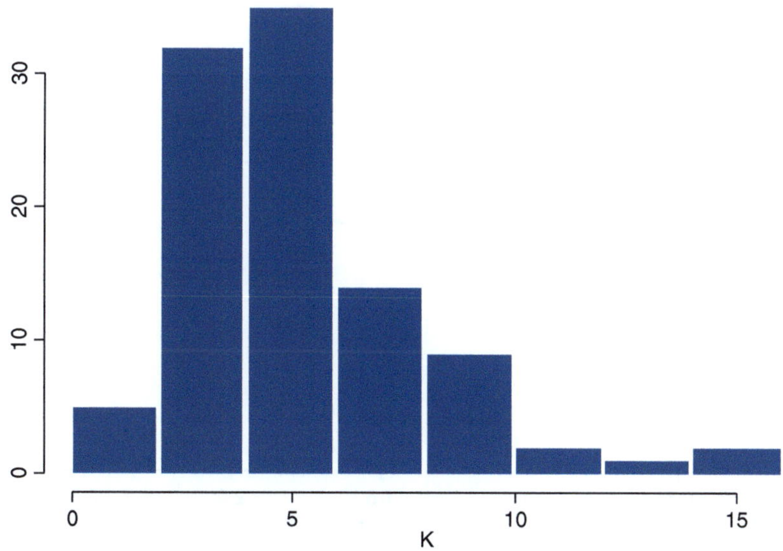

Fig. 1.28 Histogram of ratio: final wt. to initial wt. (year 2010)

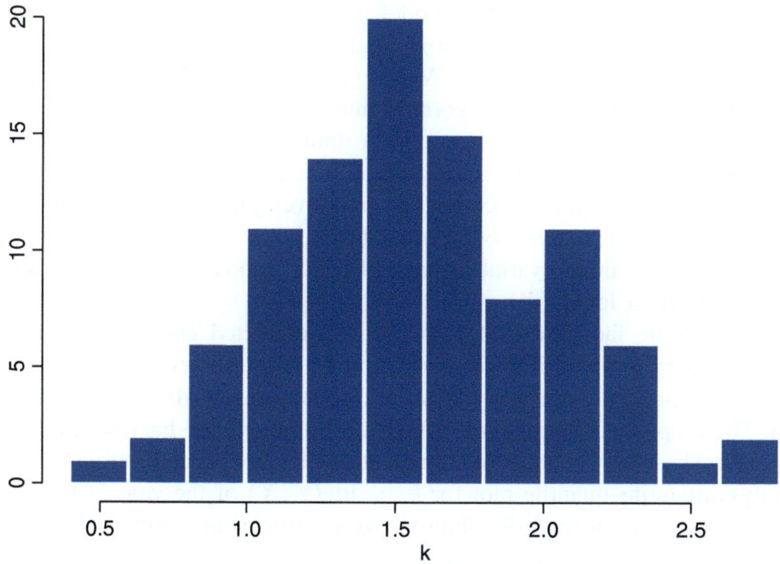

Fig. 1.29 Histogram of log(ratio)

of the model $Y = e^u X$, where $u \sim N(\mu, \sigma^2)$, estimated values of mean and variance are $(\hat{\mu} = 1.577681, \hat{\sigma^2} = 0.2116378)$, from data. Expected return of yam-yield is then about $e^{1.577681} = 4.84371$ times, in the year 2010.

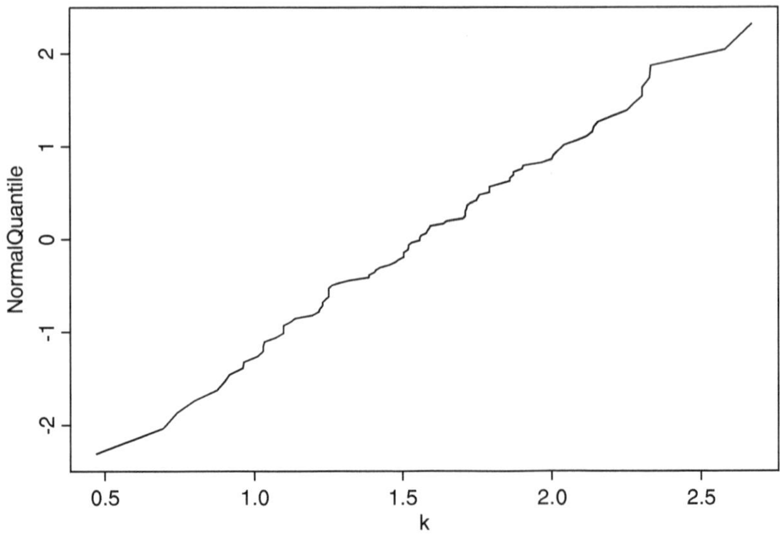

Fig. 1.30 Normal quantile vs. k quantile plot

As already pointed out, such a model may have poor performance for small values of X, near the critical seed-corm weight for germination.

The model $Y = e^u X$, where $u \sim N(\mu, \sigma^2)$ seems plausible when we consider yam-yield per unit gram of seed corm planted Y/X, is subjected to many independent causes, each of which may take a dominant role, e.g., humidity in soil and other favorable conditions during sprouting, availability of manure and soil nutrients, meteorological parameters at desired level, plant care, etc. Thus arise the possibility of a multiplicative model, where (Y/X) can be represented as product of many independent random variables each of which is positive, leading to a normal distribution for $k = \log(Y/X)$.

In view of the fact that 100 yam pits of conducted growth experiment are adjacent, leading to possibility of correlation amongst observations, estimation of parameters from such correlated Gaussian observations needs a theoretical justification. The crop over 100 pits has different lifetime and are harvested at different time points. These may be thought of realization of a process at different time points. To analyse quantile–quantile plot for $k = \log(Y/X)$ of the year 2010 shown in Fig. 1.30 indicating normal distribution, we need to study empirical distribution of observations on nonstationary Gaussian process realized on non-equispaced time points. Apart from possible correlation amongst observations these may be nonhomogeneous. Some yam plants may have short life span. Nonhomogeneity within soil structure, plant care, etc. may lead to a nonhomogeneous process. The following result for weakly correlated process with polynomially decaying correlation function is proved in Dasgupta (2013b), validating estimation of relevant parameters of limiting Gaussian process from realized data set.

Theorem. *Consider a Gaussian process* $X(t),\ 0 \le t \le T$ *with mean* $m(t)$ *and covariance kernel* $\sigma(t,u) = \sigma(t)\sigma(u)\rho(t,u)$, *where* $m(t) \to 0,\ \sigma(t) \to \sigma;\ t \to \infty$. *Assume* $X(t)$ *has the weak limit denoted by* $X(\infty)$ *and the correlation function* $|\rho(t,u)| < K|t-u|^{-\beta},\ K > 0,\ \beta > 0$. *Consider the empirical distribution function of the process based on the observations at time points* $t_1, t_2, \cdots, t_n$ *which are not necessarily equispaced. Let the time interval* $[0, T)$ *of recording the observations be subdivided into* k *subintervals and the length of each subinterval and the number of observations in each subinterval increase to* ∞. *Also let the time gap between two consecutive observations within each subinterval be homogeneous and the number* n^* *of "isolated" observations which do not fall in any one of the homogeneous subintervals, be negligible compared to* n, *i.e.,* $n^* = o(n)$. *Then the empirical distribution function of the recorded observations from the process is a strongly consistent estimate for distribution function of the limiting variable* $X(\infty)$, *as* $n \to \infty$.

Limit $n \to \infty$ in the above has to be interpreted as (large) number of yam yield harvested sequentially in a production season, and continued over different production seasons in the same or similar experimental block(s). The assumption of polynomially decaying correlation function is mild in the sense that if two yam-pits are harvested at wide time gaps, one of the plants cease to exist to influence the other yam harvested long after. Assumption $m(t) \to 0,\ \sigma(t) \to \sigma;\ t \to \infty$ are natural when error means are centered and the process stabilizes. Proof of the theorem given in Dasgupta (2013b) indicates that the conclusion remains valid in presence of a negligible number of outliers that may appear in any conducted experiment, justifying exclusion of nil yields. With an application of above theorem normal distribution on logarithmic scale for yam yield seems appropriate, as seen on approximately linear quantile–quantile plot of Fig. 1.30.

An alternative model $\log Y = a \log X + E$ may also be examined. Ratio a of logarithm of final weight (Y) to that of initial weight (X) for 97 nonnegative Y values are shown in Fig. 1.31. The histogram is negatively skew with mode at -0.5. In comparison, we may accept the former model $Y = e^u X$, where $u \sim N(\mu, \sigma^2)$.

One may bootstrap the distribution the mean of the hundred $K = Y/X$ values. With 100 sample values taken from the original K values by simple random sampling with replacement, bootstrap mean is calculated; and the histogram based on 10^6 bootstrap simulation for mean ratio is shown in Fig. 1.32. Distribution appears to be nearly normal.

Figure 1.33 shows the same for bootstrap median of ratio. Here the rugged nature of histogram is due to skew histogram of K, see Fig. 1.28.

For smoothing the ruggedness, a small perturbation of magnitude $N(0, \sigma^2)$, with $\sigma^2 = 1/100$ is added to the K values and the smoothed version of Figs. 1.32, 1.33 is shown in Figs. 1.34, 1.35, respectively. In Fig. 1.35 the mode is at 4.65, and the histogram is still rugged.

Estimation of common ratio via ratio of the total of X, Y values is known to be efficient. Figure 1.36 shows the bootstrap distribution of ratio estimate based on total of variate values. The distribution seems to be normal with mean 4.805 and

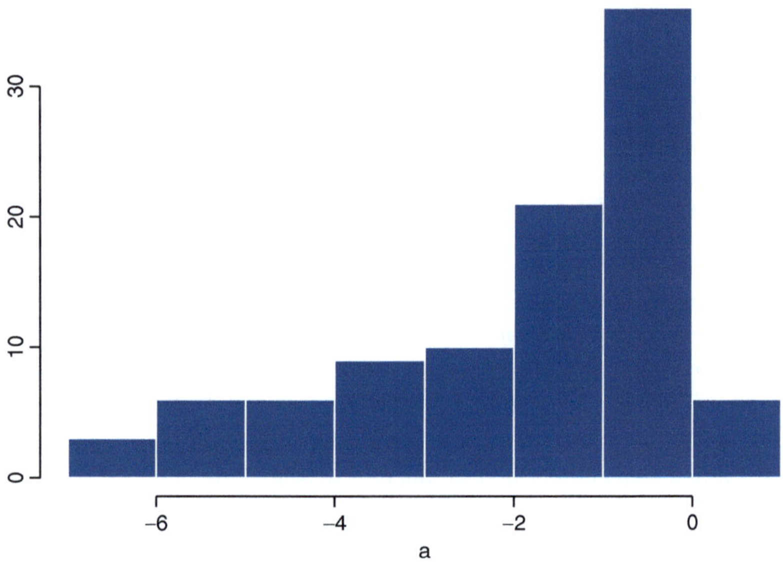

Fig. 1.31 Histogram of the ratio: log(Y)/log(X)

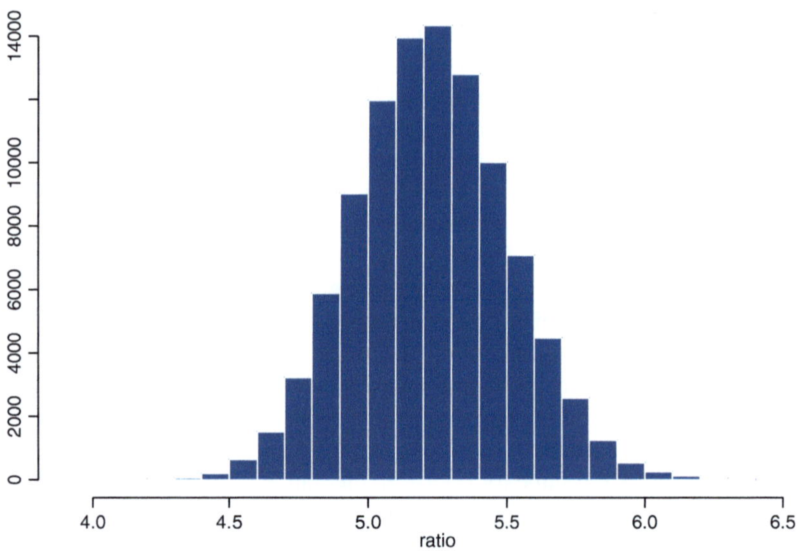

Fig. 1.32 Bootstrap histogram of mean ratio

variance 0.0425; the mode of histogram is at 4.8. The 2σ confidence interval for percentage return of yam yield is (440–520)%.

Confirmatory studies were made for growth curve, an upward trend in lowess regression is seen for yam weight vs. plant age (in days) in Fig. 1.37, exhibiting a spike in growth curve towards end; much like the yam growth curve shown in

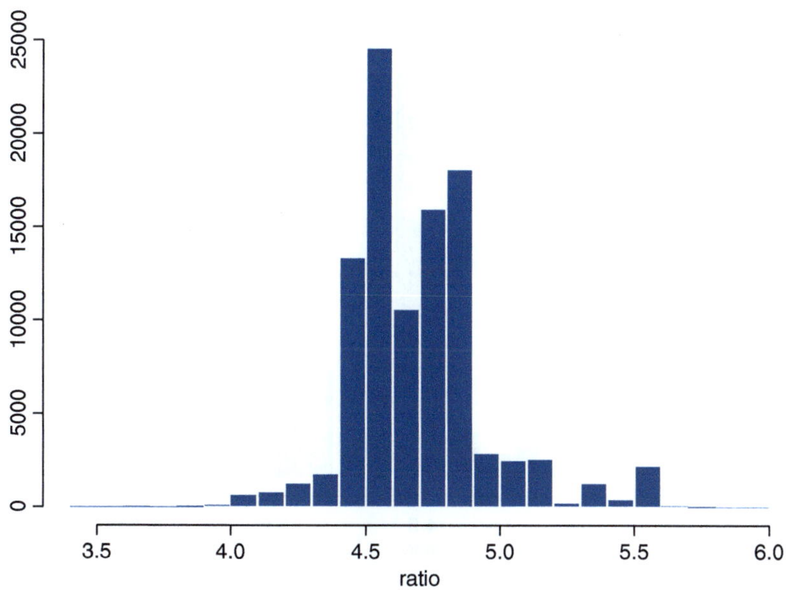

Fig. 1.33 Bootstrap histogram of median ratio

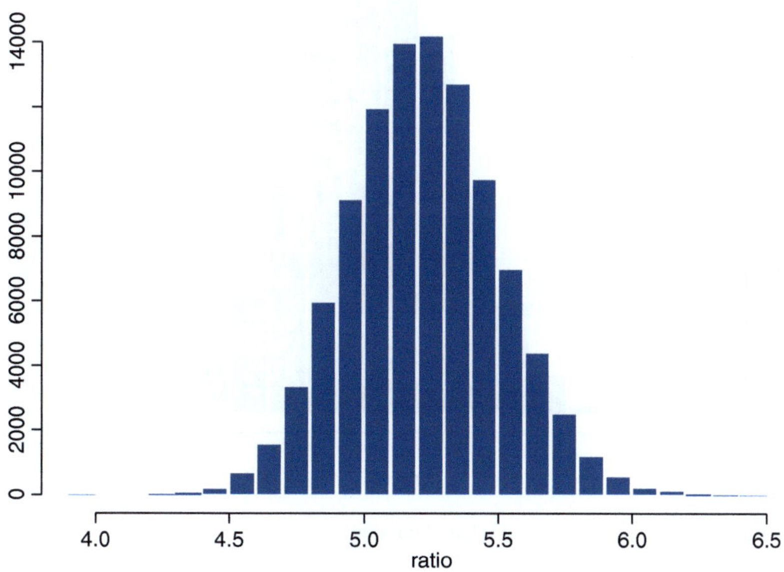

Fig. 1.34 Smooth bootstrap histogram of mean ratio

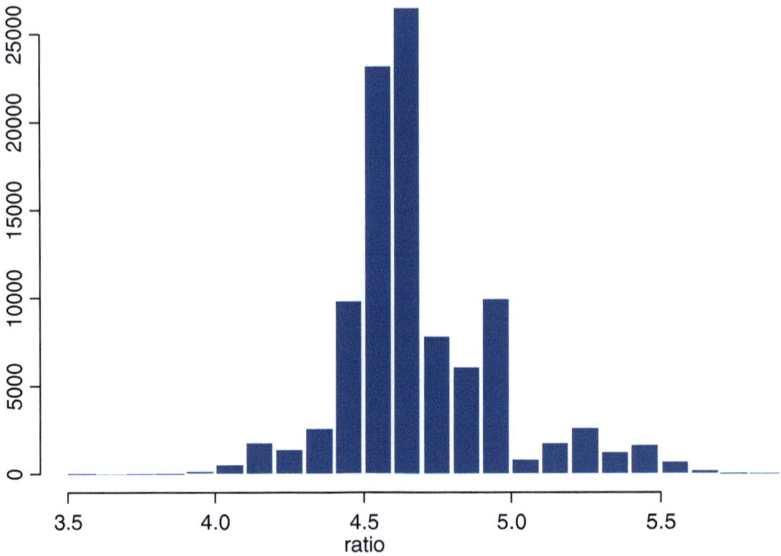

Fig. 1.35 Smooth bootstrap histogram of median ratio

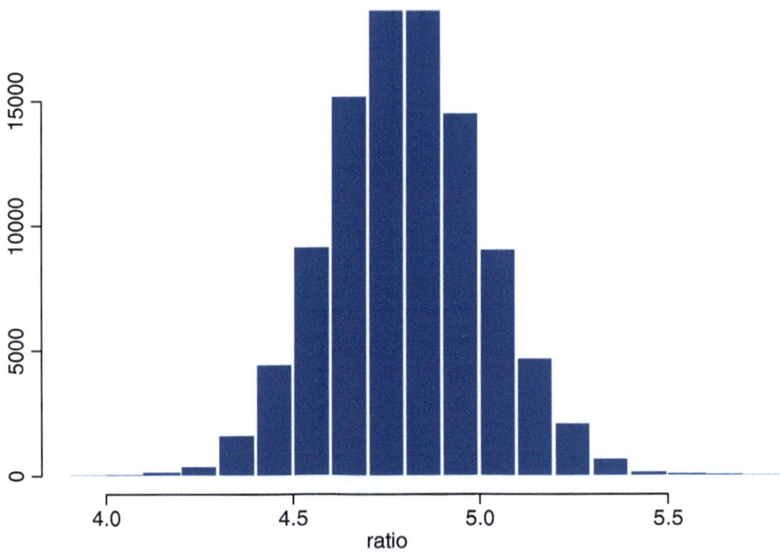

Fig. 1.36 Bootstrap histogram of ratio based on total

Fig. 1.9, indicating that slight increase in plant-life towards end results in relatively high yield.

Analysis of Data for the Year 2011. Almost same feature is also observed in Fig. 1.38 for the yam growth curve of experiment no. 2 of the year 2011.

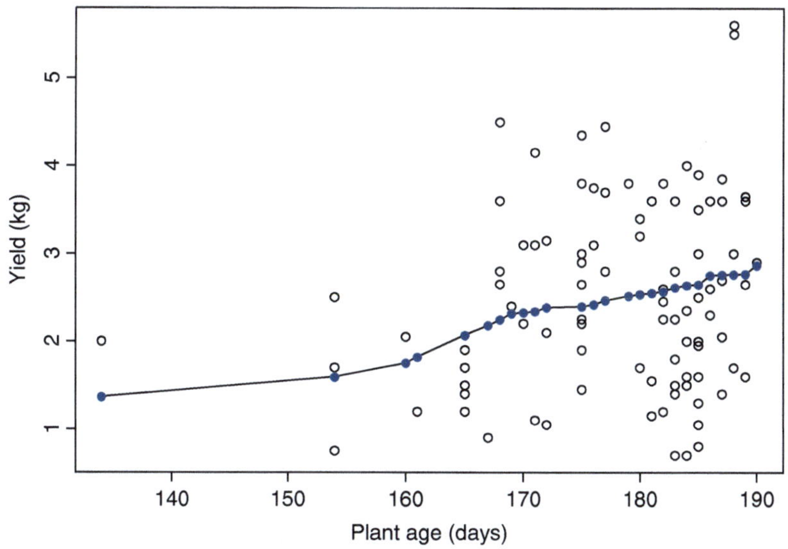

Fig. 1.37 Growth curve of Yam yield, year 2010

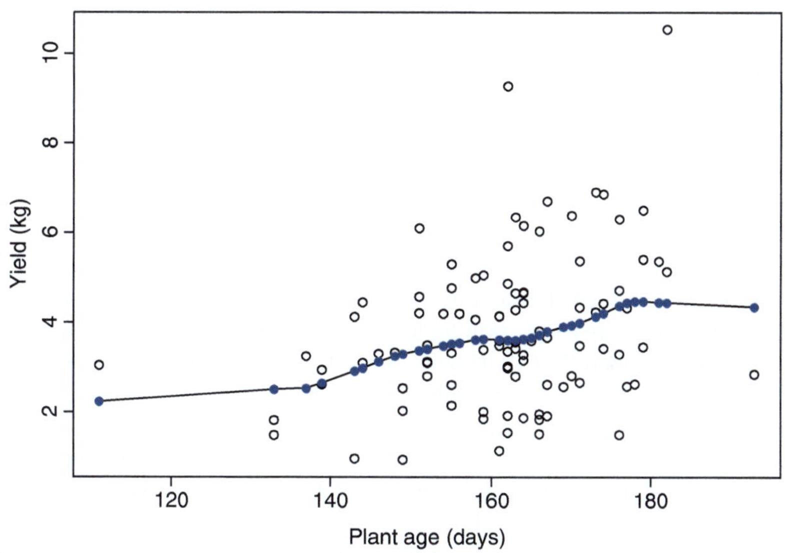

Fig. 1.38 Growth curve of Yam yield, year 2011, Expt. 2

Upward trend near spike is slightly dampened at the end by a single observation corresponding to the combination $E\epsilon$, i.e., seed weight 800 g, but the roughest and worst seed-skin condition made it not conducive for a high yield.

Sometimes it may so happen that the plants at the middle of a season may turn yellow and pale due to harsh environment that is quite common in Jharkhand region

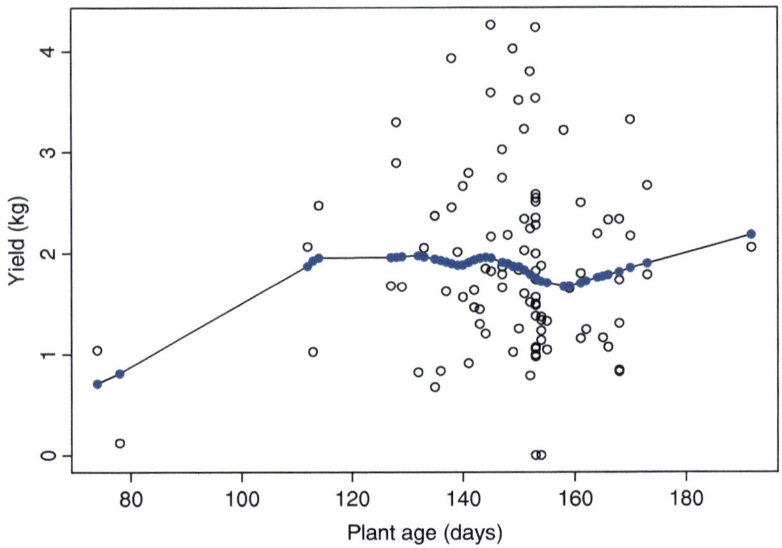

Fig. 1.39 Growth curve of Yam yield, year 2011, Expt. 4

that occasionally faces dry spell of weather. Lifeless plants may turn back to lush and green condition consuming a part of yam deposited underground when climate is conducive with several showers, see Fig. 1.39; where a sharp rise is seen in growth towards end after a little bit fall of the curve for experiment no. 4 conducted in unfertile soil, having little irrigation.

For the same experiment, arithmetic mean from non-zero entities of 100 pit observations is shown in Fig. 1.40, categorised with respect to 25 combinations of seed weight and seed skin textures. Combination $D\gamma$ (cut seed-corm of weight 650 g, with moderately rough skin texture) turns out to be optimal for yam yield in unfertile land. Compare Fig. 1.40 with Fig. 1.1, where least square estimates from linear model provides a similar result for growth experiment conducted in another unfertile piece of land in the year 2008.

Extensive sprouting like maximum number of sprouts 14, as seen in the year 2009 were not observed further in subsequent years. However, data from experiment no. 2 suggest a positive association between the number of sprouts and yam yield in Fig. 1.41, where mean yield for a fixed number of sprouts show upward trend as number of sprouts increased. We may ignore one observation towards end of figure, this plant had 8 light sprouts with small girth at base.

We have the following recommendation to farmers for cultivation in unfertile, lateritic soil texture full of gravels. The sprouted corm (cut corms will also suffice), with moderately rough surface having approximate weight 650 g is appropriate for the lateritic and gravel mixed soil of Giridih region of Jharkhand for best production. A little bit of cow dung/vermicompost and coal ash as manure at initial stage can greatly enhance the crop return to about 584% of initial seed weight.

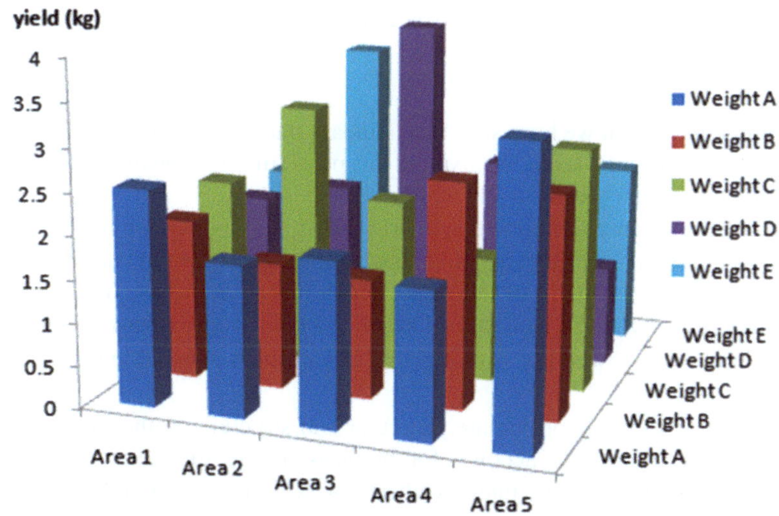

Fig. 1.40 Simple average of yield (kg) Expt. 4, 2011

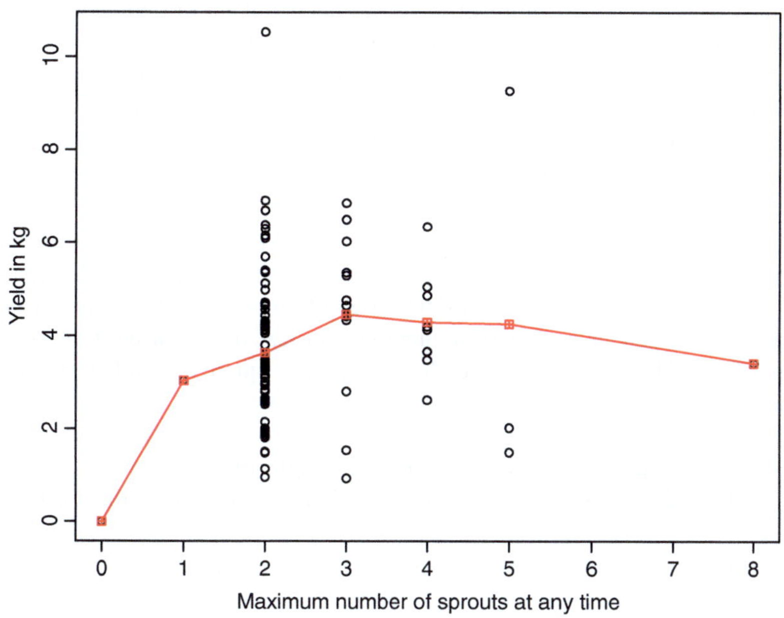

Fig. 1.41 Yield (grouped mean) vs max no. of sprouts (all areas) Expt. 2, 2011

1.3 Above-Ground Biomass as a Predictor and Model Selection

In the previous section we have seen that a linear relationship may hold amongst the variables in logarithmic scale. We explore this possibility for above-ground biomass as a predictor of yam-yield.

The volume of an approximate cylindrical stem may be written as $V_s = \pi r^2 h$, where r is the stem radius that is obtainable from the girth $(2\pi r)$ at base/middle/top of the stem, and h is height of the stem. Recall that girth at base has a very high correlation with height (h) of stem. One may obtain approximate weight of the stem as $W_s^{(1)} = V_s d$, where d is the density of yam stem obtainable via destructive testing.

Weight $W_s^{(2)}$ of the approximately circular leaf structure on the top of the stem should also be taken into account for biomass. Large spread of this umbrella like structure contributes to photosynthesis in presence of sunlight resulting in underground carbohydrate deposition of Yam. Diameter (or radius) readings of this may be taken at a point on circumference $(\theta = 0)$ through which the diameter is maximum, along with additional diameter readings at angles $\theta = \pi/3,\ 2\pi/3$; for a particular stem. Average a of these readings may be taken for diameter. Branching of leaf structure from stem in a curved manner causes depth of structure towards top of stem for accumulated biomass, i.e., high value of vertical distance (h^*) between top of leaf surface and stem before branching is an indication of high leaf mass. More curved and far the leaf structure is from the top of stem where branching starts, denser is the vegetation resulting in more leaf weight, especially at later stages of plant growth, when the matured leafs start hanging. Weight $W_s^{(2)}$ can then be approximately expressed as $W_s^{(2)} = \pi a^2 h^* d^*/4$, where d^* is leaf weight per cubic unit of length, obtainable via destructive testing. Alternatively, for a flexible leaf structure one may find area of approximately circular yam leaf structure with diameter a after straightening yam leaf lightly at the time of taking readings for a. Then multiply area by leaf weight per square unit of length (obtained via independent destructive testing) to estimate leaf weight.

A sturdy and heavy stem can hold the proportionate heavy mass on the top of it. Assume that the stem weight and the weight of leaf structure on the top are proportional to a first degree, making allowance for measurement errors; i.e., $W_s^{(1)} \propto W_s^{(2)}$.

In such a situation one may write the total vegetative mass $W = W_s^{(1)} + W_s^{(2)}$ that satisfies an approximate relation, $W \propto W_s^{(1)}$, and $W \propto W_s^{(2)}$.

In other words $W \propto \sqrt{W_s^{(1)} W_s^{(2)}}$, which leads us to conclude that W has a linear regression in logarithmic scale with the variables considered above: height of stem, girth of stem, diameter of the circular leaf structure on the top, densities d and d^*, depth h^*, etc.

To be precise, the shape of a typical yam stem is more like a truncated cone that is gradually tapered towards top from base at ground level. Volume of such truncated cone can be expressed as $V_s = \pi \times h \times (r_1^2 + r_2^2 + r_1 r_2)/3$, where r_1 and r_2 are the radius of the conical stem at base and at the top, respectively. As before, this can be approximately linearised in logarithmic scale when average girth is considered.

We have already seen that yam yield has a very high correlation with stem height in logarithmic scale. So, it is quite likely that a more precise relation may hold in linear regression of yam yield in logarithmic scale with inclusion of additional predictor variables mentioned above, as multiple regression is non-decreasing in inclusion of additional variables. Inclusion of informative variable will, however, increase the value of R^2.

Tenacity of the above assumptions made is planned to be conducted in field experiments. An affirmative test would point out that above-ground biomass is an important predictor for yield.

Under a multiplicative model that can be linearised considering logarithm, Geometric mean (G.M) is more appropriate than arithmetic mean (A.M) to represent average value. In the above, one may consider G.M. (provided none of the observations is zero), in presence of repeated measurements.

Now write the linear model in logarithmic scale as

$$\log y = a_0 + \sum_{i=1}^{p} a_i \log x_i + \epsilon \tag{1.1}$$

where y is the yield, $\{x_i, i = 1, \cdots, p\}$ are predictor variables mentioned above, $a_0, a_1, \cdots, a_p$ are constants to be determined by the method of least squares, and $E(\epsilon) = 0$; being expectation of error term. Under general conditions the least square estimates $\hat{a}_i$ are strongly consistent for $a_i, i = 0, 1, \cdots, p$; for example, see Bretona and Musiela (1987).

Thus, for the predicted value $\hat{y}$ of yield, we have

$$E(\log \hat{y}) \approx \hat{a}_0 + \sum_{i=1}^{p} \hat{a}_i \log x_i = \hat{g}(\mathbf{x}) \tag{1.2}$$

say. This provides a *lower bound* for expected yield,

$$E(\hat{y}) \geq e^{\hat{g}(\mathbf{x})} \tag{1.3}$$

by an application of Jensen's inequality, as $\log x$ is a concave function. The function $g(\mathbf{x}) = a_0 + \sum_{i=1}^{p} a_i \log x_i$ is obtained by the method of least squares.

The model $Y = e^u X$, where $u \sim N(\mu, \sigma^2)$ verified in the earlier section for yam data of the year 2010 is a special case of (1.1). Allowing an intercept term, the least square regression line for yam data of the year 2010 is

$$\log Y = 1.222024 + 0.5418401 \log X \tag{1.4}$$

In other words, based on data of the year 2010, the expected yam production at Giridih with initial seed weight X has the following lower bound.

$$E(\hat{Y}) \geq 3.39405\ X^{0.5418401} \tag{1.5}$$

Let us interpret the above result. For the lowest seed weight in conducted experiments $X = 0.2$ kg, i.e., 200g, the expected return is above 1.41 kg i.e., more than 7 times. For seed weight 350g, the expected return is above 1.92 kg i.e., more than 5.49 times, etc. For the highest seed weight viz., 800g considered in the experiments, the expected return is above 3 kg i.e., more than 3.75 times. Finally, when seed weight is 1 kg (seed weight usually do not exceed 1 kg) the expected yield is greater than 3.39 kg.

1.4 Model Sensitivity and Generalized Mahalanobis Distance

For two distribution functions F_1, F_2 having densities $f_1(x)$ and $f_2(x)$ with respect to some measure ν, the generalised Mahalanobis distance square Δ^2 is defined as

$$\Delta^2_{f_1,f_2} = -8\ \log\ d_{f_1,f_2} \tag{1.6}$$

where $d = d_{f_1,f_2} = \int \{f_1(x) f_2(x)\}^{1/2} d\nu$ is the Hellinger affinity or Bhattacharya affinity between $f_1(x)$ and $f_2(x)$; see Dasgupta (2008).

For $\alpha \in (0,1)$ one may consider a generalisation of the above affinity d as $d^{(\alpha)} = d^{(\alpha)}_{f_1,f_2} = \int f_1^{1-\alpha} f_2^{\alpha} d\nu$. Assume that the densities are separated by $\delta\theta$ in parameter space, i.e., $f_1 = f(\theta), f_2 = f(\theta + \delta\theta)$. Then, $d^{(\alpha)} = \int \left[\frac{f(\theta+\delta\theta)}{f(\theta)}\right]^{\alpha} f(\theta) d\nu$. For small $\delta\theta$, by binomial theorem with rational index one may write this as $d^{(\alpha)} \approx 1 - \frac{\alpha(1-\alpha)}{2} E(f'/f)^2 (\delta\theta)^2 = 1 - \frac{\alpha(1-\alpha)}{2} i(\theta)(\delta\theta)^2$, where $i = i(\theta) = E(f'/f)^2$ is the information in a single observation. For the class of distance $d^{(\alpha)}$, sensitivity of the variable for small change in parameter depends on information i at θ. The measure $d^{(\alpha)}, \alpha \in (0,1)$ is optimized with affinity closer to 0 at $\alpha = 1/2$, suggesting a symmetric index in f_1 and f_2, leading to Hellinger affinity.

Intrinsic accuracy of a distribution with respect to a set of parameters $(\theta_1, \cdots, \theta_p) = \Theta$ may be judged by the extent to which the distribution is altered by a small change in the value of the parameter from Θ to $\Theta + \delta\Theta$. Proceeding like the case of a single parameter θ, where we had $d^{(\alpha)} \approx 1 - \frac{\alpha(1-\alpha)}{2} i(\theta)(\delta\theta)^2$, by multivariate Taylor expansion of Hellinger affinity d with $\alpha = 1/2$ one can show

$$d_{f_\Theta, f_{\Theta+\delta\Theta}} \approx \left[1 - \frac{1}{8}(\delta\Theta) I (\delta\Theta)'\right], \quad I = (I_{ij}), I_{ij} = E\left(\frac{\partial \log f_\Theta}{\partial \theta_i} . \frac{\partial \log f_\Theta}{\partial \theta_j}\right) \tag{1.7}$$

for small $\delta\Theta$, see (5a.4.4)–(5a.4.5) of Rao (1974) where $I = I(\Theta)$ is the Fisher's information matrix. Hence from (1.6), one may write

$$\Delta^2_{f_\Theta, f_{\Theta+\delta\Theta}} \approx (\delta\Theta)I(\delta\Theta)' = \sum\sum I_{ij}\delta\theta_i\delta\theta_j \tag{1.8}$$

Thus generalized Mahalanobis distance induces a quadratic differential metric in the r.h.s of (1.8) associated with Fisher information matrix in multiparameter set-up, measuring the sensitivity of the model. A model with high sensitivity is desirable one.

The sensitivity of the variable y under the model (1.1) is given by (1.8), specified by the distribution of variables ϵ. For pits dug in nearby plots residuals ϵ may be correlated. Growth experiments on Yam at Giridih farm are conducted in a square plot consisting of 4 blocks of 5×5 Graeco-Latin squares, consisting of 100 pits in a clustered region numbered as $i = 1, \cdots, 100$. Linear model seems appropriate at logarithmic scale in observed data; one may then assume the random variables $\epsilon_i, i = 1, \cdots, 100$ in (1.1) to be multivariate normal for which Fisher information matrix in (1.8) has a nice form.

For extended Mahalanobis distance of two (asymptotically) multivariate normal distributions $N_p(\mu_1, \Sigma_1)$ and $N_p(\mu_2, \Sigma_2)$ with possibly two distinct dispersions, if the dispersion matrices are independent of parameters i.e., $\Sigma(\Theta) = \Sigma$ then (i, j)-th element of the information matrix based on difference of two sample means $(\overline{x}^{(1)} - \overline{x}^{(2)}) \rightarrow^L N_p(\mu^{(1)} - \mu^{(2)}, \frac{\Sigma_1}{n_1} + \frac{\Sigma_2}{n_2})$, assuming the ratio of sample sizes $n_1/n_2 \rightarrow 1, n = n_1 + n_2 \rightarrow \infty$, is $I_{i,j} = \Sigma^{i,j}$, where $((\Sigma^{i,j})) = \Sigma^{-1} = (\frac{\Sigma_1+\Sigma_1}{2})^{-1}$, $\Theta = \mu^{(1)} - \mu^{(2)}$. See also (3.10)–(3.13) of Dasgupta (2013a).

If the magnitude of the matrices (in terms of trace or determinant, say) are small, then this provides a sensitive model in multivariate normal set-up with possibly different dispersion matrices under null and alternative.

References

Dasgupta, R. (2008). Quality index and Mahalanobis D^2 statistics. Advances in Multivariate statistical methods. In *Proc. of ISI Platinum Jubilee Conference* (pp. 367–382.), World Scientific.

Dasgupta, R. (2013a). Optimal-time harvest of elephant foot yam and related theoretical issues. Appearing in this volume as chapter 6.

Dasgupta, R. (2013b). South pole ozone profile and lower tolerance limit. Appearing in this volume as chapter 8.

Le Bretona, A., & Musiela, M. (1987). Strong consistency of least squares estimates in linear regression models driven by semimartingales. *Journal of Multivariate Analysis, 23*(1), 77–92.

Rao, C.R. (1974). *Linear statistical inference and its applications*. New York: Wiley Eastern.

Venkatram, R., Mani, K., & Saraswathi, T. (2007). Production and marketing of Elephant Foot Yam in Salem District of Tamil Nadu. *Journal of Root Crops, 33*(2), 133–137.

Chapter 2
Some Statistical Perspectives of Growth Models in Health Care Plans

Pranab K. Sen

Abstract Growth (and wear) curve models, having genesis in epidemiology and system biology, have cropped up in every walk of life and science. In statistics, such growth curve models have led to an evolution of multivariate analysis with better performance characteristics and enhanced scope of applications in many interdisciplinary field of research. Recent advances in bioinformatics and genomic science have opened the Pandora's box with high-dimensional data models, often with relatively smaller sample sizes. Growth curve models are especially useful in such contexts. There are also other areas where growth curve model-based analyses are in high demand. In this vein, the scope and perspectives of growth models are appraised with special emphasis on some health care and health study plans.

2.1 Introduction

In exploratory studies, especially in experimental biology, developmental biology, medicine, epidemiology, socio-economics, psychology, and more recently, in biotechnology, information technology, toxico-genomics and bioinformatics, such growth models have been systematically studied under the terminology *longitudinal data models* and *repeated measurement models*; classical *growth curve models* (GCM) in simple parametric setups are regarded as precursors. Box (1950) initiated the study of *growth and wear* curves in simple biometric setups. C.R. Rao (1958, 1965) made significant contributions to GCM while Potthoff and Roy (1964) systematically integrated GCMs in the main stream of *multivariate analysis of variance* (MANOVA) and linked it to *multivariate analysis of covariance* (MANOCOVA). Rao (1959) developed procedures for parameter estimation and

P.K. Sen (✉)
Departments of Biostatistics, and Statistics & Operations Research, University of North Carolina, Chapel Hill, NC 27599-7420, USA
e-mail: pksen@bios.unc.edu

R. Dasgupta (ed.), *Advances in Growth Curve Models*, Springer Proceedings in Mathematics & Statistics 46, DOI 10.1007/978-1-4614-6862-2_2,

estimation of confidence bands for the response curve. Cole and Grizzle (1966) and Grizzle and Allen (1969) formulated some innovative statistical analysis of growth and dose response curves. Geisser (1970, 1981) annexed Bayesian methodology in GCMs. Khatri (1966) elaborated the connection of GCM and MANO(CO)VA. Timm (1981) and Zerbe and Walker (1977) studied further MANO(CO)VA of repeated measurement designs incorporating the basics of parametric GCM. This also laid down the foundation of *random-effects* and *mixed-effects* models in MANOCOVA. In some medical (and dental) research problems, often, *area under the curve* (AUC) has been used. It is also possible to relate AUC to GCM and obtain better performance of statistical tests and estimates (Preisser et al. 2011).

In most of these developments, as has been systematically accounted in Gnanadesikan et al. (1971), it has been tacitly assumed that (i) the underlying response variables are continuous, (ii) additivity of the effects hold, and (iii) the underlying probability distributions are all multivariate normal. The latter assumption is accompanied by the homogeneity of their dispersion matrices, the so-called, *homoscedasticity* condition in a general multivariate setup. Both the linearity of the model and multinormality of the errors have been critically appraised in the past 50 years, raising concern of the scope of adaptability of normal MANOCOVA models in various applications. Nonparametric (mostly based on marginal ranks) MANO(CO)VA evolved during the 1960s and reported in Puri and Sen (1971, 1985). For GCM, such nonparametrics have been incorporated by Ghosh et al. (1973) and Sen (1973, 1985), among others. The past three decades have witnessed the development of semi-parametric GCM and longitudinal data models. Both spatial and temporal variations are accounted in such models.

In epidemiology, epidemic models are earlier examples of growth models. The growth of a disease or disorder (in a population) follows another track of discrete GCM where typically the response variable is the number of infected people or their proportion in the target population. In population dynamics, such discrete GCM are commonly perceived wherein various demographic features account for explanatory or design variables. For example, for the HIV afflicted population in a *spatiotemporal* setups, discrete GCM are quite appealing, albeit the multi-normality or the linearity of effects assumption may not be reasonable. In system biology, for example, the growth of a tumor or spread of cancerous cells, growth (curve) models are very appealing, albeit they come under high-dimensional or functional data clouds. The classical fMRI models also pertain to growth models, although the commonly assumed multi-normality condition may not be generally tenable in such contexts. In many stochastic models, such as the diffusion process, birth and death process, and morbidity (illness) process, such GCM may appear, not only with some longitudinal or temporal features of the expectation parameters but also with subtle change in the shape or dispersion parameters. For example, the drift versus dispersion in generalized random walk models. From white noise to signal detection in high dimension (as related to chaos theory) is another example of this sort. Markov processes have also been atuned to GCM with appropriate growth condition on the failure rate or reliability functions. Nonhomogeneous Poisson processes and their natural extension to *doubly stochastic Poisson processes* bear growth features in a stochastic mode. Also, GCM in HIV (AIDS) models come in a completely different setup. It is therefore perceived that conventional MANO(CO)VA-related GCM may not be universally adaptable in many other fields of applications. *Beyond parametrics* in GCM is therefore a natural avenue to traverse.

There is a class of scenarios of growth (or decay) models which are characterized by evolutionary growth or decay but subject to extraneous restraints. For example, in a *branching process* model the outcome variable may lead to an extinction if it reaches the absorbing state (0). On the other hand it can explode to an infinite state under plausible conditions on the branching parameters. In some toxicological models, such growth patterns may have similarity with GCMs but are subject to suitable upper bounds due to experimental constraints. For example, when the output variable attains an upper threshold level, the system moves to a different stage, and a new process model comes in the picture. This may be regarded as a GCM annexation to the classical *change-point model* which typically relates to either a change in location (regression) or scale parameter at an unknown time-point. The problems is much more complex in this general growth model. A typical example is the growth of HIV-AIDS afflicted population following some break-through medical intervention. For growth models with a finite upper bound, the classical Gompertz (1825) model, motivated by a distribution function to fit mortality tables is a precursor to other models such as the logistic model and its ramifications (Johnson and Kotz 1970). With this genesis, logistic regression models pertain to stochastic growth curves in more general formulations. Likewise, Poisson regression models pertain to such GCM (Sen et al. 2010).

This volume has a primary emphasis on GCM in conventional agricultural setups with emphasis on the *elephant foot yam*. In modern interdisciplinary research, typically, high-dimensional data models are encountered where some times the sample size may be relatively smaller, thus giving rise to the so-called *high dimension low sample size* (HDLSS) models. This is particularly the case with bioinformatics and toxico-genomics studies. Even in many socioeconomic investigations, HDLSS models are encountered in very nonstandard setups. The scope for traditional MANOCOVA tools in HDLSS has been critically appraised in the recent past (Sen 2006, 2008). In this context, the dimension reduction can be effectively done with appropriate GCM in beyond parametrics setups (Sen et al. 2007). Nevertheless, conventional statistical tools are of very limited utility in such HDLSS-related GCM setups. This study focuses on some high-dimensional models arising in some socioeconomic research problems where GCM may have a natural appeal. The nest section is devoted to the preliminary notion on the evolution of GCM from simple parametric to beyond parametric setups, encompassing HDLSS models as well. Some of these beyond parametrics perspectives are elaborated in Sect. 2.3. The main results on GCM approach on some general socioeconomic models are disseminated in Sect. 2.4. The concluding section is devoted to some general observations and remarks.

2.2 Preliminary Notion

Typically, GCM relates to some multi-sample or blocked design models where (in a general setup), there are n observations, each observation has p characteristics and observed at q time points. For example, in an environmental health hazard study for identifying environmental dioxin pollution (Chen et al. 2012), *finger-print analysis* comparing the polychlorinated *dibenzo-p-dioxin* and *dibenzofuran* (PCDD/F) congener profile patterns of collected samples with those of potential dioxin emission source(s), has been advocated as an important tool. There are p (= 17) PCDD/F congeners comprising a fingerprint and data collected in a longitudinal setup. This typically relate to a MANOVA model, albeit the sample sizes are small, and moreover, the underlying distributions are distinctly not multivariate normal; multivariate gamma distributions appear to be more reasonable in this setup for which the dispersion matrix depends on the mean levels and shape parameters, and hence, the homogeneity of the dispersion matrices may not hold. For a stochastic p-vector $\mathbf{X}$ following a multivariate normal distribution with mean vector $\boldsymbol{\mu}$ and dispersion matrix $\boldsymbol{\Sigma}$, in a conventional setup, it is tacitly assumed that the dispersion matrix does not depend on the mean vector. This basic assumption may not generally hold in GCM where heteroscedasticity, possible collinearity and nonlinear relationship of dispersion matrix and mean vectors may mar the simplicity of the standard GCM analysis schemes. In the above cited PCDD/F model, we have nonnegative component variables which brings the relevance of *compositional data models*. If the p coordinate variables of $\mathbf{X}$ are independent gamma variables with shape parameters $\alpha_1, \cdots, \alpha_p$ respectively (all positive), and a scale parameter $\nu(> 0)$, and if we define the proportion vector as $\mathbf{Y} = (\mathbf{X}'\mathbf{1})^{-1}\mathbf{X}$, then $\mathbf{Y}$ has the Dirichlet distribution whose mean vector and the (singular) covariance matrix depend on the scale as well as shape parameters. This particular feature not only renders a singular covariance matrix but also invalidates the routine adaption of the so-called *principal component model* PCM) or *canonical correlation* analysis. Bearing in mind such examples, we first consider a simple GCM and motivate more general ones arising thereof.

Let $t_{i1}, \cdots, t_{iq}$ be the time points for the ith subject and let

$$\mathbf{Y}_i = ((Y_{ijk}))_{j=1,\cdots,p;k=1,\cdots,q},\ i = 1, \cdots, n, \tag{2.1}$$

where $Y_{ijk} = Y_{ij}(t_{ik}), k = 1, \cdots, q$. In a balanced design, $t_{ik} = t_k, \forall i = 1, \cdots, n; k = 1, \cdots, q$. In a conventional parametric setup, it is typically assumed that $\mathbf{Y}_i$ has a matrix-valued multi-normal distribution with unknown mean (matrix) $\boldsymbol{\Theta}_i$ and unknown dispersion matrix $\boldsymbol{\Gamma}$ (of order $pq \times pq$), for $i = 1, \cdots, n$. It is thus tacitly assumed that the dispersion matrix $\boldsymbol{\Gamma}$ is common for all observations; this condition as noted earlier is violated for multivariate gamma and other non-normal distributions. In a conventional GCM setup,

$$\boldsymbol{\Theta}_i = \boldsymbol{\nu}\mathbf{A}_i + \mathbf{B}\mathbf{C}_i,\ i = 1, \cdots, n, \tag{2.2}$$

where $\boldsymbol{\nu}$ is a $p \times k$ matrix of unknown (intercept) parameters, $\mathbf{A}_i$ are matrices of known design variables (constants), $\mathbf{C}_i$ is a $r \times q$ matrix of known constants, and $\mathbf{B} = ((\beta_{jl}))$ is a $p \times r$ matrix of unknown parameters. In the k sample model, the jth column of $\mathbf{A}_i$ is equal to $\mathbf{1}$ and other columns are $\mathbf{0}$, according the ith observation belongs to the jth sample or not. In more complex designs, the choice of the known $\mathbf{A}_i$ and k depends on the design matrix. Typically, $r < q$ (to facilitate dimension reduction in a GCM setup). In the balanced design case, the $\mathbf{C}_i$ are all equal. By (2.1) and (2.2), we have

$$\mathbf{Y}_i = \boldsymbol{\nu}\mathbf{A}_i + \mathbf{B}\mathbf{C}_i + \mathbf{E}_i, i = 1, \cdots, n, \tag{2.3}$$

where the $\mathbf{E}_i$ are independent and identically distributed random matrices with null mean and dispersion matrix $\boldsymbol{\Gamma}$. One may use the vec notation to convert these $\mathbf{Y}_i$ into pq-vectors and then apply the usual MANOVA tools to draw statistical conclusions on $\mathbf{B}$. Typically, r is much smaller than q, and hence, the GCM approach works out well in having a more powerful statistical analysis when the postulated model in (2.2) holds. We refer to Rao (1965) and Gnanadesikan et al. (1971) for a systematic account of these developments.

In most of the fields of application, be it in biometry or clinical trials, system biology or bioinformatics, or the vast area of modern interdisciplinary research, usual MANOVA model assumptions are mostly untenable. In multivariate normal models, the covariance matrix is functionally independent of the mean vector, but this is not generally true for other multivariate distributions. We may refer to the fingerprint analysis problem where not only multi-normality assumption may be dubious but also homogeneity of the dispersion matrices is untenable. In the univariate setup, the Box and Cox (1964) transformation has been widely used to achieve approximate linearity of the model and improve normality approximation of such transformed variables. However, such nonlinear transformations while improving the normality approximation may adversely affect the underlying additivity structures as well as the homoscedasticity assumption. In some simple univariate models, Bartlett variance stabilizing transformations work out well. But such transformations are of not much help in stabilizing the dispersion matrices. For example, in multivariate gamma distributions, the dispersion matrix may functionally depend on the mean vector and hence the homoscedasticity condition may not hold. Because of these impasses, in multivariate GCM, in beyond parametrics approaches, some alternative analysis schemes are advocated; these are to be considered in the next section.

2.3 Beyond Parametrics Formulations

Whereas in parametric GCM, conventionally, it is assumed that the error distributions are (multi-)normal, in beyond parametrics, not only this multi-normality assumption is deemphasized but also other robustness issues are appropriately appraised. In this perspective, first consider the conventional parametric models.

In the balanced design case, all the $\mathbf{C}_i$ are the same, and without loss of generality it may be assumed that they are of rank $r(\leq q)$. We make a similar assumption for the general unbalanced case as well. Let us then consider a set of $q \times q$ matrices $\mathbf{L}_i$ and partition it as

$$\mathbf{L}_i = (\mathbf{L}_{i1}, \mathbf{L}_{i2}), \ i = 1, \cdots, n, \tag{2.4}$$

where

$$\mathbf{L}_{i1} = \mathbf{C}_i'(\mathbf{C}_i\mathbf{C}_i')^{-1} \tag{2.5}$$

is of order $q \times r$ and $\mathbf{L}_{i2}$ is of order $q \times (q - r)$ for $i = 1, \cdots, n$. Let then

$$\mathbf{Z}_i = \mathbf{Y}_i\mathbf{L}_i = (\mathbf{Z}_{i1}, \mathbf{Z}_{i2}), \tag{2.6}$$

for $i = 1, \cdots, n$. As in Potthoff and Roy (1964) and Rao (1965), we note that the $\mathbf{BC}_i\mathbf{L}_{i1} = \mathbf{B}$ while we choose the $\mathbf{L}_{i2}$ in such a way that $\mathbf{C}_i\mathbf{L}_{i2} = \mathbf{0}$ are null matrices of order $r \times (q - r)$. By (2.3) and (2.6), we have on writing $\mathbf{E}_i^* = \mathbf{E}_i\mathbf{L}_i = (\mathbf{E}_{i1}^*, \mathbf{E}_{i2}^*)$ and $\mathbf{A}_i^* = \mathbf{A}_i\mathbf{L}_i$,

$$\mathbf{Z}_i = \boldsymbol{\nu}\mathbf{A}_i^* + \mathbf{B}(\mathbf{I}_r, \mathbf{0}) + (\mathbf{E}_{i1}^*, \mathbf{E}_{i2}^*), \tag{2.7}$$

for $i = 1, \cdots, n$. This perfectly fits in to a MANOCOVA model which under the multi-normality condition has been thoroughly studied in the literature (Rao 1965, and others).

In a simple nonparametric approach (Puri and Sen 1971), it is assumed that the $\mathbf{E}_i^*$ have jointly a pq variate continuous distribution, for all $i = 1, \cdots, n$. Then linear rank statistics are constructed for each of the pq coordinates of the $\mathbf{Z}_i$, $1 \leq i \leq n$ of which pr statistics relate to the case where $\mathbf{B}$ is present in addition to the partitioned part of $\boldsymbol{\nu}\mathbf{A}_i^*$, while the remaining $p(q - r)$ linear rank statistics relate to the part where $\mathbf{B}$ does not appear but the complementary part of $\boldsymbol{\nu}\mathbf{A}_i^*$ appears. If the null hypothesis of relates to $H_0 : \mathbf{B} = \mathbf{0}$, i.e., no regression on the time points, then we can proceed in two ways. From the first part, we use the R-estimators of $\mathbf{B}$ as in Jurečková and Sen (1996) and use a Wald-type test statistic. Alternatively, assuming $\mathbf{B} = \mathbf{0}$, we estimate $\boldsymbol{\nu}$ from the entire set of linear rank statistics. In the second place, we align the $\mathbf{Z}_i$ by using these R-estimators of $\boldsymbol{\nu}$, and on the aligned linear rank statistics for the $p \times r$ sub-matrix, we construct an aligned rank MANOCOVA test statistic as in Puri and Sen (1971, 1985) wherein the $p(q-r)$ linear rank statistics are treated as covariate statistics. Such tests are based on the Chatterjee and Sen (1964) *rank permutation* principle and are conditionally distribution-free under hypotheses of invariance. For large sample sizes, under these hypotheses of invariance, they have approximately chi-square distribution with appropriate degrees of freedom. Being based on the marginal ranks of the $\mathbf{Z}_i$, $i = 1, \cdots, n$, these tests are robust against plausible model departures.

A second approach is based on *rank* (or *R*-) *estimators* of $\boldsymbol{\nu}, \mathbf{B}$ from individual observations. Note that by (2.7), for each $i(= 1, \cdots, n)$, we can obtain linear estimates of $\boldsymbol{\nu}$ and $\mathbf{B}$ from $\mathbf{Z}_i$. Given these n independent estimators of $\mathbf{B}$, it may be possible to use the weighted least squares methodology to obtain a combined sample estimator of $\mathbf{B}$ and also, to estimate its dispersion matrix (of order $pr \times pr$). Tests for suitable hypotheses on $\mathbf{B}$ can be based on these estimates using the classical Wald statistics. Such tests are, however, generally not robust due to the poor robustness properties of the estimated dispersion matrix. On the other hand, based on these $\hat{\mathbf{B}}_i, i = 1, \cdots, n$ (all of the order $p \times r$), the general theory of R-estimators developed in detail in Jurečková and Sen (1996) can be incorporated to obtain robust estimators of $\mathbf{B}$ and also to test for suitable hypotheses on $\mathbf{B}$.

We illustrate this methodology with a simple situation where the n observations can be regarded as the composite of $k(\geq 2)$ samples of sizes $n_1, \cdots, n_k$, respectively, so that $n = \sum_{s=1}^{k} n_s$. For an observation from the sth sample, referred to (2.2), the $\mathbf{A}_i$ are equal to some $\mathbf{A}_s$, for $s = 1, \cdots, k$. In this setup, $t = p + k$ where the additional k relates to the individual population effects (vectors). As such, we may proceed to test for the null hypothesis that the k columns of $\boldsymbol{\nu}$ are the same, treating $\mathbf{B}$ as a nuisance parameter (matrix). Thus, using R-estimators of $\mathbf{B}$, we may use aligned rank test based on the $k \times q$ linear rank statistics. We refer to Puri and Sen (1985) and omit the details. Alternatively, if the null hypothesis relates to $\mathbf{B} = \mathbf{0}$ (i.e., no regression over time), treating $\boldsymbol{\nu}$ as nuisance, then one can use aligned rank statistics. The dimension reduction (from pq to pr when $r << q$) generally leads to increased statistical precision.

A further source of concern is the very basic assumption of linear models in GCM. It is not uncommon in toxicology and *physiologically based pharmacokinetics* (PBPK) models to have distinct nonlinear GCM where even if normality assumption can be approximately justified, the homogeneity of the error variances may not be tenable. Further, in PBPK and certain systems models, often (stochastic) differential equations (SPDE) are incorporated to explain better the underlying kinetics. In such a case, typically, the response pattern is nonlinear and multidimensional (viz., Mandal et al. 2012). Though such nonlinear systems are often approximated by linear ones (under the usual delta method), the reliability and validity of such linearization may be open to questions. In PBPK modeling, the composite response is the synergic and chain body resistance and metabolic changes through a number of organs along with their impact on the blood circulation system. As such, a multicomponent model is usually advocated, though in most of the mathematical modeling, for drawing statistical conclusions a simplistic approach is considered. A suitable growth model connecting the impact of these organs in relation to the body reaction to external stimulus will certainly be a better solution. The GCM approach has therefore a natural appeal in this context.

In HDLSS models, though it may be tempting to use *projection pursuit* for dimension reduction, its scope may be limited to distributions admitting linear structure so that the classical *principal component model* (PCM)-based statistical methodology is appropriate. In many contexts, this linear manifold is not tenable and hence this approach may not be adaptable either. In some simple models, this approach was elaborated by Sen (1973). There is scope for expansion in more general GCM (viz., Preisser et al. 2011). They treated the *Gingivitis* problem resulting from absenting from brushing or flossing of tooth. It had 7 time points consisting of the *induction* and *resolution phases* and 31 biomarkers on 22 subjects. Although the area under the individual subject and biomarker data were initially considered, the number of subjects (22) being smaller than the number of data points (186), conventional MANOVA tools were not adaptable. Moreover, the assumption of multinormality was difficult to justify. First, in the usual way in GCM, a dimension reduction was suggested, resulting in 4 response variables for each biomarker. Second, signed-rank tests were used in the univariate as well as multivariate setups (viz., Sen and Puri 1967), resulting in more robust procedures. Further, the Chen–Stein theorem (Chen 1975) was adapted to control Type I error and related measures. This produced a better inferential procedure.

In the environmental pollution problem (Chen et al. 2012), the GCM appeal was overwhelming. However, in that *fingerprint* analysis comparing the *poly-chlorinated dibenzo-p-dioxin* and *dibenzofuran* (PCDD/F) congener profile patterns of collected samples with those of potential dioxin emission sources, there were 17 PCDD/F congeners comprising a fingerprint which did not look like to have multi-normal distribution. They differ in their emission rate and exposure pattern, and hence, it was decided to have the proportion of these 17 compounds relative to their sum. This has of course given rise to a response vector on a 16-simplex (in a *compositional data model*) for which the variance-covariance matrix is intricately dependent on the mean vector and a multi-normal distribution is far from being tenable. Based on plausible assumptions, multivariate gamma-type distributions were thought to be more appropriate. That led to the so-called *Dirichlet* type distribution. A discouraging feature is that the dispersion matrix for this multivariate random vector depends not only on the mean vector but also on the shape parameters of the underlying gamma distributions. As such, conventional growth curve model-based analysis was not pursued. It turned out that the usual procedure based on multivariate ranks (Puri and Sen 1971) have much more robustness perspectives, thus performing better than the multi-normality-based likelihood ratio type tests. Thus, beyond parametrics seems to have a better appeal.

A third illustration relates to rank analysis of covariance (R-ANOCOVA) in some nonstandard data models (Sen et al. 2013). There, the R-ANOCOVA has been extended to a more general class of linear or nonlinear models (including measurement errors or misspecified models). This would make GCM for such more nonstandard cases manageable under beyond parametric schemes.

2.4 GCM in Health Care Studies

The development and management of a health care plan is a global problem, albeit drastically different from one country to another, or even within a country, from one region to another. A health care plan may either pertain to a general (overall) population or certain subclass, termed a target population, demarcated by various socioeconomic or demographic features. Some of these features are qualitative or categorical while some others are quantitative. A health care plan is designed to assess the need for welfare or financial support for the target population for needy people when inflicted with certain type of disease or disorder. Of course, to run that, it needs sustainable funding through health insurance, government support, and other resources. Therefore, it is needed to have a complete inventory of diseases and disorders which are to be covered under the health care plan. It also needs to assess the available resources to cover the cost of providing health care for the target population. Such resources not only include the financial aspects but also the availability of ambulatory care personnel and facility, medical and paramedical personnel, general awareness of the population for some of the pertinent health hazards, lifestyle of the population concerned, and a thousand and one other associated factors, some of which may not even be properly ascribable. In this respect, the *quality of life* (QoL) and general attitude towards life have an important bearing too. There is naturally a temporal factor that relates to the adequacy or deterioration of a health care plan over time as is commonly perceived in many countries (Sen 2012). The growth of population susceptible to various diseases and disorders, by sector, spatial and temporal factors, the temporal change in the enrolment and compliance to a health care plan, growth of various burdens of disease (including virus mutations which may alter the nature of some of these diseases), (mal-)nutrition, poverty and affluence and other factors have significant bearing on such health care plans.

In most of the countries in the Western Europe, the social welfare system provides a significant support to available health care plans, although such schemes are difficult to implement in developing countries, especially, the over-populated ones, including the Indian subcontinent and China. The burden of population and the vast inequality of wealth and living standards create impasses for a unified health plan that could suit equally well the people from all walks of life. In capitalistic countries, USA is no exception, a health care insurance plan is not affordable across the various sectors of the population, and no wonder, still a big number of people are deprived of equitable health care insurance and facilities. The prevalence of certain diseases or disorders can impact a health care plan drastically. For example, diabetes is a major concern in India, China, and many other countries where consumption of carbohydrates is significantly higher. In this respect, the familial or genetic effects are very much noticeable. Breast cancer is more likely for daughters of mothers who has had such affliction. The fast changing lifestyle of a major sector of population be it in the West or in the third world countries is having an impactful aftermath on many cardio-vascular diseases. Hypertension is another big concern. On top of that, HIV (AIDS) has become a global threat, and all over the world, is having a huge toll in terms of mortality and morbidity. Arthritis and gout affect a significant part of any population, especially at golden ages. For cholera, quite prevalent in the coastal areas of the Indian subcontinent, it has been observed that there has been a mutation in the microbes which can now fight back many of the drugs (salines) which were quite effective a few years ago. Arsenic contamination of ground water is a major health concern in a vast coastal area in the eastern part of India as well as the entire southern part of Bangladesh. Most of the working class people have their daily need of drinking, cooking, washing clothes and dishes too, and even bathing, intake a perceptible amount of arsenates which may not only have carcinogenic impact on their skin, hands and feet but also have impactful effect on their ingestion system. Combined with that improper disposition of human waste adds more misery to this contamination. Dementia, Parkinson's disease, and Alzheimer may be occurring at a higher incidence rate. Smoking and lung cancer may be good relation although they have not been linked causally. Environmental smoking effect is a significant health hazard, more so in metropolitan areas where automobile exhausts contribute liberally to this pollution. In any composite health care plan, the galaxy of diseases and disorders need to accounted for, although the prevalence pattern and relative cost for cure could be quite dissimilar.

The models discussed in Sect. 2.3 can be adapted for such health care plans. However, a much more complex and interacting modeling is necessary. First and foremost, let $\mathbf{D}(s,t)$ stand for the galaxy of diseases or disorders, at time $t,\ t \in T,\ s \in \mathcal{S}$, which are to be covered under the plan. Here T stands for the time domain and S stands for the domain of other spatial as well as explanatory variables. Secondly, some of the diseases or disorders are chronic and have long-range impact, while some others are relatively short duration with a (stochastically) much smaller in-disease period. Therefore, it may be better to include statistical information on the *time under treatment or service* of various diseases and disorders. In this respect, the age at onset, duration of the service and the level (ambulatory, in-house assistance or hospitalization) distributions are needed to be charted. The prevalence of various diseases or disorders may vary considerably across the demographic and economic strata of an overall population. The coverage of health plans may also depend on such socioeconomic strata. Thus, we will have a multi-dimensional stochastic vector, say, $\mathbf{W}(s,t),\ t \in T,\ s \in \mathcal{S}$ wherein all the other information are to be included as covariates. There may be a growth of prevalence of the diseases or disorders (in some cases the opposite way), and the information on available (para-)medical or clinical help and the associated cost analysis all are needed for an in-depth assessment. The (age-specific) *life expectancy* as further categorized by sex, ethnicity, and other demographic features can be viewed as a very useful piece of information in this respect. This needs development of a suitable index of *health status* of individuals covered under the health care system that can be incorporated in the formulation of a general *exposure risk* measure $\mathbf{R}(s,t),\ t \in T,\ s \in \mathcal{S}$ whose distribution over the target population constitutes an essential component of a stochastic modeling of the overall picture. Some other factors like most of the high-cost surgeries need to be attuned to a possible health plan in such a way that a complete coverage may push up the cost factor so much that on cost ground such plans may not be affordable for a greater part of the society. Therefore, sustainability and afford-ability issues are to be weighed in objectively so as to make a plan adaptable. That also needs statistical modeling.

No plans can be sustained without a complete provision of funding through health insurance, cost sharing by the patients and government or other funding. An accounting of the relative support, their potential change over time and their matching the cost of providing the health care service is therefore desirable before any undertaking can be planned. As such, we have a complex of variable, some being response variables while others as covariates or explanatory variables, and statistical appraisal of this picture is a prerequisite. This is needed to model a composite *cost-factor* analysis based on a stochastic time-dependent $\mathbf{C}(s,t), s \in \mathcal{S},\ t \in T$ which are to be attuned to the other stochastic matrices described before.

Statistically speaking, we need to have the collection of stochastic systems:

$$(\mathbf{D}(s,t), \mathbf{W}(s,t), \mathbf{R}(s,t), \mathbf{C}(s,t)),\ t \in T,\ s \in \mathcal{S}, \tag{2.8}$$

which are to be incorporated in to a growth model for a composite model. It is also necessary to account for $U(s,t), t \in T,\ s \in \mathcal{S}$, the cost for providing health care contrasted with the resources to match that factor. This is intricately related to fixation of the health insurance premiums, projection of clinical and medical personnel cost and revenue sharing from other sources. Even in USA and other developed countries in the West, the escalating health care cost is a nightmare for concerned administrations; the problem is undoubtedly much more complex in the Indian sub-continent and China. This is highly a nonlinear system, and routine use of standard MANOCOVA or GCM may be grossly inappropriate. It may be appealing to incorporate some SPDE (as in the PBPK models). However, given the usual assumptions of white noises following suitable Gaussian laws in such SPDE, it could be difficult to formulate computationally manageable methodological justifications of SPDE sans those Gaussian components.

An essential feature of these stochastic processes is that they are not stationary even in a very broad sense. Time dependence of not only the basic marginal functions but also their association structures may generally cause tremendous roadblocks to implement standard GCM models even in a component-wise formulation. Generally, these stochastic processes have some tendency to acquire some aggregative effects, resulting in usually nonlinear trends. Thus detrending is an essential task. In the presence of nonlinear trends, usual parametric models may not only be inadequate but also too irrelevant. Beyond parametrics approaches based on *wave-length* methodology and *nonparametric smoothing* are therefore advocated. That may invariably need relatively much larger sample size and could run into cost constraints. It seems that taking into account the basic extraneous factors a multidimensional, nonstationary, and non-Gaussian process with appropriate systematic factors (most relevant to the GCM) can only be dome in a more nonparametric setup with adherence to local (sub- or semi-)martingale features may lead to more meaningful resolutions. The basic issue may be can there be sufficient statistical validation and interpretation of data collection and monitoring to induce the impact of GCM in this largely exploratory field?

2.5 Concluding Remarks

It is indeed a challenge, especially in the developing countries, to collect reliable data sets pertaining to the detailed statistical perspective as listed in the preceding section. In most of the cases, there may be data sets pertaining to marginal morbidity and mortality rates due to various (competing) causes such as the major diseases or disorders but not that much of their synergic effects, and on top of that, very little information on the health care facilities, insurance coverage, actual illness and disease-free state sojourns, cost of services and individual health insurance premium, etc. In health care and health services, especially for the senior people, composite impact of more than one disease or disorder needs to be investigated. This information can only be obtained through intensive sample surveys. The sampling frame, cost of sample survey, adequacy of sample size information, possible adjustment for non-responses, and the need for follow-up sampling, all are to be formulated in a sound statistical manner. Collecting the relevant information from census or official publications is likely to be grossly incomplete. In USA and some other countries, the Bureau of Census, regularly conducts sample surveys to update the census figures and collect some additional information. Still, they are not enough to chart out the whole complex of growth models presented in Sect. 2.4. In the Indian subcontinent, possibly the State Statistical Bureaus and the Central Statistical Organization can undertake a network of sampling scheme but would probably require statistical expertise to do it in depth and in a valid way to match the need of the general objectives of health care plans and health study protocols. There has been a sustained development of statistical thinking in public health (Sen and Rao 2000) but their adaption in health care system is one step further that requires immediate attention. My feeling is that this is a more complex problem beyond the reach of these organizations present state of activities. On top of that some other public health enterprises in India may not have the expertise and resources to undertake such schemes. It is my hope that given such exploratory studies, the implementation of actual health care plans will be facilitated. A much more detailed statistical study is indeed needed and intended in the near future.

References

Box, G.E.P. (1950). Problem in the analysis of growth and wear curves. *Biometrics, 6*, 362–389.

Box, G.E.P., & Cox, D.R. (1964). An analysis of transformations. *Journal of the Royal Statistical Society, Series, 26*, 211–252.

Chatterjee, S.K., & Sen, P.K. (1964). Nonparametric tests for the bivariate two-sample problem. *Calcutta Statistical Association Bulletin, 22*, 13–50.

Chen, C.C., Sen, P.K., & Wu, K.-Y.(2012). Permutation tests for homogeneity of fingerprint patterns of dioxin congener profiles. *Environmetrics*, in press.

Chen, L.H.Y. (1975). Poisson approximation for dependent trials. *Annals of Probability, 3*, 534–545.

Cole, J.W.L., & Grizzle, J.E. (1966) Applications of multivariate analysis of variance to repeated measurements experiments. *Biometrics, 22*, 810–828.
Geisser, S. (1970). Bayesian analysis of growth curves. *Sankhya, Ser. A, 32*, 53–64.
Geisser, G. (1981). Growth curve analysis. In P.R. Krishnaiah (Ed.) *Handbook of Statistics, Vol. 1: Analysis of Variance* (pp. 89–115). Amsterdam: North Holland.
Ghosh, M., Grizzle, J.E., & Sen, P.K. (1973). Nonparametric methods in longitudinal studies. *Journal of the American Statistical Association, 68*, 29–36.
Gnanadesikan, R., Srivastava, J.N., Roy, S.N., Foulkes, E.B., & Lee, E.T. (1971). *Analysis and design of certain quantitative multiresponse experiments*. New York: Pergamon Press.
Gompertz, B. (1825). *Philosophical Transactions of the Royal Society, Ser. A, 115*, 513–580.
Grizzle, J.E., & Allen, D.M. (1969). Analysis of growth and dose-response curves. *Biometrics, 25*, 357–381.
Johnson, N.L., & Kotz, S. (1970). *Distributions in statistics: continuous univariate distributions*. New York: Wiley.
Jurečková, J., & sen, P.K. (1996). *Robust statistical procedures: asymptotics and interrelations*. New York: Wiley.
Khatri, C.G. (1966). A note on a MANOVA model applied to problems in growth curves. *Annals of the Instutute of Statistical Mathematics, 18*, 75–86.
Mandal, S., Sen, P.K., & Peddada, S.D. (2012). Statistical inference for dynamic systems governed by differential equations with application to toxicology. (under preparation).
Potthoff, R.F., & Roy, S.N. (1964). Generalized multivariate analysis of variance model useful especially for growth curve problems. *Biometrika, 51*, 313–326.
Preisser, J., Sen, P.K., & Offenbacher, S. (2011). Multiple hypothesis testing for experimental gingivitis based on Wilcoxon signed rank statistics. *Statistics in Biopharmaceutical Research, 3*, 372–384.
Puri, M.L., & Sen, P.K. (1971). *Nonparametric methods in multivariate analysis*. New York: Wiley.
Puri, M.L., & Sen, P.K. (1985). *Nonparametric methods in general linear models*. New York: Wiley.
Rao, C.R. (1958). Comparison of growth curves. *Biometrics, 14*, 1–16.
Rao, C.R. (1959). Some problems involving linear hypotheses in multivariate analysis. *Biometrika, 46*, 49–58.
Rao, C.R. (1965). Theory of least squares when the parameters are stochastic and its application to the analysis of growth curves. *Biometrika, 52*, 447–458.
Sen, P.K. (1973). Some aspects of nonparametric procedures in multivariate statistical analysis. In D.G. Kabe, & R.P. Gupta (Eds.) *Multivariate statistical analysis* (pp. 231–240). Amsterdam: North Holland.
Sen, P.K. (1985). Nonparametric procedures for some miscellaneous problems. In P.R. Krishnaiah, & P.K. Sen (Eds.) *Handbook of statistics, Vol. 4: Nonparametric methods* (pp. 699–739). Amsterdam: North Holland.
Sen, P.K. (2006). Robust statistical procedures for high-dimensional data models with applications to genomics. *Austrian Journal of Statistics, 35*, 197–214.
Sen, P.K. (2008). Kendall's tau in high-dimensional genomic parsimony. In *Institute of mathematical statistics, collection series, vol. 3* (pp. 251–266).
Sen, P.K. (2012). Development and management of national health plans: Health economics and Statistical perspectives. In Y.P. Chaubey (Ed.) *Some topics on current issues in mathematical and statistical methods*. Singapore: World Scientific Press.
Sen, P.K., Singer, J.M., & Pedroso de Lima, A.C. (2010). *Finite Sample to Asymptotic Methods in Statistics*. Cambridge University Press, New York.
Sen, P.K., Jurečková, J., & Picek, J. (2013). Rank tests for corrupted linear models. *Journal of the Indian Statistical Association, 51*(1), in press.
Sen, P.K., & Puri, M.L. (1967). On the theory of rank order tests for location in the multivariate one sample problem. *Annals of Mathematical Statistics, 38*, 1216–1228.
Sen, P.K., & Rao, C.R. (Eds.) (2000). *Handbook of statistics, vol. 18: Bioenvironmental and public health sciences*. Amsterdam: North Holland.

Sen, P.K., Tsai, M.-T., & Jou, Y.S. (2007). High-dimension low sample size perspectives in constrained statistical inference: The SARSCOVRNA genome in illustration. *Journal of the American Statistical Association, 102*, 685–694.

Timm, N.H. (1981). Multivariate analysis of variance of repeated measurements. In P.R. Krishnaiah (Ed.) *Handbook of statistics, vol. 1: Analysis of variance* (pp. 41–87). Amsterdam: North-Holland.

Zerbe, G.O., & Walker, S.H. (1977). A randomization test for comparison of groups of growth curves with different polynomial design matrices. *Biometrics, 33*, 653–659.

Chapter 3
Testing of Growth Curves with Cubic Smoothing Splines

Tapio Nummi and Nicholas Mesue

Abstract In this paper we present a novel method for testing growth curves when the analysis is based on spline functions. The new method is based on the use of a spline approximation. For the approximated spline model an exact F-test is developed. This method also applies under a certain type of correlation structures that are especially important in the analysis of repeated measures and growth data. We tested this method on the glucose data of Zerbe (J Am Stat Assoc 74:215–221, 1979) and also investigated it by simulation experiments. The new method proved to be a very powerful modeling and testing tool especially in situations, where the growth curve may not be easy to approximate using simple parametric models.

3.1 Introduction

Longitudinal research has an important role in various fields of science, for example in medicine, economics, social sciences, and engineering. The aim is to analyze the change caused, e.g., by growth, degradation, maturation, and ageing when individuals are followed over time or according to some other ordered sequence of measurements. In this paper the focus is on complete and balanced data. One of the most important statistical models for these data is the growth curve model of Potthoff and Roy (1964). The early development of this model was mainly based on the unstructured MANOVA assumption of the covariance matrix of independent random vectors (e.g., Khatri 1966 and Grizzle and Allen 1969). Later, however, more attention has been paid to modeling the covariance matrix by using parsimonious covariance structures (see, e.g., Azzalini 1987, Lee 1988 and Nummi 1997). For excellent reviews of the growth curve model we refer to the books by Kshirsagar and Smith (1995) and Pan and Fang (2002).

T. Nummi (✉) • N. Mesue
School of Health Sciences, FIN-33014 University of Tampere, Finland
e-mail: tan@uta.fi

R. Dasgupta (ed.), *Advances in Growth Curve Models*, Springer Proceedings in Mathematics & Statistics 46, DOI 10.1007/978-1-4614-6862-2_3,

Our approach is to use cubic smoothing splines to model the mean growth curve. As is very well known cubic smoothing splines are very flexible curves with interesting mathematical properties (see, e.g., Green and Silverman 1994). For an up-to-date summary of recent methods of smoothing splines and nonparametric regression we refer to Wu and Zhang (2006). Approximate inference with smoothing splines have been studied, e.g., in Eubank and Spiegelman (1990), Schimek (2000), and Cantoni and Hastie (2002). In their simulation study Liu and Wang (2004) compared six testing statistics. Nummi et al. (2011) provided a test of a regression model against spline alternative for correlated data. The main focus in these studies have been on testing the order of the polynomial model against a spline alternative. However, testing if two or more splines are equal would be very important in many applications. Nummi and Koskela (2008) introduced some results for the estimation and rough testing of growth curves when the analysis is based on spline functions. However, very little research about testing equality of smoothing splines, especially for correlated data, has been carried out so far. In this paper we focus on testing if the progression in time is equal over the set of correlated observations.

In Sect. 3.2 we introduce the basic growth model and its estimation using cubic smoothing splines. In Sect. 3.3 a spline approximation is introduced and a test for mean curves is developed. In Sect. 3.4 a computational example of Glucose data is presented and the method is also investigated by simulation experiments.

3.2 Basic Spline Growth Model and Some Properties

One of the most important statistical models for balanced complete multivariate repeated measures data is the GMANOVA (Generalized Multivariate Analysis of Variance Model) of Potthoff and Roy (1964). The model is often also refered to as the growth curve model. This model can be written as

$$\mathbf{Y} = \mathbf{TBA}' + \mathbf{E}, \tag{3.1}$$

where $\mathbf{Y} = (\mathbf{y}_1, \mathbf{y}_2, \ldots, \mathbf{y}_n)$ is the matrix of independent response vectors, $\mathbf{T}$ is a $q \times p$ within-individual design matrix, $\mathbf{A}$ is an $n \times m$ between-individual design matrix, $\mathbf{B}$ is an unknown $p \times m$ parameter matrix to be estimated, and $\mathbf{E}$ is a $q \times n$ matrix of random errors. It is assumed that the columns $\mathbf{e}_1, \ldots, \mathbf{e}_n$ of $\mathbf{E}$ are independently normally distributed as $\mathbf{e}_i \sim N(\mathbf{0}, \mathbf{\Sigma})$, $i = 1, \ldots, n$. In the original model formulation $\mathbf{\Sigma}$ was assumed to be an unstructured covariance matrix and the analyses were mainly based on the methods developed for linear models and multivariate analysis.

Often when analyzing growth data the true growth function is more or less unknown and there may not be any theoretical justification for any specific parametric form of the curve. Parametric models are then used for descriptive purposes rather than interpretative to summarize the information of development profile. A natural first choice in such situations is a low order polynomial curve.

However, in many cases these models may fail to reveal important features of the growth process and more complicated models are therefore also needed.

Our approach is to use the cubic smoothing splines to model the mean growth curve. As is very well known cubic smoothing splines are very flexible curves with interesting mathematical properties (see, e.g., Green and Silverman 1994). We can write the model (3.1) in a slightly more general form as (see also Nummi and Koskela 2008)

$$\mathbf{Y} = \mathbf{G}\mathbf{A}' + \mathbf{E}, \tag{3.2}$$

where $\mathbf{G} = (\mathbf{g}_1, \ldots, \mathbf{g}_m)$ is the matrix of smooth mean growth curves in time points $t_1, t_2, \ldots, t_q$. We assume that the covariance matrix $\mathbf{\Sigma}$ takes certain type of parsimonious structure $\mathbf{\Sigma} = \sigma^2\mathbf{R}(\theta)$ with covariance parameters θ. In sequel we refer to this model as the spline growth model (SGM). The growth curve model of Potthoff and Roy (1964) is now the special case $\mathbf{G} = \mathbf{TB}$. The smooth solution for $\mathbf{G}$ can be obtained by minimizing the penalized least squares (PLS) criterion

$$Q = \text{tr}[(\mathbf{Y} - \dot{\mathbf{G}})'\mathbf{H}(\mathbf{Y} - \dot{\mathbf{G}}) + \alpha\dot{\mathbf{G}}'\mathbf{K}\dot{\mathbf{G}}], \tag{3.3}$$

where we denote $\dot{\mathbf{G}} = \mathbf{G}\mathbf{A}'$, $\mathbf{H} = \mathbf{R}^{-1}$, and $\mathbf{K}$ is the so-called roughness matrix arising from the common roughness penalty $RP = \int g''^2$ and α is a fixed smoothing parameter. For cubic smoothing splines the roughness matrix is

$$\mathbf{K} = \nabla\mathbf{\Delta}^{-1}\nabla', \tag{3.4}$$

where the nonzero elements of banded $q \times (q-2)$ and $(q-2) \times (q-2)$ matrices ∇ and $\mathbf{\Delta}$, respectively, are

$$\nabla_{k,k} = \frac{1}{h_k}, \; \nabla_{k+1,k} = -\left(\frac{1}{h_k} + \frac{1}{h_{k+1}}\right), \; \nabla_{k+2,k} = \frac{1}{h_{k+1}} \tag{3.5}$$

and

$$\mathbf{\Delta}_{k,k+1} = \mathbf{\Delta}_{k+1,k} = \frac{h_{k+1}}{6}, \; \mathbf{\Delta}_{k,k} = \frac{h_k + h_{k+1}}{3}, \tag{3.6}$$

where $h_j = x_{j+1} - x_j$, $j = 1, 2, \ldots, (q-1)$, and $k = 1, 2, \ldots, (q-2)$. It can be shown that Q can be rewritten in an alternative form

$$Q = \text{tr}[\{\dot{\mathbf{G}} - (\mathbf{H} + \alpha\mathbf{K})^{-1}\mathbf{HY}\}'(\mathbf{H} + \alpha\mathbf{K})\{\dot{\mathbf{G}} - (\mathbf{H} + \alpha\mathbf{K})^{-1}\mathbf{HY}\} + c], \tag{3.7}$$

where c is a constant and $(\mathbf{H} + \alpha\mathbf{K})$ is a positive definite matrix. The function Q is minimized for given α and $\mathbf{H}$ when $\dot{\mathbf{G}} = (\mathbf{H} + \alpha\mathbf{K})^{-1}\mathbf{HY}$. This gives the spline estimator

$$\tilde{\mathbf{G}} = (\mathbf{H} + \alpha\mathbf{K})^{-1}\mathbf{HYA}(\mathbf{A}'\mathbf{A})^{-1}. \tag{3.8}$$

However, the covariance matrix $\mathbf{H}$ may not be known and therefore the estimator (3.8) maybe difficult to use in practical situations. Fortunately, it can be shown that in certain important special cases the general spline estimator (3.8) simplifies to simple linear functions of the original observations $\mathbf{Y}$. One obvious condition for such kind of simplification is

$$\mathbf{KR} = \mathbf{K} \tag{3.9}$$

and since now $\mathbf{K} = \mathbf{KH}$ the spline estimators $\tilde{\mathbf{G}}$ can be simplified as

$$\hat{\mathbf{G}} = (\mathbf{I} + \alpha\mathbf{K})^{-1}\mathbf{YA}(\mathbf{A}'\mathbf{A})^{-1} = \mathbf{SYA}(\mathbf{A}'\mathbf{A})^{-1}, \tag{3.10}$$

where the smoother matrix is $\mathbf{S} = (\mathbf{I} + \alpha\mathbf{K})^{-1}$. Covariance matrices satisfying the condition (3.9) have been studied in Nummi and Koskela (2008) and Nummi et al. (2011). Some important special cases of these structures useful for growth data are $\mathbf{R} = \mathbf{I}$, $\mathbf{R} = \mathbf{I} + \sigma_d^2\mathbf{11}'$, $\mathbf{R} = \mathbf{I} + \sigma_{d'}^2\mathbf{XX}'$ and $\mathbf{R} = \mathbf{I} + \mathbf{XDX}'$, where $\mathbf{X} = (\mathbf{1}, \mathbf{x})$ and $\mathbf{x}$ is a vector of q measuring times.

If we apply the result $\text{vec}(\mathbf{ABC}) = (\mathbf{C}' \otimes \mathbf{A})\text{vec}(\mathbf{B})$, where the vec operation rearranges the columns of a matrix underneath each other, we can write the basic model (3.2) in a vector form

$$\mathbf{y} = (\mathbf{A} \otimes \mathbf{I}_q)\mathbf{g},$$

where $\mathbf{y} = \text{vec}(\mathbf{Y})$ and $\mathbf{g} = \text{vec}(\mathbf{G})$. If the spline estimates are written in vector form we have

$$\hat{\mathbf{g}} = [(\mathbf{AA}')^{-1}\mathbf{A}' \otimes \mathbf{S}]\mathbf{y}$$

and the smoother of the whole data is

$$\hat{\mathbf{y}} = (\mathbf{P}_a \otimes \mathbf{S})\mathbf{y} = \mathbf{S}_*\mathbf{y}, \tag{3.11}$$

where we denote $\mathbf{P}_a = \mathbf{A}(\mathbf{A}'\mathbf{A})^{-1}\mathbf{A}'$ and $\mathbf{S}_* = (\mathbf{P}_a \otimes \mathbf{S})$. The effective degrees of freedom of the smoother can now be given as

$$edf_* = \text{tr}(\mathbf{S}_*) = \text{tr}(\mathbf{P}_a \otimes \mathbf{S}) = \text{tr}(\mathbf{P}_a)\text{tr}(\mathbf{S}) = m \times edf, \tag{3.12}$$

where $edf = \text{tr}(\mathbf{S})$ is the effective degrees of freedom of the smoother $\mathbf{S}$. It is further easy to see that the generalized cross-validation criteria for choosing the smoothing parameter α take the form

$$GCV(\alpha) = \frac{\frac{1}{nq}\sum_{i=i}^{nq}[y_i - \hat{y}_i]^2}{(1 - \frac{m \times edf}{nq})^2}, \tag{3.13}$$

where y_i and $\hat{y}_i$ are individual elements of the observed and smoothed vectors $\mathbf{y}$ and $\hat{\mathbf{y}}$, respectively.

3.3 Testing of Mean Curves

It is very well known that exact tests may be difficult to develop when making statistical inference based on smoothing splines. Our interest in this study focuses on testing if the progression in time is the same in treatment groups considered. In this study an exact test based on spline approximations for testing growth curves is developed.

3.3.1 *Spline Approximation*

It has been demonstrated by Nummi et al. (2011) that the approximation discussed in this paper is quite good for relatively smooth data. More detailed consideration of spline approximations can be found, e.g., in Hastie (1996). In a general case the smoother matrix $\mathbf{S}$ is not a projection matrix and therefore certain results, e.g. in testing, developed for general linear models are not directly applicable. Our approach is to utilize an approximation for the smoother matrix $\mathbf{S}$ with the properties of a projection matrix. As discussed by Hastie (1996) the smoother matrix can be written as

$$\mathbf{S} = \mathbf{M}(\mathbf{I} + \alpha\Lambda)^{-1}\mathbf{M}', \tag{3.14}$$

where $\mathbf{M}$ is the matrix of q orthogonal eigenvectors of $\mathbf{K}$ and Λ is a diagonal matrix of corresponding q eigenvalues. It is easily seen that $\mathbf{K}$ and $\mathbf{S}$ share the same set of eigenvectors $\mathbf{m}_1, \mathbf{m}_2, \ldots, \mathbf{m}_q$ and the eigenvalues are connected such that the eigenvalues of $\mathbf{S}$ are $\gamma = 1/(1 + \alpha\lambda)$. In sequel we assume that eigenvectors $\mathbf{m}_1, \mathbf{m}_2, \ldots, \mathbf{m}_q$ are ordered according to the eigenvalues of $\mathbf{S}$. It is well known that the sequence of eigenvectors appears to increase in complexity like a sequence of orthogonal polynomials. The first two eigenvalues of $\mathbf{S}$ are always 1. We can set $\mathbf{m}_1 = \mathbf{1}/\sqrt{n}$ and $\mathbf{m}_2 = \mathbf{t}_*$, where $\mathbf{t}_* = (\mathbf{t} - \bar{t}\mathbf{1})/S_t$, $\bar{t}$ is the mean and $S_t = \sqrt{\sum_{i=1}^{q}(t_i - \bar{t})^2}$ is the square root of the sum of squares of the time points $t_1, \ldots, t_q$. Therefore the first two eigenvectors $\mathbf{m}_1$ and $\mathbf{m}_2$ span the subspace corresponding to the straight line model. In the mixed model formulation of the spline solution (e.g. Verbyla et al. 1999) this corresponds to the fixed part of the model. It is also easily observed that if the value of the smoothing parameter α increases the fit approaches the straight line model and the fitted line (fixed part) is not influenced by any specific choice of α.

Clearly, one obvious approximation of the spline fit (3.10) is the spline model

$$\bar{\mathbf{G}} = \mathbf{P}_m\mathbf{Y}\mathbf{A}(\mathbf{A}'\mathbf{A})^{-1}, \tag{3.15}$$

where $\mathbf{P}_m = \mathbf{M}_*\mathbf{M}_*'$ and $\mathbf{M}_*$ contains the $c(\leq q)$ first eigenvectors of $\mathbf{M}$. This corresponds to minimizing the least squares (LS) criteria

$$Q' = \mathrm{tr}(\mathbf{Y} - \dot{\mathbf{G}})'(\mathbf{Y} - \dot{\mathbf{G}}), \tag{3.16}$$

where $\dot{\mathbf{G}} = \mathbf{GA}'$. Note that the smoother matrix $\mathbf{S}$ and the smoothing parameter need not be computed here. However, the number of eigenvectors c from $\mathbf{K}$ used in the approximation needs to be estimated. This is easily done by, for example, using a modified generalized cross-validation criteria

$$GCV^*(c) = \frac{\frac{1}{nq}\sum_{i=i}^{nq}[y_i - \bar{y}_i]^2}{(1 - \frac{m \times c}{nq})^2}, \tag{3.17}$$

where $\bar{y}_i$ is now computed using the formula (3.11) with $\mathbf{S}$ replaced by $\mathbf{P}_m$.

3.3.2 Constructing a Test for Mean Spline Curves

First, consider the set of fitted spline curves

$$\hat{\mathbf{Y}} = \hat{\mathbf{G}}\mathbf{A}'. \tag{3.18}$$

As discussed in the previous section we may use the approximation

$$\hat{\mathbf{Y}} = \bar{\mathbf{G}}\mathbf{A}' = \mathbf{M}_*\hat{\Omega}\mathbf{A}', \tag{3.19}$$

where we denoted $\hat{\Omega} = \mathbf{M}_*'\mathbf{YA}(\mathbf{A}'\mathbf{A})^{-1}$. All the relevant information for testing mean profiles is now in the matrix $\hat{\Omega}$, which can now be considered to be an unbiased estimate of the unknown parameter matrix of the statistical model $E(\mathbf{Y}) = \mathbf{M}_*\Omega\mathbf{A}'$. Therefore in sequel we confine in testing linear hypothesis of the form

$$H_0 : \mathbf{C}\Omega\mathbf{D} = \mathbf{0},$$

where $\mathbf{C}$ and $\mathbf{D}$ are known $\nu \times c$ and $m \times g$ matrices with ranks ν and g, respectively. Since $\text{vec}(\mathbf{ABC}) = (\mathbf{C}' \otimes \mathbf{A})\text{vec}(\Omega)$, the vector form of H_0 is given by

$$H_0 : (\mathbf{D}' \otimes \mathbf{D})\omega = \mathbf{0},$$

where $\omega = \text{vec}(\Omega)$. If we take the vector form of $\hat{\Omega}$, we get

$$\hat{\omega} = \text{vec}(\hat{\Omega}) = [(\mathbf{A}'\mathbf{A})^{-1}\mathbf{A}' \otimes \mathbf{M}_*']\mathbf{y}. \tag{3.20}$$

It is now easily seen that the covariance matrix of $\hat{\omega}$ is

$$Var(\hat{\omega}) = \sigma^2[(\mathbf{A}'\mathbf{A})^{-1} \otimes \mathbf{M}_*'\mathbf{RM}_*]. \tag{3.21}$$

If we denote $Var(\mathbf{D}' \otimes \mathbf{C})\hat{\omega} = \mathbf{W}$, it is then obvious that under the null hypothesis

$$\mathbf{W}^{-1/2}(\mathbf{D}' \otimes \mathbf{C})\hat{\omega} \sim N_{\nu g}(\mathbf{0}, \mathbf{I})$$

and

$$\sigma^{-2}Q_* = \hat{\omega}'(\mathbf{D} \otimes \mathbf{C}')\mathbf{W}^{-1}(\mathbf{D}' \otimes \mathbf{C})\hat{\omega} \sim \chi^2_{\nu g}.$$

By using the results tr($\mathbf{AZ}'\mathbf{BZC}$) $=$ (vec $\mathbf{Z}$)$'$($\mathbf{CA} \otimes \mathbf{B}$)vec $\mathbf{Z}$, it is further easy to see that Q_* can be rewritten as

$$Q_* = \mathrm{tr}\{[\mathbf{D}'(\mathbf{A}'\mathbf{A})^{-1}\mathbf{D}]^{-1}[\mathbf{C}\hat{\Omega}\mathbf{D}]'[\mathbf{CM}_*'\mathbf{RM}_*\mathbf{C}']^{-1}[\mathbf{C}\hat{\Omega}\mathbf{D}]. \tag{3.22}$$

If σ^2 is estimated by

$$\hat{\sigma}^2 = \frac{1}{n(q-c)}\mathrm{tr}\mathbf{Y}'(\mathbf{I} - \mathbf{P}_m)\mathbf{Y}, \tag{3.23}$$

it can be shown that $n(q-c) \times \hat{\sigma}^2 \sim \chi^2_{n(q-c)}$ and since Q_* and $\hat{\sigma}^2$ are independent testing can be based on the F-ratio. Then under the null hypothesis

$$F = \frac{Q_*/\nu g}{\hat{\sigma}^2} \sim F[\nu g, n(q-c)]. \tag{3.24}$$

Testing can then be based on the quantiles of the F-distribution. However, in practical situations the matrix $\mathbf{R}$ contains unknown parameters that need to be estimated and therefore the distribution of F in general case is only approximate. However, if we are only interested in progression in time we can drop the first eigenvector $\mathbf{m}_1$ corresponding to the constant term in the approximation model (see Sect. 3.3.1). Therefore we can take $\mathbf{C} = [\mathbf{0}, \mathbf{I}]$, and if we assume the uniform covariance model $\mathbf{R} = d^2\mathbf{1}\mathbf{1}' + \mathbf{I}$, it can be shown that

$$\mathbf{CM}_*'\mathbf{RM}_*\mathbf{C}' = \mathbf{C}\{d^2\mathbf{e}_1\mathbf{e}_1' + \mathbf{I}\}\mathbf{C}' = \mathbf{I}, \tag{3.25}$$

where $\mathbf{e}_1 = (1, 0, \ldots, 0)'$. Therefore the term Q_* simplifies to

$$Q_* = \mathrm{tr}[\mathbf{C}\hat{\Omega}\mathbf{D}]\{[\mathbf{D}'(\mathbf{A}'\mathbf{A})^{-1}\mathbf{D}]^{-1}[\mathbf{C}\hat{\Omega}\mathbf{D}]', \tag{3.26}$$

which does not contain unknown parameters of the covariance matrix and therefore for this special case the distribution of the F-statistic is exact. This is an important result since the uniform covariance model is quite common and a good approximation in many situations. The F-test proposed here provides means to test if the progression in time is the same over treatment groups when the models are based on spline curves. Following the same kind of considerations it would be easy to develop an exact F-statistic to test if the progression around the fitted straight line (the so-called random part in mixed model formulation) is the same over treatment groups with the more general assumption of linear correlation structure $\mathbf{R} = \mathbf{XDX}' + \mathbf{I}$.

3.4 Computational Examples

3.4.1 *Standard Glucose Tolerance Test*

As the first computational example we consider the glucose data of Zerbe (1979). In these data glucose tolerance tests were administered to 13 control and 20 obese patients. Plasma inorganic phosphate measurements determined from blood samples drawn 0, 0.5, 1, 1.5, 2, 3, 4, and 5 h after standard oral glucose dose were taken. The curves plotted for the control and obese patients are plotted in Fig. 3.1. In Fig. 3.1, two features of the plotted curves are quite obvious. First, there is a considerable variation in patient's individual levels. Secondly, the functional form of the dependency of plasma inorganic phosphate and time is quite complicated and possibly different for control and obese patients. In Zerbe (1979) a polynomial of degree of 4 was used to model this relationship.

To set up the spline growth model the between-individual design matrix $\mathbf{A}$ was first defined. For 13 control patients the rows of $\mathbf{A}$ are $(1, 0)$, $i = 1, \ldots, 13$ and for

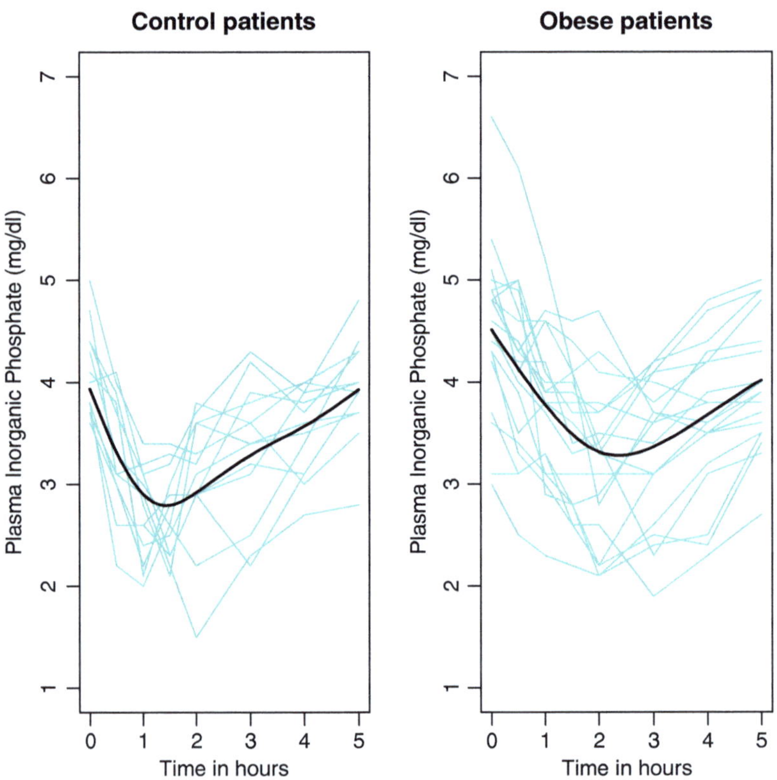

Fig. 3.1 Plasma inorganic phosphate measurements for control and obese patients

20 obese patients the rows of $\mathbf{A}$ are $(0, 1)$, $i = 14, \ldots, 33$. The minimum value of $GCV = 0.4484152$ is obtained at $\alpha = 0.09410597$. This gives the total effective degrees of freedom $edf_* = 9.310273$. The fitted curves are plotted in Fig. 3.1. It can be observed that the fitted spline curves very nicely depict the mean performance of measurements in both groups.

To test if the progression in time is the same in both groups we first determined the dimension c needed in the spline approximation. Minimizing the modified generalized cross-validation criterion gives $c = 5$. To test the null hypothesis we took

$$\mathbf{C} = [\mathbf{0}, \mathbf{I}_4] \quad \text{and} \quad \mathbf{D} = [1, -1]'.$$

Next we calculated the estimate $\mathbf{C}\hat{\Omega}\mathbf{D}$. This yields

$$\mathbf{C}\hat{\Omega}\mathbf{D} = (0.74897422,\ -0.03613939,\ 0.46201190,\ 0.54998030)'$$

and the residual variance estimate for this setup is $\hat{\sigma}^2 = 0.09408348$. For the covariance matrix $\mathbf{R}$ we assumed the uniform correlation model and therefore the exact version of the test statistics can be used. Then the value of Q_* is given as

$$Q_* = \text{tr}[\mathbf{C}\hat{\Omega}\mathbf{D}]\{[\mathbf{D}'(\mathbf{A}'\mathbf{A})^{-1}\mathbf{D}]^{-1}[\mathbf{C}\hat{\Omega}\mathbf{D}]' = 8.494923$$

and the value of the test statistics is then

$$F = \frac{Q_*/\nu g}{\hat{\sigma}^2} = \frac{8.494923/4}{0.09408348} = 22.57283.$$

If this is compared to the critical value $F_{0.95}(4, 99) = 2.447$, the null hypothesis of equal progression in mean plasma inorganic phosphate for control and obese patients is clearly rejected.

3.4.2 A Simulation Study

In order to demonstrate the advances of the methodology presented we conducted a simulation study. In this study two models were tested

$$y = 1 + 0.5 \times t + \epsilon, \tag{3.27}$$

$$y = 1 + 0.5 \times t + a \times \cos(0.4\pi t) + \epsilon, \tag{3.28}$$

with $t = 1, \ldots, 10$ and independent random errors $\epsilon_i \sim N(0, 1)$. The coefficient a takes the values 0, 0.1, 0.2, 0.3, 0.4, 0.5, 0.6, and 0.7. The first group of growth curves consists of 100 random vectors generated from model 3.27 and the second consists of 100 random vectors generated from model 3.28. So, for each value

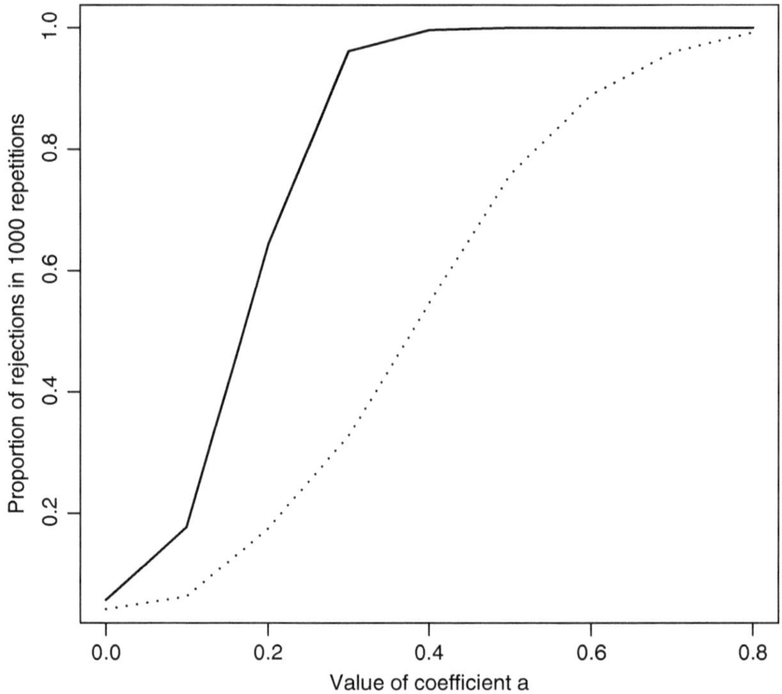

Fig. 3.2 Proportion of the rejections in 1,000 repetitions under models (3.27) and (3.28)

of a these two sets of growth curves were generated. The mean growth curves are then tested against the null hypothesis that the progression in time is the same in both groups. Two methods were utilized. The spline testing method presented in this paper and the second method utilized here was the basic parametric least squares fit of the third degree polynomial model. The power was estimated with the significance level 0.05 by counting the percentage of rejections in the 1,000 repetitions.

The results are shown in Fig. 3.2. Clearly, the spline test presented in this paper performed better than the test based on the least squares fit of the third degree polynomial. This is obviously due to the fact that the fit provided by the splines better depicts the peculiarities of the unknown growth function.

3.5 Concluding Remarks

Traditional analyses of growth curves are often based on simple parametric curves, which may not satisfactorily depict all the features of the growth process during the testing period. The method presented in this paper is based on cubic smoothing splines, which provides a very flexible modeling tool for the analysis. However, very little research on the statistical inference (especially testing) of cubic smoothing splines for correlated data has been carried out. The novel test presented in this paper seems to provide a good alternative, especially when more accurate modeling of growth process is required.

References

Azzalini, A., (1987). Growth Curves analysis for patterned covariance matrices. In: New perspectives in theoretical and applied statistics (eds puri, M., Vilaplana, J.P. & Wertz, W) New York: Wiley, 63–70

Cantoni, E., & Hastie, T. (2002). Degrees-of-freedom tests for smoothing splines. *Biometrika, 89*(2), 251–263.

Eubank, R.L., & Spiegelman, C.H. (1990). Testing the goodness of fit of a linear model via nonparametric regression techniques. *Journal of the American Statistical Association, 85*(410), 387–392.

Green, P.J., & Silverman, B.W. (1994). *Nonparametric regression and generalized linear models.* London: Chapman and Hall.

Grizzle, J.E., & Allen, D.M. (1969). Analysis of growth and dose response curves. *Biometrics, 25*, 357–381.

Hastie, T. (1996). Pseudosplines. *Jounal of the Royal Statistical Society, Series B, 58*, 379–396.

Khatri, C.G. (1966). A note on a MANOVA model applied to problems in growth curves. *Annals of the Institute of Statistical Mathematics, 18*, 75–86.

Kshirsagar, A.M., & Smith, W.B. (1995). *Growth curves.* New York: Marcel Dekker.

Lee, J.C. (1988). Tests and model solution for general growth curve model. Biometrics, *47*, 147–159.

Liu, A., & Wang, Y. (2004). Hypothesis testing in smoothing spline models. *Journal of Statistical Computation and Simulation, 74*, 581–597.

Nummi, T. (1997). Estimation in random effects growth curve model. *Journal of Applied Statistics, 24*(2), 157–168.

Nummi, T., & Koskela, L. (2008). Analysis of growth curve data using cubic smoothing splines. *Journal of Applied Statistics, 35*, 1–11.

Nummi, T., Jianxin, P., Siren, T., & Liu, K. (2011). Testing for cubic smoothing splines under dependent data. *Biometrics, 67*(3), 871–875. DOI: 10.1111/j.1541-0420.2010.01537.x.

Pan, J., & Fang, K. (2002). *Growth curve models and statistical diagnostics*, Springer series in Statistics. New York: Springer.

Potthoff, R.F., & Roy, S.N. (1964). A generalized multivariate analysis of variance model useful especially for growth curve problems. *Biometrika, 5*, 313–326.

Schimek, M.G. (2000). Estimation and Inference in partially linear models with smoothing splines. *Journal of Statistical Planning and Inference, 91*, 525–540.

Verbyla, A.P., Cullis, B.R., Kenward, M.G., & Welham, S.J. (1999). The analysis of designed experiments and longitudinal data by using smoothing splines (with discussion). *Journal of the Royal Statistical Society, Series C, 48*, 269–311.

Wu, L., & Zhang, J.T. (2006). *Nonparametric regression methods for longitudinal data analysis.* New Jersey: Wiley.

Zerbe, G.O. (1979). Randomization analysis of the completely randomized design extended to growth and response curves. *Journal of the American Statistical Association, 74*, 215–221.

Chapter 4
Nonuniform Rates of Convergence to Normality for Two-Sample U-Statistics in Non IID Case with Applications

Ratan Dasgupta

Abstract Rates of convergence to normality are studied for two-sample U-statistics in non iid case under certain conditions which ensures that all moments of the kernel exist but the moment generating function of the kernel may not exist. Applications are made to compute normal approximation zone for the tail probability, nonuniform L_p version of Berry–Esseen theorem and moment type convergences of a standardized U-statistic. The normal approximation zone goes beyond moderate deviation and extends up to large deviation. As an application, efficiency of U-statistic-based tests, when the basic observations are discretized, is studied. It is seen that sometimes test efficiency may increase after discretization. Possible explanation is provided for such intriguing phenomena. A further application is made of deviation probabilities to compare agricultural production scenarios, e.g., growth of Elephant-foot-yam, over years. Yam stem growth is a good predictor for underground yam deposition. A nonparametric robust procedure to estimate derivative of a function based on discrete data is proposed. Performance of the proposed technique that is insensitive to outliers is investigated and found to be satisfactory. The procedure of analysis adopted in yam data may be extended to the cases where variables of interest are continuous growth curves that need to be compared over different time cycles in terms of rare events.

MS subject classification: 62G20, 62P30

R. Dasgupta (✉)
Theoretical Statistics and Mathematics Unit, Indian Statistical Institute, Kolkata, India
e-mail: ratandasgupta@gmail.com

R. Dasgupta (ed.), *Advances in Growth Curve Models*, Springer Proceedings in Mathematics & Statistics 46, DOI 10.1007/978-1-4614-6862-2_4,

4.1 Introduction

The class of U-statistics is a highly fruitful extension of the usual sample mean to a natural higher level of generality. Since their inception in 1948 the basic probability theory of U-statistics has become long established, see Sen (1992) for some historical references. The asymptotics of U-statistics, defined on a sample space from some distribution, entails a hierarchy of limit results depending on the order of degeneracy of the underlying kernel, starting with asymptotic normality, followed by weighted sum of chi-squares as the limit random variable and then the higher order cases best described in terms of multiple Wiener integral.

Amongst the results on uniform rates of convergence in CLT for U-statistics based on n iidrvs, Callert and Janssen (1978) obtained the optimal rate $O(n^{-1/2})$ under the existence of the third absolute moment of the kernel; recently Bentkus et al. (2009) obtained rates for *adjusted* normal approximation involving third derivative of standard normal cdf, in terms of error bound for main linear part involving iidrvs via Hoeffding decomposition plus the variance of the remainder.

In this paper we consider nonuniform rates of convergence to normality for 2-sample U-statistic in non iid case under some exponential type moment conditions. The genesis of the present problem was launched in Ghosh and Dasgupta (1980) and has seen further developments in Dasgupta (1988, 1989, 2006, 2008, 2010).

Let $U_{n,m}$ be a two-sample U-statistic based on the independent but not necessarily identically distributed random variables $X_1, \cdots, X_n$ and $Y_1, \cdots, Y_m$ with kernel ϕ and degree (r, s) i.e.,

$$U = (n_{C_r} m_{C_s})^{-1} \sum_{\substack{1 \le i_1 < \cdots < i_r \le n \\ 1 \le j_1 < \cdots < j_s \le m}} \phi(X_{i_1}, \cdots, X_{i_r}; Y_{j_1}, \cdots, Y_{j_s}), \tag{4.1}$$

where the kernel ϕ is symmetric in its arguments X_is and Y_js. Without loss of generality let,

$$E\phi(X_{i_1}, \cdots, X_{i_r}; Y_{j_1}, \cdots, Y_{j_s}) = 0, \quad \forall\, i_1 \neq \cdots \neq i_r, \quad j_1 \neq \cdots \neq j_s. \tag{4.2}$$

An example of such a statistic is Wilcoxon 2-sample statistic. Nonuniform rates of convergence in CLT for two-sample U-statistics is studied in Dasgupta (2008) when a finite ($\geq$2) moment of the kernel ϕ exist. A Berry–Esseen bound for random index was also established therein.

We study nonuniform rates of convergence to normality for two-sample U-statistics in non iid case under certain conditions which ensures that all the moments of the kernel exist but the moment generating function of the kernel may not exist. As an application of these rates, we compute normal approximation zone for the tail probability of standardized two-sample U-statistic that go beyond the moderate deviation zone. The results in this paper obtained under assumptions that are stronger than those obtained in Dasgupta (2008), which seems to be first of this kind in literature for two-sample U-statistic. We show that the normal approximation zone is of comparable order with that for standardized sum of iid random variables obtained under similar moment assumptions. We also show that for Wilcoxon 2-sample statistic, the normal approximation zone is the same as that for standardized sum of iid random variables. Nonuniform L_p version of Berry–Esseen theorem and moment type convergences of a standardized U-statistic are also studied.

In Sect. 4.2 we outline the decomposition for two-sample U-statistics. In Sects. 4.3 and 4.4 we study convergence rates in CLT along with allied results, based on two different types of moment bound for the kernel ϕ. Let $m = O_e(n)$ and ν be the number of arguments in ϕ. For the first type of moment bound which ensures $E\exp[s\{\log_e(1+|\phi|)\}^{\nu/(\nu-1)}] < \infty$, $s > 0$, normal approximation zone is $O((\log n)^{\nu/(\nu-1)})$, $\nu > 1$. These zones are larger than moderate deviation zone, as $\nu > 1$. The second type of moment bound ensures $E\exp(s|\phi|^{1/\nu}) < \infty$, $s > 0$, and the corresponding normal approximation zone turns out to be $o(n^{1/(2(\nu+\nu+1))})$, a large deviation zone. It is known that Chernoff-type large deviation behaviour is different for U-statistics from that of iidrvs, whereas here it is shown that w.r.t. Linnik type large deviation these are equivalent (Remark 3, regarding Wilcoxon statistic), for two-sample case in iid/non iid set-up.

Bayes Risk Efficiency (BRE) considers testing problem in terms of rare events. The concept of BRE in its large deviation counterpart studied in Dasgupta (2010) revolves around comparing tiny errors. In Sect. 4.5, we assess the efficiency of a test based on U-statistic when the basic observations are discretized with possibilities of ties in data. Random (data dependent) scaling is now quite frequent in limit theorems inter alia, and one may use (random) variance calculated from sample for such purposes of scaling, as advocated in this paper. A remarkable feature revealed in our study is that sometimes test efficiency *increases* after discretization, although there is a possible loss of information when the basic observations are discretized. Statistical implications of the result are discussed. A further application is made in agricultural statistics to compare production scenarios over years in Sect. 4.6. A nonparametric technique to estimate growth rate based on lowess (locally weighted scatterplot smoothing) regression is proposed to examine whether the rate falls below a level and the crop is ready for harvest. Among all non parametric regression techniques, lowess smoothing has a special role compared to others. The procedure down weights an outlier over several iterations, so as to remain insensitive towards it. This is one of the motivations to develop a new technique for estimating derivatives based on lowess. The technique is explained by live yam data from Indian Statistical Institute, Giridih farmhouse. The proposed technique of estimating derivative of a function based on discrete data has potential of general applications. Law of iterated logarithm (LIL), which has application to develop tests of power 1, is related to normal approximation zone of order ($\log\log n$) by Tomkins theorem, is also a consequence of nonuniform bounds in CLT; requiring only a little bit more (viz., logarithmic power > 1) than the second moment of individual random variables in the set-up of triangular array of independent random variables. We indicate another application of LIL; production scenarios are proposed to be compared in terms of relative frequencies of rare events in large deviation zone and by LIL. The technique may be extended to the cases where a large number of observations are continuously/sequentially recorded over time. Proof of Proposition 2, a technical result on equivalence of moment bound and m.g.f. of the transformed random variable is given in Appendix.

4.2 Decomposition of 2-Sample U-Statistic and Estimate of Remainder

For completeness and reader's convenience, we indicate the steps of decomposition, as shown in Dasgupta (2008), for $r = 2$ and $s = 2$. It is possible to generalize for other values of (r, s), see also Ghosh and Dasgupta (1980). Define

$$\psi^{i_1}_{i_2,i_3,i_4}(x_{i_1}) = \phi^{i_1}_{i_2,i_3,i_4}(x_{i_1}) = E[\phi(X_{i_1}, X_{i_2}, Y_{i_3}, Y_{i_4})|X_{i_1} = x_{i_1}], \tag{4.3}$$

similarly define $\psi^{i_3}_{i_1,i_2,i_4}(y_{i_3})$.

$$\phi^{i_1,i_2}_{i_3,i_4}(x_{i_1}, x_{i_2}) = E[\phi(X_{i_1}, X_{i_2}, Y_{i_3}, Y_{i_4})|X_{i_1} = x_{i_1}, X_{i_2} = x_{i_2}], \tag{4.4}$$

similarly, define $\phi^{i_3,i_4}_{i_1,i_2}(y_{i_3}, y_{i_4})$.

$$\psi^{i_1,i_2}_{i_3,i_4}(x_{i_1}, x_{i_2}) = \phi^{i_1,i_2}_{i_3,i_4}(x_{i_1}, x_{i_2}) - \psi^{i_1}_{i_2,i_3,i_4}(x_{i_1}) - \psi^{i_2}_{i_1,i_3,i_4}(x_{i_2}), \tag{4.5}$$

similarly, define $\psi^{i_3,i_4}_{i_1,i_2}(y_{i_3}, y_{i_4})$.

$$\phi^{i_1,i_2,i_3}_{i_4}(x_{i_1}, x_{i_2}, y_{i_3}) = E[\phi(X_{i_1}, X_{i_2}, Y_{i_3}, Y_{i_4})|X_i = x_{i_1}, X_{i_2} = x_{i_2}, Y_{i_3} = y_{i_3}]. \tag{4.6}$$

$$\psi^{i_1,i_2,i_3}_{i_4}(x_{i_1}, x_{i_2}, y_{i_3}) = \phi^{i_1,i_2,i_3}_{i_4}(x_{i_1}, x_{i_2}, y_{i_3}) - \psi^{i_1,i_2}_{i_3,i_4}(x_{i_1}, x_{i_2}) \tag{4.7}$$
$$- \psi^{i_1,i_3}_{i_2,i_4}(x_{i_1}, y_{i_3}) - \psi^{i_2,i_3}_{i_1,i_4}(x_{i_2}, y_{i_3}) - \psi^{i_1}_{i_2,i_3,i_4}(x_{i_1}) - \psi^{i_2}_{i_1,i_3,i_4}(x_{i_2}) - \psi^{i_3}_{i_1,i_2,i_4}(y_{i_3}).$$

In the same fashion similar terms can be defined.

Finally, let

$$\begin{aligned}
&\psi^{i_1,i_2,i_3,i_4}(x_{i_1}, x_{i_2}, y_{i_3}, y_{i_4}) \qquad (4.8)\\
&= \phi(x_{i_1}, x_{i_2}, y_{i_3}, y_{i_4}) - \psi^{i_1,i_2,i_3}_{i_4}(x_{i_1}, x_{i_2}, y_{i_3}) - \psi^{i_1,i_2,i_4}_{i_3}(x_{i_1}, x_{i_2}, y_{i_4})\\
&- \psi^{i_1,i_3,i_4}_{i_2}(x_{i_1}, y_{i_3}, y_{i_4}) - \psi^{i_2,i_3,i_4}_{i_1}(x_{i_2}, y_{i_3}, y_{i_4})\\
&- \psi^{i_1,i_2}_{i_3,i_4}(x_{i_1}, x_{i_2}) - \psi^{i_1,i_3}_{i_2,i_4}(x_{i_1}, y_{i_3}) - \psi^{i_1,i_4}_{i_2,i_3}(x_{i_1}, y_{i_4})\\
&- \psi^{i_2,i_3}_{i_1,i_4}(x_{i_2}, y_{i_3}) - \psi^{i_2,i_4}_{i_1,i_3}(x_{i_2}, y_{i_4}) - \psi^{i_3,i_4}_{i_1,i_2}(y_{i_3}, y_{i_4})\\
&- \psi^{i_1}_{i_2,i_3,i_4}(x_{i_1}) - \psi^{i_2}_{i_1,i_3,i_4}(x_{i_2}) - \psi^{i_3}_{i_1,i_2,i_4}(y_{i_3}) - \psi^{i_4}_{i_1,i_2,i_3}(y_{i_4}).
\end{aligned}$$

Write,

$$\overline{\psi}^{(1)}(X_{i_1}) = \frac{1}{[(n-1)\ {}^{m}C_2]} \sum_{\substack{i_2 = 1,\cdots,n \\ i_2 \neq i_1 \\ 1 \leq i_3 < i_4 \leq m}} \psi^{i_1}_{i_2,i_3,i_4}(X_{i_1}),$$

$$\overline{\psi}^{(1)}(Y_{i_3}) = [(m-1)\ {}^{n}C_2]^{-1} \sum_{\substack{i_4 = 1,\cdots,m \\ i_4 \neq i_3 \\ 1 \leq i_1 < i_2 \leq n}} \psi^{i_3}_{i_1,i_2,i_4}(Y_{i_3}),$$

$$\overline{\psi}^{(2)}(X_{i_1}, X_{i_2}) = ({}^{m}C_2)^{-1} \sum_{1 \leq i_3 < i_4 \leq m} \psi^{i_1,i_2}_{i_3,i_4}(X_{i_1}, X_{i_2}),$$

$$\overline{\psi}^{(2)}(X_{i_1}, Y_{i_3}) = [(n-1)(m-1)]^{-1} \sum_{\substack{i_2 = 1,\cdots,n;\ i_2 \neq i_1 \\ i_4 = 1,\cdots,m;\ i_4 \neq i_3}} \psi^{i_1,i_3}_{i_2,i_4}(X_{i_1}, Y_{i_3}),$$

$$\overline{\psi}^{(3)}(X_{i_1}, X_{i_2}, Y_{i_3}) = (m-1)^{-1} \sum_{i_4 = 1,\cdots,m;\ i_4 \neq i_3} \psi^{i_1,i_2,i_3}_{i_4}(X_{i_1}, X_{i_2}, Y_{i_3}),$$

$$\overline{\psi}^{(3)}(X_{i_1}, Y_{i_3}, Y_{i_4}) = (n-1)^{-1} \sum_{i_2 = 1,\cdots,n;\ i_2 \neq i_1} \psi^{i_1,i_3,i_4}_{i_2}(X_{i_1}, Y_{i_3}, Y_{i_4}),$$

define other terms similarly.

Then, with

$$V_1 = \frac{2}{n} \sum_{i_1=1}^{n} \overline{\psi}^{(1)}(X_{i_1}) + \frac{2}{m} \sum_{i_3=1}^{m} \overline{\psi}^{(1)}(Y_{i_3}),$$

$$V_2 = \binom{n}{2}^{-1} \sum_{1 \leq i_1 < i_2 \leq n} \overline{\psi}^{(2)}(X_{i_1}, X_{i_2}) + \binom{m}{2}^{-1} \sum_{1 \leq i_3 < i_4 \leq m} \overline{\psi}^{(2)}(Y_{i_3}, Y_{i_4})$$
$$+ \frac{4}{mn} \sum_{\substack{i_1 = 1,\cdots,n \\ i_3 = 1,\cdots,m}} \overline{\psi}^{(2)}(X_{i_1}, Y_{i_3}),$$

$$V_3 = (m\ {}^{n}C_2)^{-1} \sum_{\substack{1 \leq i_1 < i_2 \leq n \\ i_3 = 1,\cdots,m}} \overline{\psi}^{(3)}(X_{i_1}, X_{i_2}, Y_{i_3})$$
$$+ (n\ {}^{m}C_2)^{-1} \sum_{\substack{1 \leq i_3 < i_4 \leq m \\ i_1 = 1,\cdots,n}} \overline{\psi}^{(3)}(X_{i_1}, Y_{i_3}, Y_{i_4}) \text{ and}$$

$$V_4 = (m_{C_2} n_{C_2})^{-1} \sum_{\substack{1 \le i_1 < i_2 \le n \\ 1 \le i_3 < i_4 \le m}} \psi^{i_1,i_2,i_3,i_4}(X_{i_1}, X_{i_2}, Y_{i_3}, Y_{i_4}),$$

one has,

$$\begin{aligned} U &= V_1 + V_2 + V_3 + V_4 \\ &= V_1 + R_{n,m}; \text{ where } R_{n,m} = V_2 + V_3 + V_4. \end{aligned} \tag{4.9}$$

In the above representation V_1 is the main part. To be precise, one should write the main part as $V_1 = 2n^{-1} \sum_{i_1=1}^{n} \overline{\psi}_{1i_1}^{(1)}(X_{i_1}) + 2m^{-1} \sum_{i_3=1}^{m} \overline{\psi}_{2i_3}^{(1)}(Y_{i_3})$, which is a weighted sum of independent differently distributed random variables with different functions $\overline{\psi}_{1i_1}^{(1)}(.)$ and $\overline{\psi}_{2i_3}^{(1)}(.)$ for which application of standard theory is possible. In fact, we use the set-up of triangular array for random variables where variables in each array are independent. The arrays may themselves be dependent.

Although to simplify complex notations in non iid case we suppress the lower suffixes $1i_1$ and $2i_3$ in $\overline{\psi}$, the presence of these is to be understood from the corresponding random variables of 1st type and 2nd type X_{i_1} and Y_{i_3}, respectively, present within the first brackets; it *should not be misunderstood* as if $\overline{\psi}_{1i_1}^{(1)}(.) = \overline{\psi}_{2i_3}^{(1)}(.) = \overline{\psi}_{1i_1}^{(1)}$. The same liberty is taken in notation of other types of conditional expectation, and even there the meaning of the notations is precise in presence of the random variables X_{i_1}, Y_{i_3}, etc., within ψ. The quantities $\psi, \overline{\psi}$ etc., are obtained via conditional expectation of kernel ϕ. These retain the independent structure if the associated random variables in $\psi, \overline{\psi}$ etc., are non overlapping. Results on triangular array of independent random variables thus become applicable. By Jensen's inequality for conditional expectation and C_{2q} inequality, the moment bounds of ψ or $\overline{\psi}$ are ultimately connected to ϕ, and these are of same order as that for ϕ.

The remaining parts in (4.9) viz., V_2, V_3, and V_4 are comparatively negligible than the main part V_1. The following moment estimates of V_2, V_3, V_4, and the remainder $R_{n,m}$ hold, see Dasgupta (2008).

Proposition 1. *Let* (4.2) *holds and for an integer* $q \ge 1$,

$$\delta_q = \sup_{\substack{m \ge 2 \\ n \ge 2}} [\tbinom{n}{2}\tbinom{m}{2}]^{-1} \sum_{\substack{1 \le i_1 < i_2 \le n \\ 1 \le i_3 < i_4 \le m}} E|\phi(X_{i_1}, X_{i_2}, Y_{i_3}, Y_{i_4})|^{2q} < \infty. \tag{4.10}$$

Then,

$$E(V_2)^{2q} \le (n^{-2q} + m^{-2q} + (mn)^{-q})\, L^q (2q)!\delta_q, \tag{4.11}$$

$$E(V_3)^{2q} \le (m^{-q}n^{-2q} + n^{-q}m^{-2q})\, L^q (3q)!\delta_q, \tag{4.12}$$

$$E(V_4)^{2q} \le (mn)^{-2q}\ L^q(4q)!\delta_q, \tag{4.13}$$

where $L(>1)$ *is a constant independent of* m, n *and* q. *Finally, from* (4.11) *to* (4.13), *for* $R_{n,m}$ *defined in* (4.9), *one has*

$$ER^{2q}_{n,m} \le n^{-2q}\ L^q(\nu q)!\ \delta_q, \tag{4.14}$$

under the assumption $m = O_e(n)$, *where* ν *is the number of arguments in* ϕ.

In the decomposition (4.3)–(4.9), $\nu = r + s = 4$.

4.3 Rates of Convergence and Deviation Probabilities: Part 1

The representation (4.9) permits us to compute the convergence rates and the deviation probabilities of two-sample U-statistics. Consider $m = O_e(n)$. Then from (4.9),

$$U = V_1 + R_{n,m},\ \text{where}\ V_1 = \frac{2}{n}\sum_{i=1}^{n}\overline{\psi}^{(1)}(X_i) + \frac{2}{m}\sum_{i=1}^{m}\overline{\psi}^{(1)}(Y_i)$$

is the weighted sum of $(m+n)$ independent random variables.

Assume that the kernel ϕ satisfies

$$\sup_{n\ge1,m\ge1}\ (n_{c_2}m_{c_2})^{-1}\sum_{\substack{1\le i_1<i_2\le n\\ 1\le j_1<j_2\le m}} E\mid\phi(X_{i_1},X_{i_2},Y_{j_1},Y_{j_2})\mid^{2q} = \delta_q \le Le^{w_o q^\nu} \tag{4.15}$$

$\forall q > 1$, and for some $L > 1$, where $w_o > 0$, $\nu > 1$. In Appendix we show that condition (4.15) is equivalent to the following:

$$\sup_{n\ge1,m\ge1}\ (n_{c_2}m_{c_2})^{-1}\sum_{\substack{1\le i_1<i_2\le n\\ 1\le j_1<j_2\le m}} E\exp[s\{\log_e(1+|\phi|)\}^{\nu/(\nu-1)}] < \infty \tag{4.16}$$

where $\phi = \phi(X_{i_1},X_{i_2},Y_{j_1},Y_{j_2})$, and $s = w_o^{-1/(\nu-1)}$.

This ensures existence of m.g.f. for $\nu/(\nu-1)$th power of a logarithmic function of ϕ. The assumption implies ϕ has a wider moment-bound compared to a random variable with finite m.g.f. The bound on moment-growth for ϕ is of such a high order that, $\nu/(\nu-1)$th power of the *tamed variable* $\log(1+\phi)$ *admits m.g.f.*, instead of kernel ϕ. Although finiteness of all moments of ϕ is ensured, this assumption is very mild compared to stringent assumption of finite m.g.f. for ϕ.

Now write,

$$U_n^* = U_{n,m}^* = [\mathrm{var}(V_1)]^{-1/2} U_{n,m} = [\mathrm{var}(V_1)]^{-1/2} V_1 + R_{n,m}^* \tag{4.17}$$

where, $V_1 = \frac{2}{n}\sum_{i=1}^{n} \overline{\psi}^{(1)}(X_i) + \frac{2}{m}\sum_{i=1}^{m} \overline{\psi}^{(1)}(Y_i)$; $R_{n,m}^* = [\mathrm{var}(V_1)]^{-1/2} R_{n,m}$, $\sigma^{*2} = \mathrm{var}(V_1) = 4(\sum_{i=1}^{n} E[\overline{\psi}^{(1)}(X_i)]^2/n^2 + \sum_{i=1}^{m} E[\overline{\psi}^{(1)}(Y_i)]^2/m^2) = O_e(\frac{1}{n+m})$, $\sigma_n^2 = \sigma_{n,m}^2 = (n+m)^2\sigma^{*2} = (n+m)^2\mathrm{var}(V_1) = O_e(n+m) = O_e(n)$, provided

$$\inf_{n\geq 1} n^{-1}\sum_{i=1}^{n} E[\overline{\psi}^{(1)}(X_i)]^2 > 0, \quad \inf_{m\geq 1} m^{-1}\sum_{i=1}^{m} E[\overline{\psi}^{(1)}(Y_i)]^2 > 0. \tag{4.18}$$

Let $L > 1$ be a generic constant. The first term in the r.h.s. of (4.17) is then standardized sum of independent random variables and the second term is a remainder with $E(R_{nm}^*)^{2q} \leq n^{-q}\ L^q(\nu q)!\ \delta_q$. Now, $e^{w^* q^\nu} >> L^q(\nu q)!$, $w^* > 0$, $\nu > 1$. Thus,

$$E(R_{nm}^*)^{2q} \leq n^{-q}\ Le^{w\, q^\nu}, \quad w > w_o, \tag{4.19}$$

for a different (large) choice of L. The next theorem states the normal approximation zone for tail probability of the standardized two-sample U-statistic.

Theorem 1. *Let* $m = O_e(n)$. *Under the assumptions* (4.2), (4.15)/(4.16) *and* (4.18), *for the standardized* U*-statistic* U_n^* *defined in* (4.17), *one has*
$1 - P(U_n^* \leq t_n) \sim \Phi(-t_n) \sim P(U_n^* \leq -t_n)$ *for*
$t_n^2 \leq \alpha(\log n)^{\nu/(\nu-1)} + M,\ M > 0,\ t_n \to \infty$,
where $\alpha = (2-\epsilon)w_o^{-1/(\nu-1)}(\nu-1)\nu^{-\nu/(\nu-1)}$, $s = w_o^{-1/(\nu-1)}$, $\epsilon > 0$ *is arbitrary small,* w_o *and* ν *are defined in* (4.15) *and* $M > 0$ *may be arbitrary large.*

Proof. Write,

$$\begin{aligned}|P(U_n^* \leq t) - \Phi(t)| \leq\ & |P(\mathrm{var}(V_1))^{-1/2}V_1 \leq t \pm a_n(t)) - \Phi(t \pm a_n(t))| \\ & + |\Phi(t \pm a_n(t)) - \Phi(t)| + P(|R_{nn}^*| > a_n(t)).\end{aligned} \tag{4.20}$$

Under (4.16), first term of the r.h.s. of (4.20) may be approximated from Theorem 2.1 of Dasgupta (1989); the condition (4.5) therein is satisfied for independent random variables $\overline{\psi}^{(1)}(X_i)$ and $\overline{\psi}^{(1)}(Y_j)$ with $g(x) = \exp[s'\{\log_e(1+|x|)\}^{\nu/(\nu-1)}]$, $s' < s$, since from Jensen's inequality for conditional expectation and C_{2q} inequality, the $2q$th moment bounds for $\overline{\psi}^{(1)}(X_i)$ and $\overline{\psi}^{(1)}(Y_j)$ are of same order as that for ϕ. Also recall the steps of truncation of random variables in Theorem 2.1 of Dasgupta (1989). There were n independent random variables in nth array, and the variables were truncated at $rs_n|t|$, where s_n^2 is the sum of variances of variables in that array, $s_n^2 = O_e(n)$. In the present case of U-statistics we have sum of $(n+m)$ independent random variables in V_1 with variance of the sum as $\sigma^{*2} = \text{var}(V_1) = O_e(\frac{1}{n+m})$. So, an appropriate truncation point of the random variables $\frac{2}{n}\overline{\psi}^{(1)}(X_i)$ and $\frac{2}{m}\overline{\psi}^{(1)}(Y_j)$ in V_1 would be $r\sigma^*|t|$.

In other words the truncation points for $\overline{\psi}^{(1)}(X_i)$ and $\overline{\psi}^{(1)}(Y_j)$ would be $r\sigma_n|t|$, as $\sigma_n^2 = (n+m)^2\sigma^{*2} = O_e(n+m)$. Since the value of constant $r \in (0, 1/2)$ is adjustable, O_e terms arising out of truncation can also be adjusted appropriately in the set-up.

Take $a_n = n^{-\gamma}$, $\gamma > 0$, small. Then following the steps of (4.23) of Dasgupta (1989), for $1 \le t^2 \le 2(\log t + \log g(r\sigma_n t))$, one gets

$$
\begin{aligned}
|P(\text{var}(V_1))^{-1/2}V_1 \le t \pm a_n(t)) - \Phi(t \pm a_n(t))| &\le b\exp(-(t \pm n^{-\gamma})^2/2) \\
&\times |t \pm n^{-\gamma}|^{-1}|\exp(O(|t \pm n^{-\gamma}|^3\, n^{-1/2})) - 1| + \sum_{i=1}^{n} P(\frac{2}{n}\overline{\psi}^{(1)}(X_i) > r\lambda\sigma^*|t|) \\
&+ \sum_{j=1}^{m} P(\frac{2}{m}\overline{\psi}^{(1)}(Y_j) > r\lambda\sigma^*|t|), \\
&\le b\exp(-t^2/2)|t|^{-1}|\exp(O(|t|^3\, n^{-1/2})) - 1| + \sum_{i=1}^{n} P(\overline{\psi}^{(1)}(X_i) > r\lambda\sigma_n|t|) \\
&+ \sum_{j=1}^{m} P(\overline{\psi}^{(1)}(Y_j) > r\lambda\sigma_n|t|), \qquad (4.21)
\end{aligned}
$$

for some $\lambda, 0 < \lambda < 1$, and let $b(>0)$ be a generic constant. The factor λ appears as we considered $t \pm a_n(t)$ in the above inequality. Next,

$$
|\Phi(t \pm a_n(t)) - \Phi(t)| \le b\; a_n(t)e^{-t^2/2} = bn^{-\gamma}e^{-t^2/2}. \qquad (4.22)
$$

Finally, consider
$P(|R^*_{nn}| > a_n(t)) \le a_n^{-2q}(t)\, n^{-q}\; Le^{wq^{\nu}} = P^*$, say.
The minimum of the above is attained when,
$\frac{d}{dq}\log P^* = -2\log a_n - \log n + w\nu q^{\nu-1} = 0 \quad \Rightarrow q = [(2\log a_n + \log n)/(w\nu)]^{1/(\nu-1)}$.
So, $\log P^* = w(1-\nu)q^{\nu} + \log L = w(1-\nu)[(2\log a_n + \log n)/(w\nu)]^{\nu/(\nu-1)} + \log L$.
Then
$\log P^* = w(1-\nu)[(1-2\gamma)(\log n)\,/(w\nu)]^{\nu/(\nu-1)} + \log L$. That is,

$$
P(|R^*_{nn}| > a_n(t)) = O[e^{-w(\nu-1)\{(1-2\gamma)(\log n)\,/(w\nu)\}^{\nu/(\nu-1)}}]. \qquad (4.23)
$$

For normal approximation zone, one requires the terms in r.h.s of (4.20) to be $o(\Phi(-t))$, where $\Phi(-t) \sim \frac{1}{\sqrt{2\pi}} t^{-1}e^{-t^2/2}$, $t \to \infty$. The restriction from the first term on the r.h.s. of (4.21) states $t = o(n^{1/6})$.

From second and third term on the r.h.s. of (4.21) the restriction is $t^2 \leq 2(\log t + \log g(r\lambda\sigma_n t))$.

A similar condition $1 \leq t^2 \leq 2(\log t + \log g(r\sigma_n t))$ is required for validity of (4.21). Now, $t^2 \leq 2(\log t + \log g(r\sigma_n t)) \approx 2\log t + 2s'[\log(r\sigma) + \frac{1}{2}\log n + \log t]^{\nu/(\nu-1)}, r > 0, \sigma > 0 \Rightarrow t^2 \leq 2s'(\log n)^{\nu/(\nu-1)}(1 + o(1))$, and $\log t \leq \frac{\nu}{2(\nu-1)} \log\log n\ (1 + o(1))$. Hence the conditions from (4.21) simplify to

$$t^2 \leq 2s'\left[\frac{1}{2}\log n + \frac{\nu}{2(\nu-1)}\log\log n)\right]^{\nu/(\nu-1)} + \frac{\nu}{\nu-1}\log\log n + M, \quad (4.24)$$

$s' < s = w_o^{1/(1-\nu)}$ and $M > 0$.

Next the restriction from (4.22) for normal approximation is $t = o(n^{\gamma})$.

Finally, with a small choice of $\gamma(> 0)$, restriction from (4.23) is

$$t_n^2 \leq 2w_1^{-1/(\nu-1)}(\nu-1)\nu^{-\nu/(\nu-1)}(\log n)^{\nu/(\nu-1)}, w_1 > w > w_o,\ \nu > 1. \quad (4.25)$$

It therefore follows that (4.24) and (4.25) determine the normal approximation zone. The leading term in (4.24) is $t^2 \leq 2s'(\frac{1}{2}\log n)^{\nu/(\nu-1)}$. Comparing this with (4.25) with $w_1 \approx w_o = s^{1-\nu}$, $s' \approx s = w_o^{1/(1-\nu)}$, and observing that $\nu^{\nu} \geq (\nu-1)^{\nu-1}$, $\nu > 1$ it follows that the zone (4.25) is more restrictive, and hence determines the normal approximation zone. The theorem follows by taking w_1 arbitrary close to w_o in (4.25).

Remark 1. The leading term for t_n^2 in Theorem 1 is of order $(\log n)^{\nu/(\nu-1)}$. The normal approximation zone for standardized sample sum of iid random variables is also of same order, as given by (4.24); see also Theorem 2.3 of Dasgupta (1989). These zones are larger than moderate deviation zone, as $\nu > 1$. Moderate deviation results hold when some finite moment (>2) exists. In (4.15)/(4.16) we assumed existence of *all* the moments for kernel ϕ, and hence the resulting zone gets extended beyond moderate deviation.

Denote $G_n(t) = P(U_n^* \leq t)$. The next theorem provides an overall nonuniform bound in the CLT for standardized U-statistics.

Theorem 2. *Under the assumptions of Theorem 1, there exists a constant $b > 0$, depending on w and ν such that the following holds.*

$$|G_n(t) - \Phi(t)| \leq b\, n^{-\frac{1}{2}+\epsilon_n} e^{-c(\log(1+|t|))^{\nu/(\nu-1)}},\ \nu > 1,\ -\infty < t < \infty, \quad (4.26)$$

where $c = w(\nu-1)\{2/(w\nu)\}^{\nu/(\nu-1)} > 0$, and $\epsilon_n = (2c)^{-(\nu-1)/\nu}(\log n)^{-1/\nu} = O((\log n)^{-1/\nu}) \to 0$, as $n \to \infty$.

Remark 2. Observe that in the nonuniform bound (4.26), the part depending on $|t|$ decreases at a faster rate than any polynomial power of $|t|$. Further observe that the uniform bound of the rate approaches to the optimal bound $O(n^{-1/2})$, as the excess $\epsilon_n = O((\log n)^{-1/\nu}) \to 0, n \to \infty$.

Proof of Theorem 2. Let $a_n(t) = n^{-\frac{1}{2}+\epsilon_n}|t|$, then proceeding like (4.21), we have for the region $1 \le t^2 \le 2(\log t + \log g(r\sigma_n t))$,

$$|P(\text{var}(V_1))^{-1/2}V_1 \le t \pm a_n(t)) - \Phi(t \pm a_n(t))| \tag{4.27}$$
$$\le b\exp(-t^2/2)|t|^{-1}|\exp(O(|t|^3\, n^{-1/2})) - 1| + \sum_{i=1}^{n} P(\overline{\psi}^{(1)}(X_i) > r\lambda\sigma_n|t|)$$
$$+\sum_{j=1}^{m} P(\overline{\psi}^{(1)}(Y_j) > r\lambda\sigma_n|t|),$$

for some $\lambda, 0 < \lambda < 1$. Also,

$$|\Phi(t \pm a_n(t)) - \Phi(t)| \le bn^{-\frac{1}{2}+\epsilon_n}|t|e^{-t^2/2}. \tag{4.28}$$

Next, proceeding like (4.23)

$$P(|R^*_{nn}| > a_n(t)) = O[e^{-w(\nu-1)[(2\log a_n(t)+\log n)/(w\nu)]^{\nu/(\nu-1)}}] \tag{4.29}$$
$$\le be^{-w(\nu-1)[2(\epsilon_n \log n+\log t)/(w\nu)]^{\nu/(\nu-1)}}$$
$$\le be^{-c[(\epsilon_n \log n)^{\nu/(\nu-1)}+(\log t)^{\nu/(\nu-1)}]}, \quad c = w(\nu-1)\{2/(w\nu)\}^{\nu/(\nu-1)}$$
$$= O[n^{-1/2}e^{-c(\log(1+|t|))^{\nu/(\nu-1)}}], \text{ as } \epsilon_n = (2c)^{(1-\nu)/\nu}(\log n)^{-1/\nu}.$$

In the complementary zone, i.e., for $t^2 > 2(\log t + \log g(r\sigma_n t))$, from Theorem 2.5 of Dasgupta (1989), we have in place of (4.27), the following:

$$|P(\text{var}(V_1))^{-1/2}V_1 \le t \pm a_n(t)) - \Phi(t \pm a_n(t))| \le O(|t|g(r\lambda\sigma_n|t|))^{-1+\epsilon_{n,t}}$$
$$+\sum_{i=1}^{n} P(\overline{\psi}^{(1)}(X_i) > r\lambda\sigma_n|t|) + \sum_{j=1}^{m} P(\overline{\psi}^{(1)}(Y_j) > r\lambda\sigma_n|t|), \tag{4.30}$$

for some $\lambda,\ 0 < \lambda < 1$, and $\epsilon_{n,t} \to 0$, as $n \to \infty$.

For the remaining zone $t^2 < 1$, an uniform bound $O(n^{-\frac{1}{2}+\epsilon_n})$ is available for $|G_n(t) - \Phi(t)|$ letting $a_n = n^{-\frac{1}{2}+\epsilon_n}$, i.e., taking $|t| = 1$ in $a_n(t)$ and proceeding as before and then using the inequality $||F(X+Y) - \Phi|| \le ||F(X) - \Phi|| + (\sqrt{2\pi})^{-1}a_n + P(|Y| > a_n)$.

From the above uniform bound and from (4.27) to (4.30), we may now have an overall nonuniform bound for $|G_n(t) - \Phi(t)|$. The order of the last two terms in the r.h.s. of (4.27)/(4.30) is $O(|t|g(r\lambda\sigma_n|t|))^{-1}$, where $g(x) = \exp[s'\{\log_e(1 + |x|)\}^{\nu/(\nu-1)}]$, $\sigma_n = O_e(n^{1/2})$, $s' < s = w_o^{1/(1-\nu)}$, the coefficient of $(\log(1 + |t|))^{\nu/(\nu-1)}$ in the exponent on r.h.s. of (4.29) dominates. Hence the theorem.

As a consequence of Theorem 2, the following two theorems on nonuniform L_p version of Berry–Esseen theorem and moment type convergence are immediate, when one uses the representation, $|Eh(U_n^*) - Eh(T)| \leq \int_0^\infty h'(t)|P(|U_n^*| \leq t) - P(|T| \leq t)|dt$, for $h(x) = x^2 g(x)$; see also Theorems 2.7 and 2.8 of Dasgupta (1989).

Theorem 3. *Under the assumptions of Theorem 2,*

$$||e^{c(\log(1+|t|))^{\nu/(\nu-1)}}(1 + |t|)^{-q/p}(G_n(t) - \Phi(t))||_p = O(n^{-\frac{1}{2}+\epsilon_n}). \tag{4.31}$$

for $p \geq 1$ *and any* $q > 1$.

Theorem 4. *Under the assumptions of Theorem 2 and for a non-negative even function g with*

$$\frac{d}{dx}[x^2 g(x)] = O((1 + x)^{-q} e^{c(\log(1+|x|))^{\nu/(\nu-1)}}),\ \forall x > 0,\ \text{and}\ q > 1; \tag{4.32}$$

the following holds for standardized U-statistic U_n^* *and a* $N(0,1)$ *variable T.*

$$|E(U_n^{*2} g(U_n^*)) - E(T^2 g(T))| = O(n^{-\frac{1}{2}+\epsilon_n}). \tag{4.33}$$

4.4 Rates of Convergence and Deviation Probabilities: Part 2

Next consider a different moment bound for the kernel ϕ in place of (4.15)/(4.16).

Assume that

$$\sup_{n\geq 1, m\geq 1} (n_{c_2} m_{c_2})^{-1} \sum_{\substack{1 \leq i_1 < i_2 \leq n \\ 1 \leq j_1 < j_2 \leq m}} E\,|\,\phi(X_{i_1}, X_{i_2}, Y_{j_1}, Y_{j_2})\,|^{2q} = \delta_q \leq L^q e^{\nu q \log q} \tag{4.34}$$

$\forall q > 1$, where $L > 0$, $\nu > 1$. The above condition is implied by

$$\sup_{n\geq 1, m\geq 1} (n_{c_2} m_{c_2})^{-1} \sum_{\substack{1 \leq i_1 < i_2 \leq n \\ 1 \leq j_1 < j_2 \leq m}} E \exp(s|\phi|^{1/\nu}) < \infty, \tag{4.35}$$

where $0 < s < s_o = \nu e^{-1} L^{-1/\nu}$ and $\phi = \phi(X_{i_1}, X_{i_2}, Y_{j_1}, Y_{j_2})$.

This follows along the lines of Dasgupta (2006), see Proposition 2.1 and Remark 2.2 therein. Although this is a stronger assumption than (4.15)/(4.16), it is still a milder restriction compared to assuming existence of m.g.f. for ϕ, which corresponds to $\nu = 1$.

The following theorem provides a normal approximation zone under this moment bound. For the special case $\phi = \phi(X_{i_1}, X_{i_2}, Y_{j_1}, Y_{j_2})$, one has $\nu = 4$.

Theorem 5. *Let* $m = O_e(n)$. *Under the assumptions* (4.2), (4.34)/(4.35) *and* (4.18), *for the standardized* U*-statistic* U_n^* *defined in* (4.17), *one has* $1 - P(U_n^* \le t_n) \sim \Phi(-t_n) \sim P(U_n^* \le -t_n)$, $t_n \to \infty$, *for* $t_n = o(n^{1/(2(\nu+\nu+1))})$.

Proof. Let $a_n(t) = n^{-\gamma}$, and proceed like (4.20)–(4.23) to have similar expressions for the present moment bound with $g(x) = \exp(s|x|^{1/\nu})$.

Over the zone $1 \le t^2 \le 2(\log t + \log g(r\sigma_n t))$ ($\Rightarrow 1 \le t^2 \le 2s(r\sigma_n t)^{1/\nu} \Rightarrow 1 \le t^2 \le \epsilon n^{\nu/(2\nu-1)}$, for some $\epsilon > 0$), one has

$$
\begin{aligned}
&|P(\mathrm{var}(V_1))^{-1/2} V_1 \le t \pm a_n(t)) - \Phi(t \pm a_n(t))| \\
&\qquad \le b\ \exp(-t^2/2)|t|^{-1}|\exp(O(|t|^3\, n^{-1/2})) - 1| + \\
&\qquad + \sum_{i=1}^{n} P(\overline{\psi}^{(1)}(X_i) > r\lambda\sigma_n|t|) + \sum_{j=1}^{m} P(\overline{\psi}^{(1)}(Y_j) > r\lambda\sigma_n|t|), \qquad (4.36)
\end{aligned}
$$

for some $\lambda,\ 0 < \lambda < 1$.

Order of last two terms in the r.h.s. of (4.36) is $O(|t|g(r\lambda\sigma_n t))^{-1} = o(t^{-1}e^{-t^2/2})$, for $t^2 \le \epsilon n^{\nu/(2\nu-1)}$, for some $\epsilon > 0$, as $g(x) = \exp(s|x|^{1/\nu})$. Again,

$$
|\Phi(t \pm a_n(t)) - \Phi(t)| \le b\ a_n(t) e^{-t^2/2} = b n^{-\gamma} e^{-t^2/2}. \qquad (4.37)
$$

Next, in view of $E(R_{nm}^*)^{2q} \le n^{-q}\ L^q (\nu q)!\ \delta_q$, we have from (4.34)

$$
E(R_{nm}^*)^{2q} \le n^{-q}\ L^q e^{(\nu+\nu)q \log q},\ \forall q > 1. \qquad (4.38)
$$

$$
P(|R_{nm}^*| > a_n(t)) \le a_n^{-2q}(t) n^{-q}\ L^q e^{(\nu+\nu)q\log q} = P^*,\ \text{say}. \qquad (4.39)
$$

Then, $\log P^* = -2q \log a_n - q \log n + q \log L + (\nu+\nu) q \log q$.
$\frac{d}{dq} \log P^* = -2 \log a_n - \log n + \log L + (\nu+\nu) + (\nu+\nu) \log q = 0$
$\Rightarrow (\nu+\nu)\log q = 2\log a_n + \log n - \log L - (\nu+\nu) \Rightarrow q = (a_n^2 n L^{-1})^{1/(\nu+\nu)} e^{-1}$.

Thus, minimizing the r.h.s. of (4.39) with the above value of q, one gets

$$
P(|R_{nm}^*| > a_n(t)) \le P^* = e^{-(\nu+\nu)q} = e^{-(\nu+\nu)(a_n^2 n L^{-1})^{1/(\nu+\nu)} e^{-1}}. \qquad (4.40)
$$

Now the restriction from (4.37) for normal approximation zone is $t = o(n^{\gamma})$. From (4.40) the restriction is $t \le \epsilon (a_n^2 n)^{1/(2(\nu+\nu))} = \epsilon\{n^{1-2\gamma}\}^{1/(2(\nu+\nu))}$, for some

$\epsilon > 0$. Equating the powers of n from these two restrictions one gets $\gamma = 1/(2(v + \nu + 1))$. The restrictions from (4.36) are $t = o(n^{1/6})$ and $t \leq \epsilon n^{\nu/(2(2\nu-1))}$, for some $\epsilon > 0$; these turn out to be broader zone than $t = o(n^{\gamma}),\ \gamma = 1/(2(v + \nu + 1))$. Thus the former two restrictions with $\gamma = 1/(2(v + \nu + 1))$ determine the zone. Hence the theorem.

Remark 3. The kernel of Wilcoxon 2-sample statistic is an indicator function and therefore bounded. Consequently $\delta_q = O(L^q)$, and v may be taken to be 0 in the calculations (4.38)–(4.40). For random variables having finite m.g.f., restriction from (4.36) for the main part of projection is $t = o(n^{1/6})$, see also Dasgupta (1989). The number of arguments ν in the kernel ϕ for Wilcoxon 2-sample statistics is 2. Thus from the above calculations, normal approximation zone for Wilcoxon 2-sample statistic is $t = o(n^{\gamma}) = o(n^{1/6})$, as $\gamma = 1/(2(v + \nu + 1)) = 1/6$, with $v = 0,\ \nu = 2$. This zone, in general, is the best possible zone even for iid random variables, see Theorem 2.3 of Dasgupta (1989).

The next theorem provides an overall nonuniform bound for $|G_n(t) - \Phi(t)|$.

Theorem 6. *Under the assumptions of Theorem 5, there exists a constant $b > 0$, depending on β, ν and δ such that the following holds:*

$$|G_n(t) - \Phi(t)| \leq b\, n^{-\frac{1}{2}} (\log n)^{\delta} e^{-\beta |t|^{(1/v)\wedge(2/(v+\nu))}}, \quad -\infty < t < \infty, \tag{4.41}$$

where $\beta > 0$, may be arbitrary large and $\delta > (v + \nu)/2$; may be arbitrary near to $(v + \nu)/2$.

Proof. To obtain an overall nonuniform bound in this case consider a different choice $a_n(t) = n^{-\frac{1}{2}}(\log n)^{\delta}|t|,\ \delta > 0$, to be chosen later. Proceeding like (4.36), we have for the region $1 \leq t^2 \leq 2(\log t + \log g(r\sigma_n t))$,

$$\begin{aligned} &|P(\mathrm{var}(V_1))^{-1/2} V_1 \leq t \pm a_n(t)) - \Phi(t \pm a_n(t))| \\ &\leq b \exp(-t^2/2)|t|^{-1} |\exp(O(|t|^3\, n^{-1/2})) - 1| + \sum_{i=1}^{n} P(\overline{\psi}^{(1)}(X_i) > r\lambda\sigma_n|t|) \\ &\qquad + \sum_{j=1}^{m} P(\overline{\psi}^{(1)}(Y_j) > r\lambda\sigma_n|t|), \end{aligned} \tag{4.42}$$

for some $\lambda,\ 0 < \lambda < 1$. Also,

$$|\Phi(t \pm a_n(t)) - \Phi(t)| \leq b n^{-\frac{1}{2}} (\log n)^{\delta} |t| e^{-t^2/2}. \tag{4.43}$$

Next, proceeding like (4.40)

$$P(|R^*_{nm}| > a_n(t)) \le e^{-(v+\nu)(a_n^2(t)nL^{-1})^{1/(v+\nu)}e^{-1}} \tag{4.44}$$
$$\le bn^{-1/2}e^{-\beta|t|^{2/(v+\nu)}},$$

where $\beta > 0$, may be taken arbitrarily large, selecting $\delta > (v+\nu)/2$.

In the complementary zone, i.e., for $t^2 > 2(\log t + \log g(r\sigma_n t))$, from Theorem 2.5 of Dasgupta (1989), we have in place of (4.36), the following:

$$|P(\mathrm{var}(V_1))^{-1/2}V_1 \le t \pm a_n(t)) - \Phi(t \pm a_n(t))| \le O(|t|g(r\lambda\sigma_n|t|))^{-1+\epsilon_{n,t}}$$
$$+\sum_{i=1}^{n} P(\overline{\psi}^{(1)}(X_i) > r\lambda\sigma_n|t|) + \sum_{j=1}^{m} P(\overline{\psi}^{(1)}(Y_j) > r\lambda\sigma_n|t|), \tag{4.45}$$

for some $\lambda,\ 0 < \lambda < 1$, and $\epsilon_{n,t} \to 0$, as $n \to \infty$.

From (4.42) to (4.45), we may have an overall nonuniform bound for $|G_n(t) - \Phi(t)|$. The order of the last two terms in the r.h.s. of (4.42)/(4.45) is $O(|t|g(r\lambda\sigma_n|t|))^{-1} \le bn^{-1/2}e^{-\beta|t|^{1/\nu}}$, as $g(x) = \exp(s|x|^{1/\nu})$.

For the zone $t^2 < 1$, an uniform bound $O(n^{-\frac{1}{2}}(\log n)^{\delta}),\ \delta > (v+\nu)/2$ holds for $|G_n(t) - \Phi(t)|$, letting $a_n = n^{-\frac{1}{2}}(\log n)^{\delta}$ and proceeding like Theorem 2.

Hence the theorem. Observe that the part involving $|t|$ in (4.41) decreases at a faster rate than that of Theorem 2.

As a consequence of Theorem 6, the following two theorems is immediate.

Theorem 7. *Under the assumptions of Theorem 5, for any $p > 1$*

$$||e^{\beta|t|^{(1/\nu)\wedge(2/(v+\nu))}}(G_n(t) - \Phi(t))||_p = O(n^{-\frac{1}{2}}(\log n)^{\delta}), \tag{4.46}$$

where $\beta(>0)$ may be arbitrary large; and $\delta > (v+\nu)/2$, may be taken arbitrary near to $(v+\nu)/2$.

Theorem 8. *Under the assumptions of Theorem 5, and for a non-negative even function g with*

$$\frac{d}{dx}[x^2 g(x)] = O(e^{\beta|x|^{(1/\nu)\wedge(2/(v+\nu))}}),\ \forall x > 0, \tag{4.47}$$

where $\beta(>0)$ may be arbitrary large, the following holds for standardized U-statistic U_n^* and a $N(0,1)$ variable T.

$$|E(U_n^{*2}g(U_n^*)) - E(T^2 g(T))| = O(n^{-\frac{1}{2}}(\log n)^{\delta}). \tag{4.48}$$

Remark 4. The condition $m = O_e(n)$ may be relaxed. One needs to assume that (n, m) are nondecreasing in each coordinate, while both the coordinates tend to infinity. The moment bound of remainder components in (4.11)–(4.13) depends on m as well as n, sometimes in an isolated manner. The results on nonuniform rates in CLT and allied subsequent results then hold with n replaced by $m \wedge n$.

4.5 Application I: Efficiency and Discretization

The null distribution of Wilcoxon 2-sample statistic is distribution free when the samples are drawn from a continuous distribution F, as $X > Y \Leftrightarrow F(X) > F(Y)$. In the following we study the effect of discretization on the efficiency of U-statistics considering the parent distribution to be $U(0, 1)$ relative to discretized version of $U(0, 1)$.

There are possibilities of ties while sampling from a discrete distribution. The kernel ϕ associated with Wilcoxon two-sample statistics is an indicator function taking the values 0 and 1 when there are no ties. In case of a tie between x, y values it assumes the mid value $1/2$.

Consider $m = n$. The mean and variance of the U-statistic U_n and the corresponding discretized version U_n' are as follows:

$$E(U_n) = n^2/2 = E(U_n'). \tag{4.49}$$

$$V(U_n) = n^2(2n+1)/12,\ V(U_n') = n^2\{2n+1-\epsilon/n(2n-1)\}/12, \tag{4.50}$$

where $\epsilon = \sum_k (a_k^3 - a_k)$; a_k being the number of observations in k block of ties when all the x, y observations are pooled together and ordered.

Discretization and subsequent presence of ties in observations in U-statistics are known to slow down the convergence to limiting normal distribution. We would like to study the performance of U_n and U_n', as test statistics in terms of type I and type II errors for moderately large values of n.

For standardized U-statistics U_n^* and $U_n'^*$, normal approximation for tail probability on the zone $t = o(n^{1/6})$ is possible in the limit as $n \to \infty$, see Remark 3. This is a large deviation zone. Proposition 1 of Dasgupta (2010) extends the result of Rubin and Sethuraman (1965) from moderate deviations to large deviations of the form $an^{\alpha/2}$, $a > 0$, $\alpha > 0$.

For a statistic $T = T_n = T_n(x)$, consider a large deviation zone as critical region $w = w(a, n, \alpha) = \{||T_n|| > an^{\alpha/2}\}$ for testing $H_o : E(T) = 0$ vs. $H_1 : E(T) \neq 0$. In iid set-up of p-dimensional normal random vectors $X_1, \cdots, X_n$; $E(X) = \zeta$ with $T_n = \sqrt{n}\ \overline{X}_n$, one may write (see Dasgupta 2010, Rubin and Sethuraman 1965), the type I risk of the above critical region in Bayesian framework as $\pi = \pi_n \approx K(n^{-1}\log n)^{(p+\lambda-2)/2} = Kd_n$ say, where $\lambda = \lambda_n = a^2 n^\alpha/\log n$, $a > 0$, $\alpha > 0$. Here λ depends on $W = W(\zeta) \propto ||\zeta||^\lambda$, the weight function, defined as prior density times the loss due to accepting $H_o : \zeta = 0$ when H_o is false. The type II error is $\log n$ times type I error. Efficiency of a test statistic may then be interpreted in terms of the slope K. The slope K affects the decrease in error, lower value corresponds to the superior test statistic. The ratio of two type I errors corresponding to two test statistics approximately equals to the ratio of two slopes. Similar relationship holds for type II errors.

One may compare two test statistics for testing H_o in terms of type I and type II errors as done subsequently, for large deviation critical region w, from a general point of view, not necessarily Bayesian.

To assess the relative performance of U_n^* and $U_n'^*$, we compare the type I and type II errors for large deviation type critical regions. Since there are $2n$ independent observations in Hájek's projection of U-statistic, $2n$ takes the role of n. The truncation point considered in the critical region $(2n)^{1/6.01}$ satisfies $t = o(n^{1/6})$ of Remark 3. The framework of the theory has been used to fix the set-up of simulation with finite sample sizes. We simulate the exact tail probabilities based on a large number of random simulations to compare U_n^* and $U_n'^*$, as there are situations when theory of normal approximation and simulations may not quite agree, and related constants have a role for moderately large sample size. Observe that, $(2n)^{1/6.01} \approx 3.156, (3.481)$ for $n = 500, (950)$, in the range of simulation. These truncation points fall beyond 3σ limit, satisfying the criteria of rare events as in large deviation.

In the following table, the probabilities $P(|T_n| > (2n)^{1/6.01})$ are simulated for $T_n = U_n^*,\ U_n'^*$, where the observations x, y are from Uniform (0, 1); in the discrete version U', observations x, y are rounded up after first decimal place. For different values of $n \in [500, 950]$, the type I error π for U_n^* appear in second column, and the power $(1 - \beta)$ of the test under the alternative hypothesis $H_1 : x \sim U(0,1),\ y \sim U(0.1, 1.1)$ are given in third column of Table 4.1. Fourth and fifth columns refer to the same for discretized version of Wilcoxon two-sample statistic, when the basic observations (x, y) are rounded up at first decimal place. The sixth column refer to ratio (π/π') of type I errors, for U_n^* and $U_n'^*$. The seventh column refers to the ratio (β/β') for type II errors. The results given in Table 4.1 are based on 5×10^4 random simulations performed in S-PLUS with assigned seed value 1111.

The behaviour of type I and type II errors reflect the performance of U_n and U_n' for moderately large values of n, placing U_n' in a superior position.

We further explain Table 4.1. For $n = 500$, the first entry in second column viz., .00174 is the simulated type I error of U_n^* estimated as the relative frequency of the cases where $(|U_n^*| > (2n)^{1/6.01})$, out of 5×10^4 simulations. The next value .00158 in the first row refers to the same for $U_n'^*$. The next entry .98294 is the power of the critical region $\{|U_n^*| > (2n)^{1/6.01}\}$, simulated under alternative hypothesis $H_1 : x \sim U(0,1),\ y \sim U(0.1, 1.1)$. The fifth entry .98338 refers to the same for discretized version $U_n'^*$. The sixth entry 1.10127 is the ratio of two type I errors .00174 and .00158. The seventh entry 1.02647 $(= (1-.98294)/(1-.98338))$ refers to ratio of type II errors of $U_n'^*$ and U_n^*.

The last two columns represent the efficiency of discretized version of U-statistic $U_n'^*$ compared to U_n^*, as ratio of errors, over different values of n.

Simulation supplements theory, especially for examining the closeness to limiting behaviour in a sample of finite size. Estimate of the ratio of slopes are directly computed via simulation.

It is surprising to observe that discretized version $U_n'^*$ perform relatively better than the original U_n^*. Mean and median of K/K' values corresponding to type I errors are 1.070844 and 1.08255, respectively. The values corresponding to type II

Table 4.1 Simulated errors and estimated slope ratio (seed 1111)

	Type I error		Power		$K/K' \approx$	
n	π	π'	$1-\beta$	$1-\beta'$	π/π'	β/β'
500	.00174	.00158	.98294	.98338	1.10127	1.02647
550	.00116	.00110	.99070	.99090	1.05455	1.02198
600	.00114	.00114	.99396	.99456	1.0	1.11029
650	.00100	.00094	.99676	.99690	1.06383	1.04516
700	.00088	.00090	.99836	.99858	0.97778	1.15493
750	.00066	.00070	.99894	.99902	0.94286	1.08163
800	.00070	.00060	.99936	.99954	1.16667	1.39130
850	.00066	.00056	.99966	.99970	1.17857	1.13333
900	.00038	.00034	.99986	.99992	1.11765	1.75
950	.00042	.00038	.99988	.99990	1.10526	1.2

Table 4.2 Simulated errors and estimated slope ratio (seed 123)

	Type I error		Power		$K/K' \approx$	
n	π	π'	$1-\beta$	$1-\beta'$	π/π'	β/β'
500	.00188	.00180	.98226	.98216	1.04444	0.99439
550	.00128	.00128	.98910	.98952	1.0	1.04008
600	.00116	.00122	.99362	.99380	0.95082	1.02903
650	.00104	.00096	.99662	.99642	1.08333	0.94413
700	.00108	.00106	.99782	.99788	1.01887	1.02830
750	.00090	.00080	.99892	.99884	1.125	0.93103
800	.00078	.00076	.99932	.99940	1.02632	1.13333
850	.00076	.00080	.99946	.99948	0.95	1.03846
900	.00066	.00056	.99974	.99980	1.17857	1.3
950	.00044	.00040	.99996	.99996	1.1	1.0

errors are 1.191509 and 1.12181, respectively. The procedure was repeated thrice with assigned seed values as 123, 15 and 57; for different data realization and the results are given in Tables 4.2–4.4, revealing similar features.

From Table 4.2, mean and median of K/K' values corresponding to type I errors are 1.047735 and 1.03538, respectively. The values corresponding to type II errors are 1.043875 and 1.028665, respectively.

From Table 4.3, mean and median of K/K' values corresponding to type I errors are 1.086323 and 1.04533, respectively. The values corresponding to type II errors are 1.065862 and 1.04301, respectively.

From Table 4.4, mean and median of K/K' values corresponding to type I errors are 1.067166 and 1.06936, respectively. The values corresponding to type II errors are 1.1156 and 1.03703, respectively.

The values remain stable over different simulations as seen from Tables 4.1 to 4.4.

For type I errors, pooled mean and median estimates of efficiency K/K', combining values from Tables 4.1 to 4.4, are 1.068017 and 1.05919, respectively. The values corresponding to type II errors are 1.104212 and 1.04301, respectively.

Table 4.3 Simulated errors and estimated slope ratio (seed 15)

	Type I error		Power		$K/K' \approx$	
n	π	π'	$1-\beta$	$1-\beta'$	π/π'	β/β'
500	.00172	.00170	.98166	.98238	1.01177	1.04086
550	.00132	.00132	.99016	.99044	1.0	1.02929
600	.00106	.00104	.99434	.99446	1.01923	1.02166
650	.00136	.00120	.99640	.99668	1.13333	1.04516
700	.00088	.00090	.99826	.99822	0.97778	1.08434
750	.00090	.00076	.99886	.99898	1.18421	1.11765
800	.00060	.00056	.99940	.99950	1.07143	1.2
850	.00062	.00046	.99968	.99974	1.34783	1.23077
900	.00048	.00048	.99984	.99982	1.0	0.88889
950	.00038	.00034	.99990	.99990	1.11765	1.0

Table 4.4 Simulated errors and estimated slope ratio (seed 57)

	Type I error		Power		$K/K' \approx$	
n	π	π'	$1-\beta$	$1-\beta'$	π/π'	β/β'
500	.00172	.00148	.98170	.98258	1.16216	1.05052
550	.00118	.00118	.98962	.98960	1.0	0.99808
600	.00102	.00092	.99396	.99374	1.10870	0.96486
650	.00092	.00084	.99644	.99654	1.09524	1.04516
700	.00102	.00100	.99796	.99802	1.02	1.02890
750	.00064	.00068	.99888	.99882	0.94118	0.94915
800	.00052	.00058	.99920	.99930	0.89655	1.14286
850	.00066	.00060	.99960	.99966	1.1	1.17647
900	.00060	.00046	.99982	.99990	1.30435	1.8
950	.00048	.00046	.99984	.99984	1.04348	1.0

Combining all the entries of K/K' for type I and type II errors from Tables 4.1 to 4.4, the grand pooled mean and median estimate for efficiency are mean $= 1.086114$, median $= 1.04516$.

Median being a robust estimator one may conclude that an estimate of efficiency for the discrete version to be $e = 1.04516$.

Remark 5. The result in favour of discretization may be explained as follows. Under the null hypothesis there are possibilities of more frequent ties in combined observations compared to that under alternative hypothesis; when the basic observations are discretized. Ties in discretized version of U affects the variance of the distribution in such a manner that in two different situations i.e., depending on whether the null hypothesis is true or false, the distribution is squeezed or spread, respectively, near the tail, after the distribution of U' is scaled by $\text{Var}^{1/2}(U_n')$, as specified in (4.50). As a result type I error shrinks and power is enlarged for test based on U' compared to those for U, when critical region is taken to be a large deviation zone. The result indicates that discretization of continuous data may not be always detrimental to the interest of a statistician who may be concerned about the size and power of a test.

4.6 Application II: Yam-Yield Comparison and Growth Rate Estimation

Elephant foot yam is a tuber crop with good market value. The above-ground biomass or vegetative growth, along with main and auxiliary stems growth of Elephant foot yam can be frequently recorded over time, and these are good indicators for underground yam growth. Timely application of growth nutrients has positive effect on yam. Denote the unknown mean and dispersion of growth by $\mu_i(t)$ and $\sigma_i(t)$, respectively, over productive time t for two different years $i = 1, 2$. Suppose that growth observations $x_i(t), i = 1, 2$ are taken over discretized time points $t = t_j, j = 1, \cdots, m$ in productive growth periods e.g., March–November for yam, over years $i = 1, 2$. Yam-stems also have good market value as edible vegetable. The stems are about to turn pale and stooping when the underground yam is mature for harvesting. If the growth rates of stems fall below a preassigned level, farmer may like to harvest both yam and stems for marketing. We shall estimate growth rate and compare the production scenarios over years.

Let the estimates of $(\mu_i(t), \sigma_i(t))$ based on the observations $x_i(t), t = t_j, j = 1, \cdots, m$ be $(\hat{\mu}_i(t), \hat{\sigma}_i(t)), i = 1, 2$. To compare growth rates in different years we compute for $j \neq k$, the approximate values of rate given by $(\hat{\mu}_i(t_k) - \hat{\mu}_i(t_j))/(t_k - t_j) = \hat{\mu'}_i(t_j(k))$, say. Similarly define $\hat{\sigma'}_i(t))$. Let $y_{j(i)}, z_{j(i)}$ be lowess estimates of rate of change of the mean μ and standard deviation σ at time $t = t_j$ for the years $i = 1, 2$; based on $\hat{\mu'}, \hat{\sigma'}$ values, respectively. Comparison of growth rate is also relevant in economic time series, change in atmospheric carbon dioxide concentration, demography etc. The lowess estimates are insensitive to outliers.

We explain the technique by an example. Tests on growth models are being conducted in Indian Statistical Institute's Giridih farm at Jharkhand, India. One of the goals in growth experiments is to check the effect of seed weight and surface texture on yam produced. Growth of main stem is a potential indicator of underground yam deposition. In Fig. 4.1 we plot the main stem growth $x(t)$ of an yam plant, from a seed corm with weight 350 g and having moderately rough surface texture, over different time points $t = t_1, \cdots, t_{25}$. Figure 4.1 also provides the lowess estimate $\hat{\mu}(t_j)$ of the growth by locally smoothing the points $x(t_j)$. Squared residuals of basic points from response curve provide estimates of dispersion. Variance estimate may be obtained by lowess smoothing of the above, following a similar procedure adopted for estimating mean. Computation of derivative for mean component is described below. Same may be applied for derivative of standard deviation. Consistency and other properties of such lowess estimates are studied in Dasgupta (2013). Estimated growth rate for time point t_j is weighted average of $\hat{\mu'}_i(t_j(k))$, with smoothing weight w_k for kth point; a choice is to let w_k decrease proportionally to power or exponential of the distance between jth and kth time point, e.g., $w_k^{(1)}(t_j) \propto |t_j - t_k|^{-1.5}$, or $w_k^{(2)}(t_j) \propto e^{-|t_j - t_k|}$, w_k is normalised; sum of the weights being unity. Thus distant points are given less weight. The averaged points so obtained are lowess smoothed to have a robust estimate of the growth

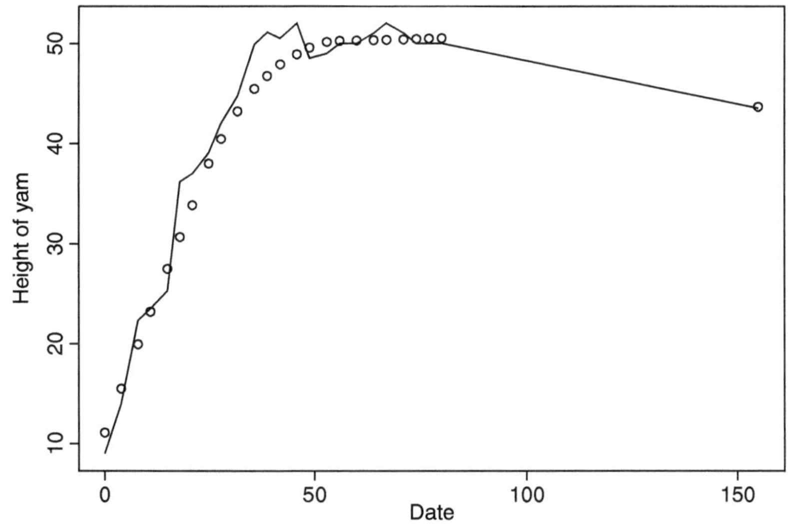

Fig. 4.1 Lowess fit for height curve of yam plant 9, stem 1

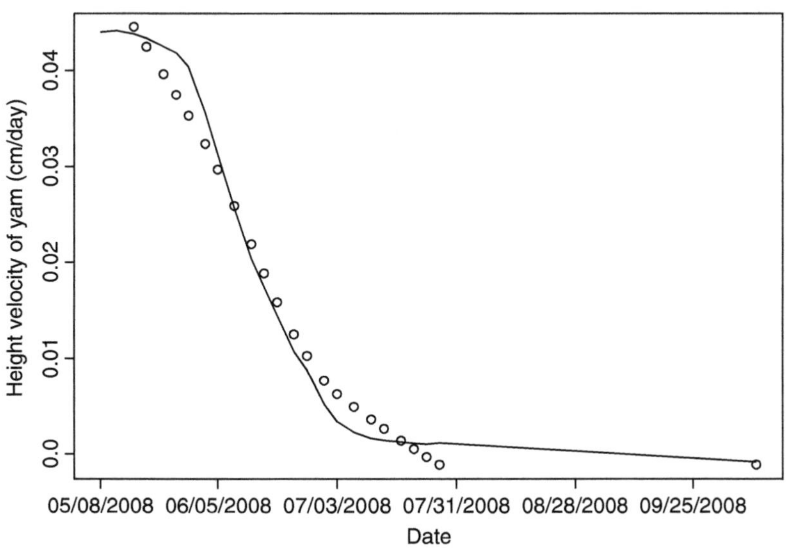

Fig. 4.2 Height velocity curve with local averaging through x^(−1.5)

rate curve. The resultant rate curve is insensitive to outliers, as lowess regression is so. In next two figures we plot estimated growth rate for different j, as the weighted average of growth rates $\hat{\mu}'(t_j(k))$ averaged over k, as basic points. These are connected by lines and the lowess estimate y_j of growth rate with weights $w^{(1)}$ and $w^{(2)}$ are shown as points in Figs. 4.2 and 4.3, respectively.

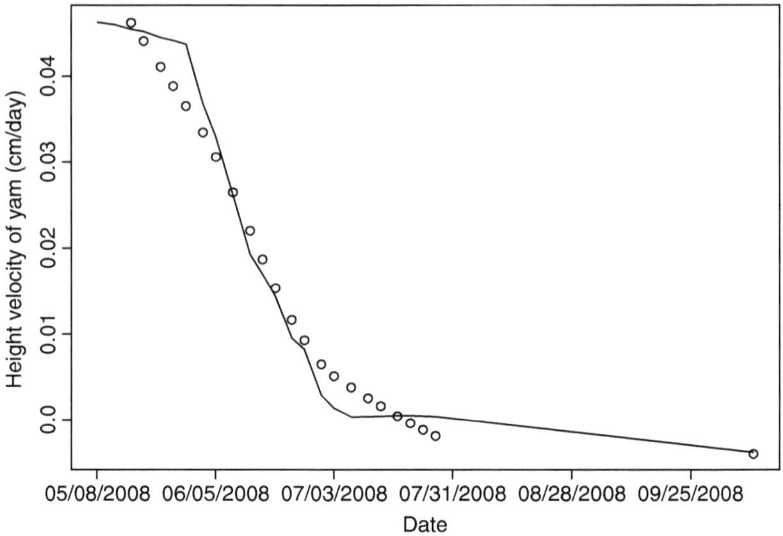

Fig. 4.3 Height velocity curve with local averaging through exp(−x)

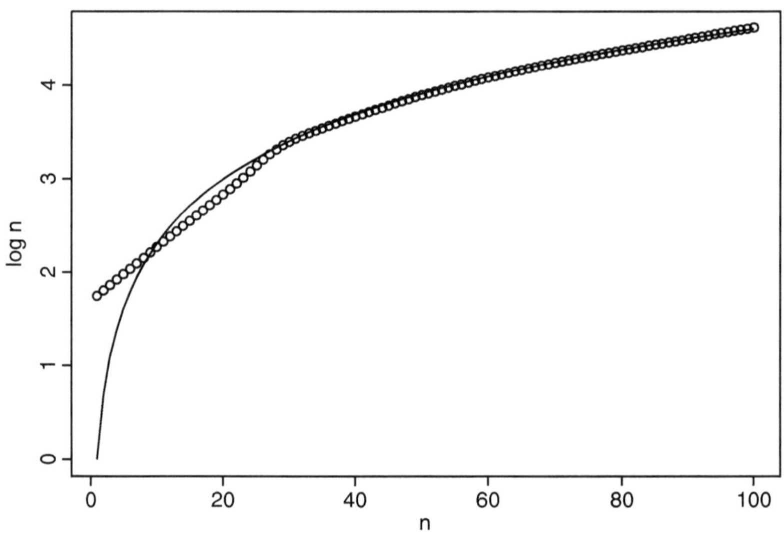

Fig. 4.4 Lowess fit for f(x) = log x

In a similar fashion derivative values z, i.e., rate of change of standard deviation σ over time may be estimated first by estimating $\sigma(t)$ based on the residuals $(x-\hat{\mu})$.

We check performance of the proposed technique in estimating the growth rate of the function $f(x) = \log x$. In Fig. 4.4 we plot the points $y = f(x) = \log x, x \in \{1, \cdots, 100\}$, along with the lowess estimate. In Fig. 4.5 we plot estimated derivative

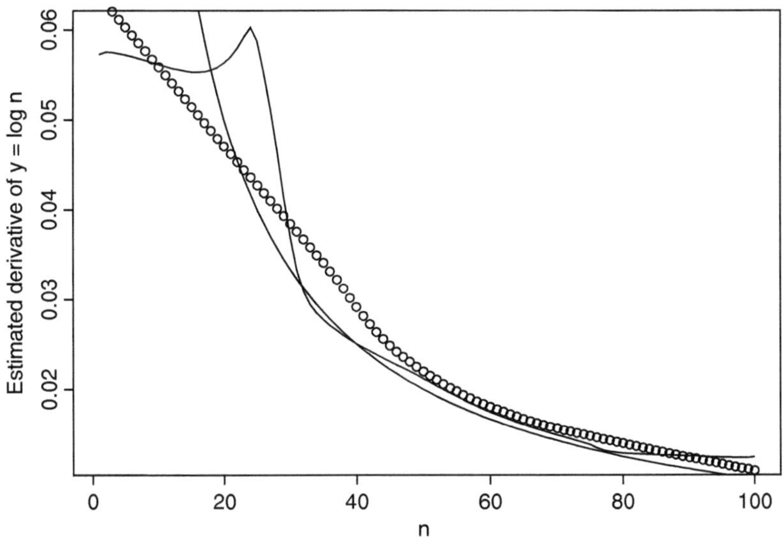

Fig. 4.5 Estimated derivative with local averaging through x^(−1.5)

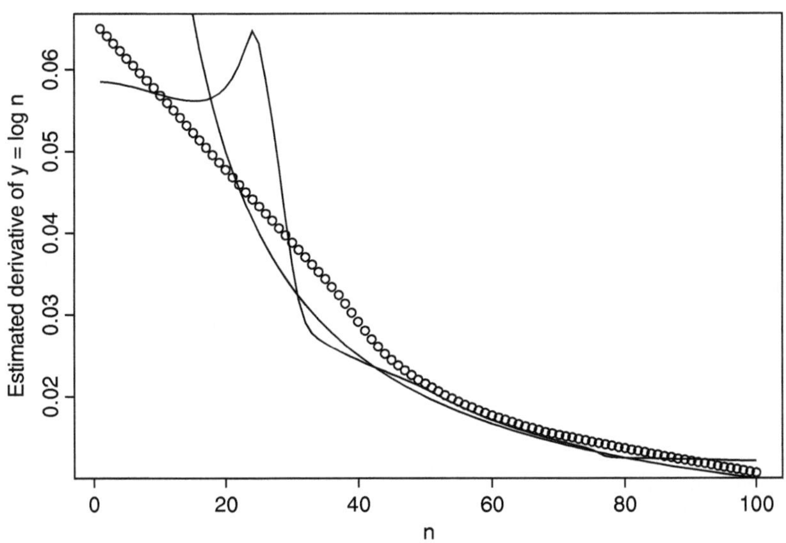

Fig. 4.6 Estimated derivative with local averaging through exp(−x)

via averaged values of $\hat{\mu}'$ with weight function $w^{(1)}$ along with the smooth lowess curve. The curve is close to the corresponding theoretical function $f'(x) = 1/x$. Figure 4.6 repeats the same as that of Fig. 4.5 with exponentially decaying weight function $w^{(2)}$.

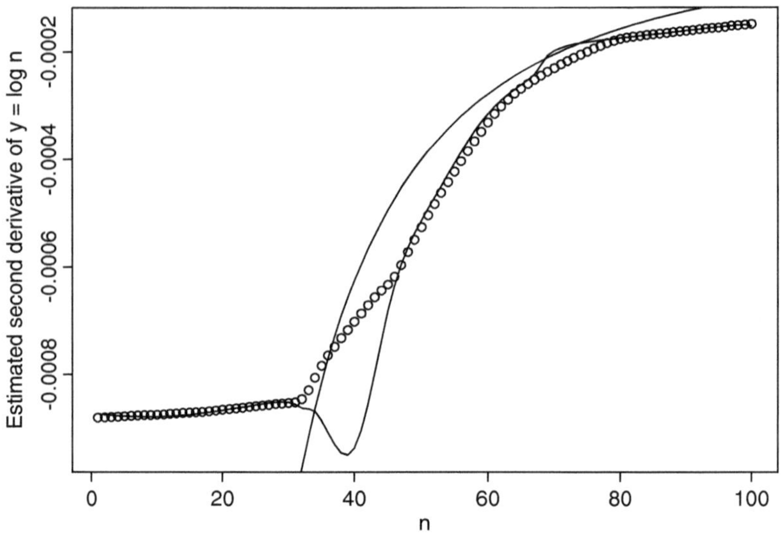

Fig. 4.7 Estimated 2nd derivative with local averaging through x^(−1.5)

The method may be used to estimate the higher order derivatives as well, in a robust manner. As for example, take the lowess values obtained from the program for first derivative as the input values at second stage to estimate the second derivative. In Fig. 4.7 we plot the curves corresponding to second derivative of $f(x) = \log x$, along with theoretical curve $f''(x) = -1/x^2$. The assigned weight function is $w^{(1)}$. Performance of the proposed technique is seen to be satisfactory, see also Dasgupta (2013).

The quantities $(y_{j(i)}, z_{j(i)})$ so obtained may be plotted over years $i = 1, 2$ to see whether there are significant variations over years.

Consider 2-sample Wilcoxon U-statistics based on y and/or z values to compare their variation over 2 years in a nonparametric manner. Lowess may induce dependence in resultant variables. However, we may ignore this when the fraction of data used in smoothing is small and weight functions decrease fast towards tail. Wilcoxon statistic is robust against outliers. Rare event of falling in the large deviation zone of the standardized U-statistics will indicate production scenarios are indeed different over years.

The values of standardized U based on y are computed progressively over $t = 1, \cdots, n$; as $n \uparrow m$, where m is large. Similar calculations are made for z-based U-statistics. We then compute the proportions of standardized U_n falling within the large deviation zone $\{u : u > (2n)^{1/6.01}\}$ as considered in Sect. 4.5, corresponding to y and z for sufficiently large values of n in a neighbourhood of m, i.e., with $n = m - k, \cdots, m$; for some integer $k > 0$. Cluster of standardized U in large deviation zone is a rare event under the null hypothesis H_o of no variation of production scenarios. We shall see that under alternative $H_1 = H_o^c$ the sequence of standardized U-statistics diverge and will cross the large deviation boundary $(2n)^{1/6.01}$, under H_1.

A test for H_o of similar production scenario in 2 years is given by standardized U_n and U_n'. From replicated independent observations from different farm productions in a year calculate the proportion of standardized U_n or U_n' falling in critical zone $\{u : u > (2n)^{1/6.01}\}$, this has to be compared with normal approximation $\Phi(-(2n)^{1/6.01})$ of large deviation probability, with ratio of the probabilities converging to 1. One may reject H_o if the relative error exceeds 1%, say.

To see the convergence rate for the ratio of probabilities to unity, note that for Wilcoxon test ϕ is indicator function with third semi-invariant zero under H_o. Consequently for the main part V_1 the normal approximation zone is up to $o(n^{3/8})$, and the speed of convergence for normal approximation at $t = o(n^{1/6})$ is quite fast, $P((\text{var}(V_1))^{-1/2}V_1 \geq t)/\Phi(-t) = O(\exp(o(n^{-1/3}))) = 1 + o(n^{-1/3})$ from (3.1) of Dasgupta (2010); truncation for bounded random variables is not needed, so the last term in (3.1) therein may be disregarded. However, the contributions from other two components in (4.37) and (4.40) of this paper provide added constants $(2\pi)^{-1/2}$ and $e^{2/e}$, respectively, in $o(1)$ term in the above approximation. Observe that $L = 1$ for bounded kernel of indicator function, and $\nu = 0, v = 2$.

One may analyse the data on weights of yams that are harvested sequentially in a production session. The stems are sold as edible vegetable before it becomes hard and pale over time with no market value. Yams with retarded growth of stems, which are going to turn pale, are in the first group for digging out. Thus, there is a natural sequence for weights of yam arrival to be stored in farm. Let these sequence for the year $i = 1, 2$ be $w_{j(i)}, i = 1, 2; j = 1, \cdots, m$. The number of yams produced m is usually quite large in a year for a yam farm. Data of yam weight being readily available over 2 different years, one may obtain an estimate of $P(w_{(1)} < w_{(2)})$ and associated standard error from the observed value of Wilcoxon U-statistic and its variance. Study of "progressive fluctuation" during yam production time period of U_n values may facilitate comparison of production scenarios, and this can be accomplished by the LIL.

Let $\{T_n, n \geq 1\}$ be a sequence of random variables. We say that $\{T_n, n \geq 1\}$ satisfies the LIL, if $\text{Var}(T_n) > 0$ for almost all $n \geq 1$ and $\overline{\lim}_{n\to\infty} \frac{T_n}{\sqrt{2\text{Var}(T_n)\log\log \text{Var}(T_n)}} = 1$, $\underline{\lim}_{n\to\infty} \frac{T_n}{\sqrt{2\text{Var}(T_n)\log\log \text{Var}(T_n)}} = -1$.

The main component V_1 in Hoeffding decomposition is dominant and behaves like independent random variables (with zero mean) for which LIL holds.

Bounded LIL for one sample U-statistics for a kernel of arbitrary order is given by Adamczak and Latala (2008), also see the references therein.

Any marked and systematically sharp divergence of the quantities $\{U^*(n) = \frac{U_n - n^2/2}{\sqrt{2\text{Var}(U_n)\log\log \text{Var}(U_n)}} : n \geq 1\}$, when plotted over n, may be interpreted as an indication that H_o is not true. This large sample analysis is robust and takes into account the longitudinal variation, the time effect during a production season, e.g., March to November in a calendar year for Elephant foot yam production.

In Fig. 4.8, we show $U^*(n)$ values of standardised U_n appearing in LIL expression above for $n = 1, \cdots, 1000$; the basic kernel is $I(X < Y)$, where $X_1, \cdots, X_n$ are iid Uniform (0, 1) and $Y_1, \cdots, Y_n$ are iid Uniform (0.1, 1.1) under H_1. The

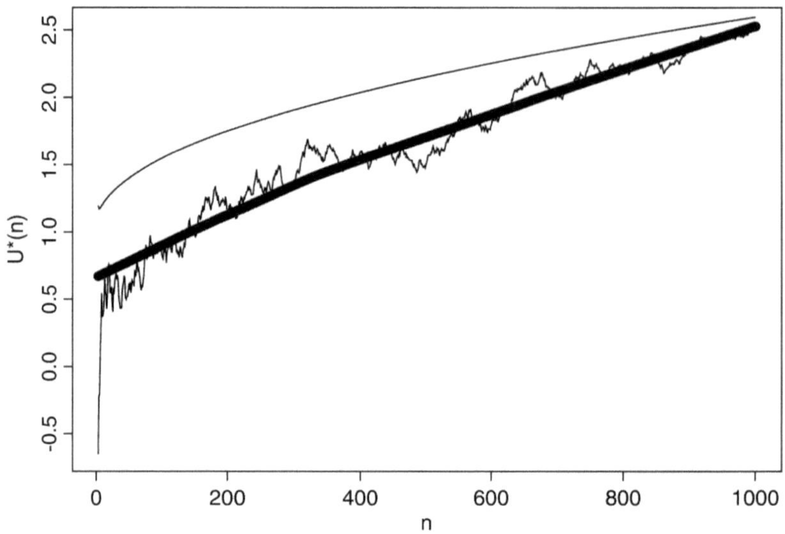

Fig. 4.8 Trajectory of U*(n) under non null case (shift = .1)

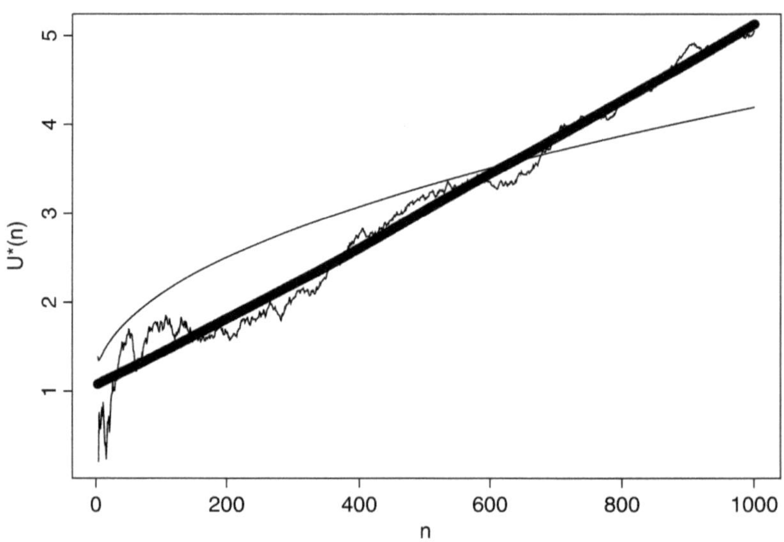

Fig. 4.9 Trajectory of U*(n) under non null case (shift = .2)

lowess curve, which smooth out the irregularity of the fluctuations, appears as a broad line.

The theoretical path $E_{H_1}(U^*(n)) + 1 \propto n^{1/2}$, for n large is shown by the thin and smooth curved line.

Figure 4.9 plots the similar quantities with additional shift for the distribution of Y, here $Y_1, \cdots, Y_n$ are taken as iid Uniform (0.2, 1.2) under H_1. The figures explain

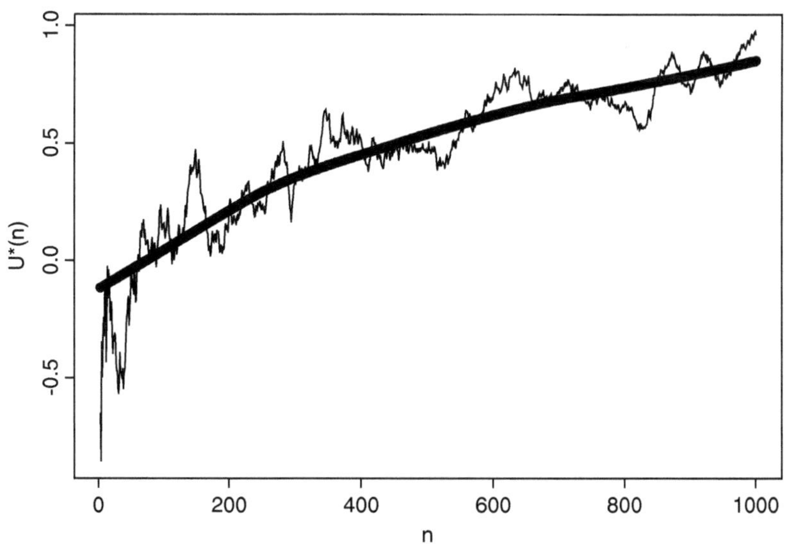

Fig. 4.10 Trajectory of U*(n) under null case

the divergence of the curves for increasing n. These also explain the effect of shift in alternative hypothesis H_1.

In Fig. 4.10, we plot the $U^*(n)$ values when H_o is true; the lowess curve appears to be stable towards 1.

In a particular year the number of yams produced is quite large, so the proposed comparison is possible for two different years in a particular geographical region. The same analysis for comparison may be done for above-ground biomass.

The simulations are done by SPLUS with assigned seed values 15, 57 and 143 for Figs. 4.8–4.10, respectively.

Similar analysis of tail comparison of the distribution of standardized U-statistics may be adopted when the response is a continuous curve over time, or when a large number of sequential observations in a time cycle are made available for comparison in two different time segments, e.g., comparison for meteorological data over years.

Appendix. We now state a result providing a moment-bound for a random variable X when m.g.f. of the transformed random variable $\{\log_e(1+|X|)\}^{\nu/(\nu-1)}$, $\nu > 1$ exist. The result is similar to A1 of Dasgupta (1988).

Proposition 2. $E\exp[s\{\log_e(1+|X|)\}^{\nu/(\nu-1)}] < \infty,\ \nu > 1,\ s > 0 \Rightarrow E|X|^m \leq Le^{w_o m^\nu}, \forall m,$ *for some* $L > 1$, *where* $w_o = s^{1-\nu} \Rightarrow E\exp[s'\{\log_e(1+|X|)\}^{\nu/(\nu-1)}] < \infty,\ 0 < s' < s(\nu-1)\nu^{-\nu/(\nu-1)}(< s)$.

Proof. Write, $\exp[s\{\log_e(1+|x|)\}^{\nu/(\nu-1)}] = (1+|x|)^{s\{\log_e(1+|x|)\}^{1/(\nu-1)}}$ and $E|X|^m < E(1+|X|)^m = E(1+|X|)^m I[s\{\log_e(1+|X|)\}^{1/(\nu-1)} \geq m] + E(1+|X|)^m I[s\{\log_e(1+|X|)\}^{1/(\nu-1)} < m]$.

The first term of the above in r.h.s. is finite from the assumption made. Now, $s\{\log_e(1+|x|)\}^{1/(\nu-1)} < m \Rightarrow (1+|x|)^m < e^{w_o m^\nu}$.

To show the second part of the implication write $P(|X| > t) \leq t^{-m}E|X|^m < t^{-m}Le^{w_o m^\nu} = P^*$, say. Minimizing P^* w.r.t. m one gets, $P(|X| > t) = O(e^{-(\nu-1)w_o\{\log(1+t)/(w_o\nu)\}^{\nu/(\nu-1)}})$. Now for a differentiable function $h \geq 0, h(0) = 0$, use the relation $Eh(X) = \int_0^\infty h'(t)P(|X| > t)dt$, and take $h(x) = \exp[s'\{\log_e(1+|x|)\}^{\nu/(\nu-1)}] - 1$, to get the result.

References

Adamczak, R., & Latala, R. (2008). The LIL for canonical U-statistics. *Annals of Probability, 36*, 1023–1058.

Bentkus, V., Jing, B. -Y., & Zhou, W. (2009). On normal approximations to U-statistics. *Annals of Probability, 37*, 2174–2199.

Callert, H., & Janssen, P. (1978). The Berry Esseen theorem for U-statistics. *Annals of Statistics, 6*, 417–421.

Dasgupta, R. (1988). Non-uniform rates of convergence to normality for strong mixing processes. *Sankhya A, 51*(Pt. 2), 436–451.

Dasgupta, R. (1989). Some further results on nonuniform rates of convergence to normality. *Sankhya A, 51*, 144–167.

Dasgupta, R. (2006). Nonuniform rates of convergence to normality. *Sankhya, 68*, 620–635.

Dasgupta, R. (2008). Convergence rates of two sample U-statistics in non iid case. *Calcutta Statistical Association Bulletin, 60*, 81–97.

Dasgupta, R. (2010). Bootstrap of deviation probabilities with applications. *Journal of Multivariate Analysis, 101*(9), 2137–2148.

Dasgupta, R. (2013). Optimal-time harvest of Elephant foot yam and related theoretical issues. *Advances in Growth Curve Models, 46*, 101–129.

Ghosh, M., & Dasgupta, R. (1980). Berry–Esseen theorems for U-statistics in the non iid case. In *Colloquia Mathematica Societatis Janos Bolyai, 32 Nonparametric statistical inference*, Hungary, pp. 293–313.

Rubin, H., & Sethuraman, J. (1965). Bayes risk efficiency. *Sankhya A, 27*, 347–356.

Sen, P. K. (1992). Introduction to Hoeffding (1948): A class of statistics with asymptotically normal distribution. In S. Kotz & N. L. Johnson (Eds.), *Breakthroughs in statistics* (Vol. I, pp. 299–307). Berlin: Springer.

Chapter 5
Correlated Bivariate Linear Growth Models: Optimal Designs for Slope Parameter Estimation

P.S.S.N.V.P. Rao and Bikas K. Sinha

Abstract Considered is a linear growth model involving each of two continuous measurable characteristics Y and Z. Along an equispaced time scale, experimental units (eus) are recruited/selected and measurements are recorded for both the characteristics. An eu may be enrolled at a time point i and may be kept under study continuously up to a time point j without any further "recall". We assume the total duration of the study to be $(2k + 1)$ units of time and set the time points as $\{-k, -(k-1), \cdots, -2, -1, 0, 1, 2, \cdots, k-1, k\}$ so that $-k \leq i \leq j \leq k$. Denote by t an arbitrary time point and by Y_t and Z_t random realizations of the characteristics Y and Z at time point t. Generally, four types of correlation structures are readily involved: $\text{Corr}(Y_t, Z_t)$; $\text{Corr}(Y_t, Y_r)$, $r \neq t$, $\text{Corr}(Z_t, Z_r)$, $r \neq t$, $\text{Corr}(Y_t, Z_r)$, $r \neq t$.

We assume a mean model: $E(Y_t) = \alpha + \beta t$ and $E(Z_t) = \gamma + \delta t$. Our purpose is to suggest an optimal design for most efficient joint estimation of the slope parameters β and δ in the above model with suitable covariance structures. Our study is closely based on that of Abt et al. (Optimal designs in growth curve models: Part I: Correlated model for linear growth: Optimal designs for slope parameter estimation and growth prediction. J Stat Plan Infer 64:141–150, 1997). Essentially we suggest using either a single pair of time points $-k$ and k, each with 50% recruit, or the stretch of *all* time points $(-k, -(k-1), \ldots, -1, 0, 1, \ldots, k-1, k)$, depending on the nature of correlations. This is based on A- and D-optimality criteria. Extensions to other mean models (such as linear–quadratic and quadratic–quadratic) are wide open.

P.S.S.N.V.P. Rao (✉)
Applied Statistics Unit, Indian Statistical Institute, Kolkata 700108, India
e-mail: pssnvprao@gmail.com

B.K. Sinha
Bayesian and Interdisciplinary Research Unit, Indian Statistical Institute, Kolkata 700108, India
e-mail: sinhabikas@yahoo.com

R. Dasgupta (ed.), *Advances in Growth Curve Models*, Springer Proceedings in Mathematics & Statistics 46, DOI 10.1007/978-1-4614-6862-2_5,

5.1 Introduction and Literature Review

This study is a sequel to [Abt–Liski–Mandal–Sinha, *Journal of Statistical Planning and Inference,1997, 64, pp. 141–150*] (Abt et al. 1997) and [Abt–Gaffke–Liski–Sinha, *Journal of Statistical Planning and Inference,1998, 67, pp. 287–296*] (Abt et al. 1998). These two references are abbreviated henceforth as ALMS (1997) and AGLS (1998) in this order.

Not to obscure the essential steps of reasoning in the complicated set-up we intend to present here, in this section, we will briefly review the results in the above two references involving one continuous response variable in the context of linear and quadratic regression models.

Let Y denote the response variable and let n_{ij} denote the number of experimental units recruited at time point i and followed through time points till j. For any single eu in this category, $\mathbf{Y_{ij}}$ denotes the column vector of order $(j - i + 1)$ given by $\mathbf{Y}_{ij} = (Y_i : Y_{i+1} : \cdots : Y_{j-1} : Y_j)'$. Further, whenever $j > i$, observations on the same eu are naturally correlated and two different correlation structures were studied in the above two papers. These correspond to (i) intraclass correlation and (ii) autocorrelation.

Again, at each of the $(j - i + 1)$ time points from i through j we have n_{ij} observations. Thus the total number of observations N is given by $N = \sum\sum n_{ij}(j - i + 1)$.

The design matrix $\mathbf{X_{ij}}$ is given by

$$\mathbf{X_{ij}} = \begin{pmatrix} \mathbf{1_{ij}} & \mathbf{t_{ij}} \end{pmatrix},$$

where $\mathbf{1_{ij}}$ denotes the column vector of order $(j - i + 1)$ with each element 1 and $\mathbf{t_{ij}}$ is also a column vector of the same order given by $\mathbf{t}'_{ij} = (i : (i + 1) : (i + 2) : \cdots : (j - 1) : j)$.

The approach to search for an optimal design is through introduction of a continuous version of the allocation design. This is done by assigning a positive mass ξ_{ij} corresponding to the pair of time points (i, j) such that $\sum\sum \xi_{ij}(j - i + 1) = 1$.

For a linear growth model $E(Y_t) = \alpha + \beta t$, with usual assumptions on the errors, ALMS (1997) investigated the nature of optimal designs for most efficient estimation of the β parameter under intraclass correlation structure. The idea was to dwell in the framework of continuous design theory and maximize information on β by appropriate choice of ξ. Towards this, the role of symmetry of the factor space was fully explored. A design ξ is said to be symmetric whenever $\xi_{ij} = \xi_{-j,-i}$ for all possible choices of $i \leq j$. Otherwise, it is said to be "asymmetric". Any asymmetric design ξ can be "symmetrized" by redefining it as $\xi^{(*)}_{(ij)} = \xi^{(*)}_{(-j,-i)} = [\xi_{(ij)} + \xi_{(-j,-i)}]/2$ whenever for a pair $i \leq j, \xi_{ij} \neq \xi_{-j,-i}$. This is done for all such pairs with positive mass. In ALMS (1997), it was argued that any asymmetric design can be "improved" by its symmetric counterpart in the sense of "information domination" for the β-parameter. Next, only four possible types of "core" symmetric designs were identified and it was resolved that only two of them, viz., $\xi^{(1)} : \xi_{(-k,-k)} = 1/2 = \xi_{(kk)}$ and $\xi^{(2)} : \xi_{(-k,k)} = 1/(2k + 1)$ are worth the consideration. Finally, the performances of these two symmetric designs were compared and it was resolved that whenever $\rho < (2k - 1)/(3k - 1)$, $\xi^{(1)}$ is better than the other. Note that for $\rho = 0$, it is well known that the design which places 50% of the observations at each of the extreme points performs the best for estimation of the slope parameter. ALMS (1997) then proceeded in other directions of optimality studies For a single measurable characteristic like Y, between two observations at two different time periods on the same experimental unit, there is likely to be a correlation and this is denoted above by ρ.

In AGLS (1998), the investigations were continued for quadratic growth model viz., $E(Y_t) = \alpha + \beta t + \gamma t^2$, with both the error structures mentioned above. The optimality criteria functions studied were more general as in traditional optimality studies. The role of symmetry of the factor space and invariance of the optimality criteria functions were once again exploited. It was argued that only point and mass symmetric designs are of relevance and two "core" symmetric designs were identified viz., $\xi^{(1)} : \xi_{(-k,-k)} = 1/2 = \xi_{(kk)}$ and $\xi^{(2)} : \xi_{(-k,k)} = 1/(2k + 1)$. Moreover, it was also observed that the single-point design $\xi^{(0)}$ with the entire mass at 0 could be suitably combined with either or both of the above two. In its most general form, non-negative "weights" assigned to the three core symmetric designs are denoted by $w^{(1)}, w^{(2)}, w^{(0)}$, respectively, so that $w^{(1)} + w^{(2)} + w^{(0)} = 1$.

With special reference to joint estimation of β and γ in the above quadratic growth model and with respect to the D-optimality criterion, here are some numerical results for the intraclass correlation model when $k = 5$:

(a) $0 < \rho < 0.3 : w^{(1)} = 0.6667, w^{(0)} = 0.3333$
(b) $\rho = 0.45 : w^{(1)} = 0.0890, w^{(2)} = 0.9110$
(c) $\rho = 0.60 : w^{(1)} = 0.0104, w^{(2)} = 0.9896$
(d) $\rho \geq 0.75 : w^{(2)} = 1.00$.

For higher values of k, the first choice extends for values of ρ up to 0.45. As we see, the last choice takes over for higher values of ρ. It may also be noted that for a linear growth model and for higher values of ρ, the last choice here also coincides with the optimal design $\xi^{(2)}$ suggested above.

AGLS (1998) dealt with other optimality criteria and the other error structure as well. Theoretical derivations, followed by numerical computations, provided necessary clue as to the nature of optimal designs.

Though we will exclusively deal with the linear growth model for joint estimation of the slope parameters of two continuous measurable characteristics, the above review for a quadratic growth model points out the difficulty level involved in handling optimality issues.

5.2 Bivariate Response Model and Optimality Issues

We follow an approach similar to that of ALMS (1997) with slight changes in the notations. Considered is a linear growth model involving each of two continuous measurable characteristics Y and Z. Along an equispaced time scale, experimental units [eus] are recruited/selected and measurements are recorded for both the characteristics. An eu may be enrolled at a time point i and may be kept under study continuously up to a time j without any further "recall". Let n_{ij} be the number of such eus between the time points i and j, that is, n_{ij} eus are recruited at time point i and followed through time points till j. Further, for $u = 1, 2, \cdots, n_{ij}$, let the column vectors $\mathbf{Y}_{ij}^{(u)}$ and $\mathbf{Z(u)_{ij}}$, each of order $(j - i + 1) \times 1$, denote the measurements of characteristics Y and Z, respectively, of the eu u during the time period between i and j. We assume the total duration of the study to be $(2k + 1)$ units of time and set the time points as $\{-k, -(k-1), \cdots, -2, -1, 0, 1, 2, \cdots, k-1, k\}$ so that $-k \leq i \leq j \leq k$.

An experimental design, for a given total number of *bivariate* observations (N), is defined as the combination of the triplets $\{(i, j); n_{ij}\}$ subject to their satisfying the condition: $N = \sum\sum(j - i + 1)n_{ij}$.

We assume the model: for the observations taken at time points i through j

$$E(\mathbf{Y}_{ij}^{(u)}) = \alpha\mathbf{e}_{ij} + \beta\mathbf{t}_{ij} \quad and \quad E(\mathbf{Z}_{ij}^{(u)}) = \gamma\mathbf{e}_{ij} + \delta\mathbf{t}_{ij},$$

where $\mathbf{t}_{ij}$ denotes the column vector $(i \quad i+1 \quad \cdots \quad j)'$ and $\mathbf{e}_{ij}$ denotes the column vector $(1 \quad 1 \quad \cdots \quad 1)'$, having the following variance–covariance structures:

$$Var(\mathbf{Y}_{ij}) = \mathbf{\Sigma}_{ij}^{(Y)}, \quad Var(\mathbf{Z}_{ij}) = \mathbf{\Sigma}_{ij}^{(Z)} \quad and \quad Cov(\mathbf{Y}_{ij}, \mathbf{Z}_{ij}) = \mathbf{\Sigma}_{ij}^{(YZ)}.$$

The dispersion matrix of $(\mathbf{Y}_{ij}, \mathbf{Z}_{ij})$, say $\mathbf{W}_{ij}$, is given by

$$\mathbf{W}_{ij} = \begin{pmatrix} \mathbf{\Sigma}_{ij}^{(Y)} & \mathbf{\Sigma}_{ij}^{(YZ)} \\ \mathbf{\Sigma}_{ij}^{(YZ)} & \mathbf{\Sigma}_{ij}^{(Z)} \end{pmatrix} = \begin{pmatrix} \mathbf{P} & \mathbf{Q} \\ \mathbf{Q} & \mathbf{R} \end{pmatrix}, \ say$$

for simplicity of notation and let the inverse of $\mathbf{W}_{ij}$ be represented as

$$\mathbf{W}_{ij}^{-1} = \begin{pmatrix} \mathbf{P} & \mathbf{Q} \\ \mathbf{Q} & \mathbf{R} \end{pmatrix}^{-1} = \begin{pmatrix} \mathbf{A} & \mathbf{B} \\ \mathbf{B} & \mathbf{C} \end{pmatrix}.$$

Now we make the following structural assumptions for the variance–covariance matrices:

$$\mathbf{\Sigma}_{ij}^{(Y)} = \sigma_Y^2\{(1-\rho_Y)\mathbf{I} + \rho_Y\mathbf{J}\}; \quad \mathbf{\Sigma}_{ij}^{(Z)} = \sigma_Z^2\{(1-\rho_Z)\mathbf{I} + \rho_Z\mathbf{J}\} \quad and$$

$$\mathbf{\Sigma}_{ij}^{(YZ)} = \sigma_Y \sigma_Z \{(\rho_* - \rho_{**})\mathbf{I} + \rho_{**}\mathbf{J}\}.$$

Notice that under these variance–covariance structure assumptions, all the matrices $\mathbf{A}, \mathbf{B}$ and $\mathbf{C}$ are of the form $u\mathbf{I} + v\mathbf{J}$ for some scalars u and v. So let us assume

$$\mathbf{A} = a_1\mathbf{I} + a_2\mathbf{J}; \quad \mathbf{B} = b_1\mathbf{I} + b_2\mathbf{J} \quad \textit{and} \quad \mathbf{C} = c_1\mathbf{I} + c_2\mathbf{J}.$$

The design matrix $\mathbf{X}_{ij}$ is given by

$$\mathbf{X}_{ij} = \begin{pmatrix} \mathbf{e}_{ij} & \mathbf{t}_{ij} & \mathbf{0} & \mathbf{0} \\ & & & \\ \mathbf{0} & \mathbf{0} & \mathbf{e}_{ij} & \mathbf{t}_{ij} \end{pmatrix},$$

where the notations are already explained before. Then the information matrix $\mathbf{I}_{ij}$ for the parameters α, β, γ and δ is given by $n_{ij}\mathbf{X}'_{ij}\mathbf{W}_{ij}^{-1}\mathbf{X}_{ij}$, that is,

$$\mathbf{I}_{ij}\begin{pmatrix} \alpha \\ \beta \\ \gamma \\ \delta \end{pmatrix} = n_{ij}\mathbf{X}'_{ij}\mathbf{W}_{ij}^{-1}\mathbf{X}_{ij} = n_{ij}\begin{pmatrix} \mathbf{e}'_{ij}\mathbf{A}\mathbf{e}_{ij} & \mathbf{e}'_{ij}\mathbf{A}\mathbf{t}_{ij} & \mathbf{e}'_{ij}\mathbf{B}\mathbf{e}_{ij} & \mathbf{e}'_{ij}\mathbf{B}\mathbf{t}_{ij} \\ \mathbf{t}'_{ij}\mathbf{A}\mathbf{e}_{ij} & \mathbf{t}'_{ij}\mathbf{A}\mathbf{t}_{ij} & \mathbf{t}'_{ij}\mathbf{B}\mathbf{e}_{ij} & \mathbf{t}'_{ij}\mathbf{B}\mathbf{t}_{ij} \\ \mathbf{e}'_{ij}\mathbf{B}\mathbf{e}_{ij} & \mathbf{e}'_{ij}\mathbf{B}\mathbf{t}_{ij} & \mathbf{e}'_{ij}\mathbf{C}\mathbf{e}_{ij} & \mathbf{e}'_{ij}\mathbf{C}\mathbf{t}_{ij} \\ \mathbf{t}'_{ij}\mathbf{B}\mathbf{e}_{ij} & \mathbf{t}'_{ij}\mathbf{B}\mathbf{t}_{ij} & \mathbf{t}'_{ij}\mathbf{C}\mathbf{e}_{ij} & \mathbf{t}'_{ij}\mathbf{C}\mathbf{t}_{ij} \end{pmatrix}.$$

Rewrite this as

$$\mathbf{I}_{ij}\begin{pmatrix} \beta \\ \delta \\ \alpha \\ \gamma \end{pmatrix} = n_{ij}\begin{pmatrix} \mathbf{t}'_{ij}\mathbf{A}\mathbf{t}_{ij} & \mathbf{t}'_{ij}\mathbf{B}\mathbf{t}_{ij} & \mathbf{t}'_{ij}\mathbf{A}\mathbf{e}_{ij} & \mathbf{t}'_{ij}\mathbf{B}\mathbf{e}_{ij} \\ \mathbf{t}'_{ij}\mathbf{B}\mathbf{t}_{ij} & \mathbf{t}'_{ij}\mathbf{C}\mathbf{t}_{ij} & \mathbf{t}'_{ij}\mathbf{B}\mathbf{e}_{ij} & \mathbf{t}'_{ij}\mathbf{C}\mathbf{e}_{ij} \\ \mathbf{e}'_{ij}\mathbf{A}\mathbf{t}_{ij} & \mathbf{e}'_{ij}\mathbf{B}\mathbf{t}_{ij} & \mathbf{e}'_{ij}\mathbf{A}\mathbf{e}_{ij} & \mathbf{e}'_{ij}\mathbf{B}\mathbf{e}_{ij} \\ \mathbf{e}'_{ij}\mathbf{B}\mathbf{t}_{ij} & \mathbf{e}'_{ij}\mathbf{C}\mathbf{t}_{ij} & \mathbf{e}'_{ij}\mathbf{B}\mathbf{e}_{ij} & \mathbf{e}'_{ij}\mathbf{C}\mathbf{e}_{ij} \end{pmatrix}.$$

Next, towards formation of a symmetrized version of a pair of allocations n_{ij} and n_{-j-i}, covering time points i to j and $-j$ to $-i$, respectively, with $i \leq j$, we add the corresponding information matrices. This yields

$$\tilde{\mathbf{I}}_{ij}\begin{pmatrix} \beta \\ \delta \\ \alpha \\ \gamma \end{pmatrix} = \mathbf{I}_{ij}\begin{pmatrix} \beta \\ \delta \\ \alpha \\ \gamma \end{pmatrix} + \mathbf{I}_{-j-i}\begin{pmatrix} \beta \\ \delta \\ \alpha \\ \gamma \end{pmatrix}$$

$$= \begin{pmatrix} (n_{ij}+n_{-j-i})\mathbf{t}'_{ij}\mathbf{A}\mathbf{t}_{ij} & (n_{ij}+n_{-j-i})\mathbf{t}'_{ij}\mathbf{B}\mathbf{t}_{ij} & (n_{ij}-n_{-j-i})\mathbf{t}'_{ij}\mathbf{A}e_{ij} & (n_{ij}-n_{-j-i})\mathbf{t}'_{ij}\mathbf{B}\mathbf{e}_{ij} \\ (n_{ij}+n_{-j-i})\mathbf{t}'_{ij}\mathbf{B}\mathbf{t}_{ij} & (n_{ij}+n_{-j-i})\mathbf{t}'_{ij}\mathbf{C}\mathbf{t}_{ij} & (n_{ij}-n_{-j-i})\mathbf{t}'_{ij}\mathbf{B}\mathbf{e}_{ij} & (n_{ij}-n_{-j-i})\mathbf{t}'_{ij}\mathbf{C}\mathbf{e}_{ij} \\ (n_{ij}-n_{-j-i})\mathbf{e}'_{ij}\mathbf{A}\mathbf{t}_{ij} & (n_{ij}-n_{-j-i})\mathbf{e}'_{ij}\mathbf{B}\mathbf{t}_{ij} & (n_{ij}+n_{-j-i})\mathbf{e}'_{ij}\mathbf{A}\mathbf{e}_{ij} & (n_{ij}+n_{-j-i})\mathbf{e}'_{ij}\mathbf{B}\mathbf{e}_{ij} \\ (n_{ij}-n_{-j-i})\mathbf{e}'_{ij}\mathbf{B}\mathbf{t}_{ij} & (n_{ij}-n_{-j-i})\mathbf{e}'_{ij}\mathbf{C}\mathbf{t}_{ij} & (n_{ij}+n_{-j-i})\mathbf{e}'_{ij}\mathbf{B}\mathbf{e}_{ij} & (n_{ij}+n_{-j-i})\mathbf{e}'_{ij}\mathbf{C}\mathbf{e}_{ij} \end{pmatrix}.$$

The above expressions as well as the expressions below follow by noting that (i) all the matrices $\mathbf{A}, \mathbf{B}, \mathbf{C}$ are of the form $u\mathbf{I} + v\mathbf{J}$, and (ii) for a matrix $\mathbf{S}_{(j-i+1)\times(j-i+1)}$ of the form $\mathbf{S} = s_1\mathbf{I} + s_2\mathbf{J}$,
$\mathbf{t}'_{-j-i}\mathbf{S}\mathbf{t}_{-j-i} = \mathbf{t}'_{ij}\mathbf{S}\mathbf{t}_{ij} = s_1\mathbf{t}'_{ij}\mathbf{t}_{ij} + s_2(\mathbf{t}'_{ij}\mathbf{e}_{ij})^2$,
$\mathbf{e}'_{-j-i}\mathbf{S}\mathbf{e}_{-j-i} = \mathbf{e}'_{ij}\mathbf{S}\mathbf{e}_{ij} = (j-i+1)[s_1 + (j-i+1)s_2]$
and $\mathbf{t}'_{-j-i}\mathbf{S}\mathbf{e}_{-j-i} = -\mathbf{t}'_{ij}\mathbf{S}\mathbf{e}_{ij} = -\mathbf{t}'_{ij}\mathbf{e}_{ij}[s_1 + (j-i+1)s_2]$.

As a result, for the symmetrized version of the above pair of allocations, that is, for the design satisfying $\tilde{n}_{ij} = \tilde{n}_{-j-i} = (n_{ij} + n_{-j-i})/2$, the above information matrix $\tilde{\mathbf{I}}_{ij}$ reduces to a block diagonal matrix with each block a square matrix of order 2, the one corresponding to the parameters β and δ being given by (writing $\mathbf{t}$ for $\mathbf{t}_{ij}$ and $\mathbf{e}$ for $\mathbf{e}_{ij}$)

$$(n_{ij} + n_{-j-i})\begin{pmatrix} a_1\mathbf{t}'\mathbf{t} + a_2(\mathbf{t}'\mathbf{e})^2 & b_1\mathbf{t}'\mathbf{t} + b_2(\mathbf{t}'\mathbf{e})^2 \\ b_1\mathbf{t}'\mathbf{t} + b_2(\mathbf{t}'\mathbf{e})^2 & c_1\mathbf{t}'\mathbf{t} + c_2(\mathbf{t}'\mathbf{e})^2 \end{pmatrix}.$$

Now, in a very general sense, we will partition the total information matrix $\mathbf{I}$, and writing $\xi_{ij} = n_{ij}/N$, we have

$$\mathbf{I}\begin{pmatrix} \beta \\ \delta \\ \alpha \\ \gamma \end{pmatrix} = \sum\sum \xi_{ij}\mathbf{I}_{ij}\begin{pmatrix} \beta \\ \delta \\ \alpha \\ \gamma \end{pmatrix} = \begin{pmatrix} \mathbf{I}_1 & \mathbf{I}_2 \\ \mathbf{I}_2 & \mathbf{I}_3 \end{pmatrix}.$$

Therefore, the information matrix for the slope parameters β and δ is given by

$$\mathbf{I}\begin{pmatrix} \beta \\ \delta \end{pmatrix} = (\mathbf{I}_1 - \mathbf{I}_2\mathbf{I}_3^{-1}\mathbf{I}_2).]$$

The following results, which are easy to verify, will be used in the sequel.

Lemma 1. *Let* $\mathbf{P}_{n\times n} = p_1\mathbf{I} + p_2\mathbf{J}$ *and* $\mathbf{R}_{n\times n} = r_1\mathbf{I} + r_2\mathbf{J}$. *Then*

(a) $\mathbf{P} \pm \mathbf{R} = (p_1 \pm r_1)\mathbf{I} + (p_2 \pm r_2)\mathbf{J}$.

(b) $\mathbf{PR} = p_1r_1\mathbf{I} + (p_1r_2 + p_2r_1 + np_2r_2)\mathbf{J}$.

(c) $\mathbf{P}^{-1} = (1/p_1)\mathbf{I} - \{p_2/p_1(p_1 + np_2)\}\mathbf{J}$.

Let $\mathbf{P}$ *and* $\mathbf{R}$ *be as defined above and let* $\mathbf{Q}_{n\times n} = q_1\mathbf{I} + q_2\mathbf{J}$. *Recall that the dispersion matrix of* $\mathbf{Y}_{ij}, \mathbf{Z_{ij}}$ *is denoted by* $\mathbf{W}_{ij}$ *and under the assumed structure of the variance–covariance matrices, it has the representation:*

$$\mathbf{W}_{ij} = \begin{pmatrix} \mathbf{P} & \mathbf{Q} \\ \mathbf{Q} & \mathbf{R} \end{pmatrix},$$

where n is to be replaced by $(j - i + 1)$.

Lemma 2. *For a positive definite variance–covariance structure, it readily follows that (i) $p_1 r_1 - q_1^2 > 0$ and (ii) $(p_1 + np_2)(r_1 + nr_2) - (q_1 + nq_2)^2 > 0$.*

The proof rests on computations of (i) generalized variance of a pair $(Y_1 - Y_2, Z_1 - Z_2)$; and (ii) generalized variance of $(\sum Y, \sum Z)$.

Set $D_1 = (p_1 r_1 - q_1^2)$, $D_2 = (p_2 r_2 - q_2^2)$ and $D_3 = [(p_1 + np_2)(r_1 + nr_2) - (q_1 + nq_2)^2]$. Recall

$$\mathbf{W}_{ij}^{-1} = \begin{pmatrix} \mathbf{A} & \mathbf{B} \\ \mathbf{B} & \mathbf{C} \end{pmatrix}.$$

Then it is not difficult to verify that

$$A_{n\times n} = a_1 I + a_2 J,\ B_{n\times n} = b_1 I + b_2 J \ \ and \ \ C_{n\times n} = c_1 I + c_2 J,$$

where $a_1 = (r_1/D_1), b_1 = (-q_1/D_1), c_1 = (p_1/D_1)$ and

$$\begin{aligned}
a_2 &= [(q_1 + nq_2)(q_2 r_1 - q_1 r_2) + (r_1 + nr_2)(q_1 q_2 - p_2 r_1)]/(D_1 D_3),\\
b_2 &= -[(q_1 + nq_2)(q_1 q_2 - p_2 r_1) + (p_1 + np_2)(q_2 r_1 - q_1 r_2)]/(D_1 D_3)\\
&= -[(q_1 + nq_2)(q_1 q_2 - p_1 r_2) + (r_1 + nr_2)(p_1 q_2 - p_2 q_1)]/(D_1 D_3),\\
c_2 &= [(q_1 + nq_2)(p_1 q_2 - p_2 q_1) + (p_1 + np_2)(q_1 q_2 - p_1 r_2)]/(D_1 D_3).
\end{aligned}$$

Further,

$$\begin{aligned}
a_1 c_1 - b_1^2 &= 1/(p_1 r_1 - q_1^2) = (1/D_1),\\
a_2 c_2 - b_2^2 &= (p_2 r_2 - q_2^2)/(D_1 D_3) = D_2/(D_1 D_3)
\end{aligned}$$

and

$$a_1 c_2 + a_2 c_1 - 2b_1 b_2 = [2q_2(q_1 + nq_2) - r_2(p_1 + np_2) - p_2(r_1 + nr_2)]/(D_1 D_3).$$

Lemma 3. *Under the assumptions that $p_2, r_2 > q_1 + q_2, q_2$, with as, bs and cs defined as in the above, (i) $a_2 < 0$; (ii) $c_2 < 0$; (iii) $(a_2 c_2 - b_2^2) > 0$.*

Proof. Note that by Lemma 2, $D_1 > 0$ and $D_3 > 0$. Further, in view of the assumptions made, $D_2 > 0$. So, claim (iii) trivially follows from the representation in terms of D_1, D_2, D_3 above. Also it is enough to verify (i) and, that too, only in terms of the numerator of a_2. Note that the coefficient of n in the expression for a_2 is given by $q_2(q_1 q_2 - p_2 r_1) + p_2(q_2 r_1 - q_1 r_2) = -q_1(r_2 p_2 - q_2^2) = -q_1 D_2 < 0$. Hence, with $n = 1$, we simplify the numerator of a_2 as $(q_1 + q_2)(q_2 r_1 - q_1 r_2) + (r_1 + r_2)(q_1 q_2 - p_2 r_1) = q_1(q_2 r_1 - q_1 r_2) + r_1 q_2^2 + r_1(q_1 q_2 - p_2 r_1) - r_1 r_2 p_2 = -q_1^2 + r_1[q_1^2 + 2q_1 q_2 + q_2^2 - p_2(r_1 + r_2)] = -q_1^2 - r_1[p_2 - (q_1 + q_2)^2] < 0$ since, by the assumption made above, $p_2 > q_1 + q_2 > (q_1 + q_2)^2$. Hence the claim.

We are now in a position to establish a general result.

Theorem. *For estimation of the slope parameters, given any general asymmetric design, its symmetrized version performs better wrt both D- and A-optimality criteria. That is, the latter provides a larger value of the determinant of the information matrix and also it provides least value of the trace of the inverse of the information matrix.*

Proof. The arguments are given in two steps as follows:

Step 1. Note that the information matrix $\mathbf{I}$ for all the four model parameters is a pd matrix and so are its main block diagonal components viz., $\mathbf{I}_1$ and $\mathbf{I}_3$. In its most general form, the design comprises both asymmetric and symmetric components. For every symmetric component, the contribution towards $\mathbf{I}_2$ is a null matrix whereas for every asymmetric component, it is a non-null matrix. On the other hand, the contribution towards $\mathbf{I}_1$ and $\mathbf{I}_3$ are positive from both types of components. These are all readily verified from the various versions of the information matrix displayed before.

Step 2. Next we invoke symmetrization in each asymmetric component of the design. This we can do one at a time or all at a time. It's only a matter of slowing down or hastening the process of "improved" performance of the resulting design which will necessarily have all symmetric components at the end. That is precisely our claim in the theorem. We have discussed the process of symmetrization before. It is readily seen that under symmetry, information matrix for the slope parameters dominates the same without symmetry in the sense of D- and A-optimality. Therefore, even if one component is left as asymmetric in our choice of a design, the performance of the design can be improved by symmetrization. In other words, a design with all components symmetrized simultaneously will do better wrt both the optimality criteria. Hence the claim.

Henceforth, we consider only symmetric designs and deal with the information matrix for the slope parameters β and δ which is given by

$$\mathbf{I}\begin{pmatrix}\beta\\ \delta\end{pmatrix} = \begin{pmatrix} a_1\mathbf{t}'\mathbf{t} + a_2(\mathbf{t}'\mathbf{e})^2 & b_1\mathbf{t}'\mathbf{t} + b_2(\mathbf{t}'\mathbf{e})^2 \\ b_1\mathbf{t}'\mathbf{t} + b_2(\mathbf{t}'\mathbf{e})^2 & c_1\mathbf{t}'\mathbf{t} + c_2(\mathbf{t}'\mathbf{e})^2 \end{pmatrix}.$$

At this stage, it may be noted that we are primarily interested in the estimation of the slope parameters β and δ and consequently, towards comparison of different designs, the scale parameters σ_Y and σ_Z may be ignored from the analysis to follow.

Following ALMS (1997), we will now confine to three competing types of symmetric designs and further establish that those of Type (III) can be improved upon by a typical member of Type (II) or of Type (I) wrt the D- and A-optimality criteria. This time we need to argue closely.

Recall Type I symmetric designs which are of the form $\xi^{(1,t)} : \xi_{(-t,-t)} = 1/2 = \xi_{(tt)}; t = 1, 2, \ldots, k$ and those of Type II are of the form $\xi^{(2,t)} : \xi_{(-t,t)} = 1/(2t + 1), t = 1, 2, \ldots, k$. On the other hand, third type of symmetric designs exhibits three sub-structures, viz.,

$$\xi^{(3,s,t)}_{(I)} : \xi_{(-t,-s)} = \xi_{(s,t)} = 1/[2(t-s+1)], 0 < s < t \le k;$$

$$\xi^{(3,0,t)}_{(II)} : \xi_{(-t,0)} = \xi_{(0,t)} = 1/[2(t+1)], 0 < t \le k;$$

$$\xi^{(3,-s,t)}_{(III)} : \xi_{(-t,s)} = \xi_{(-s,t)} = 1/[2(t+s+1)], 0 < s \ne t \le k.$$

For each of these sub-structures, computation of $\mathbf{t}'\mathbf{e}$ is routine and it is non-zero.

Next, note that the information matrix of a symmetric design which is a 2×2 matrix can be conveniently expressed as $\left[(\mathbf{t}'\mathbf{t})\begin{pmatrix} a_1 & b_1 \\ b_1 & c_1 \end{pmatrix} + [(\mathbf{t}'\mathbf{e})^2]\begin{pmatrix} a_2 & b_2 \\ b_2 & c_2 \end{pmatrix}\right]$, and, most importantly, the second part involves a positive multiplier $[(\mathbf{t}'\mathbf{e})^2]$. However, we have shown above in Lemma 3 that the associated matrix has negative diagonal elements and a positive determinant value. Therefore, this is a negative definite matrix. And this is true of every member of Type (III) family of component designs. Hence, in the sense of Loewner domination, the first part dominates the whole. In other words, wrt both D- and A-optimality, the second part can be removed. We now show that the contribution from the first part is dominated by one member of Type (II) viz., $\xi^{(2,k)} : \xi_{(-k,k)} = 1/(2k+1)$ or one member of Type (I) viz., $\xi^{(1,k)} : \xi_{(-k,-k)} = 1/2 = \xi_{(kk)}$. That will demonstrate that it is enough to confine to component designs which are mixtures of Types (I) and (II).

For this tail-end argument, note that we only need to compare $(\mathbf{t}'\mathbf{t})/2N(\mathbf{t})$ for various choices of the vector $\mathbf{t}$ among three sub-types of Type (III) designs, where $N(\mathbf{t})$ is the number of one-sided support points of the design under consideration. It follows that among all such choices, $\mathbf{t} = (-k, -(k-1))$ [together with its counterpart $-\mathbf{t} = (k-1, k)$] maximizes $(\mathbf{t}'\mathbf{t})/2N(\mathbf{t})$ and this simplifies to $[k^2 + (k-1)^2]/4$ while the contribution from $\xi^{(2,k)} : \xi_{(-k,k)} = 1/(2k+1)$ towards $\mathbf{I}_1$ amounts to $k(k+1)/3$. Next we observe that $k(k+1)/3$ exceeds $[k^2 + (k-1)^2]/4$ whenever $k \le 4$. For $k \ge 5$, it will be seen that the Type (III) component is dominated by the chosen member of Type (I).

For the design $\xi^{(1,k)} : \xi_{(-k,-k)} = 1/2 = \xi_{(kk)}$, it follows that the information matrix for β and δ is given by

$$\widetilde{\mathbf{I}^{(I)}}_{kk}\begin{pmatrix} \beta \\ \delta \end{pmatrix} = (k^2/d_1)\begin{pmatrix} 1 & -\rho_* \\ -\rho_* & 1 \end{pmatrix},$$

where $d_1 = (1 - \rho_*^2)$.

We want to derive conditions for matrix domination involving Type (I) design information matrix and first part of chosen Type (III) design information matrix. The conditions are

(i) $4D_1k^2 \ge [k^2 + (k-1)^2]r_1(1-\rho_*^2)$;
(ii) $4D_1k^2 \ge [k^2 + (k-1)^2]p_1(1-\rho_*^2)$;
(iii) $[4D_1k^2 - [k^2 + (k-1)^2]r_1(1-\rho_*^2)][4D_1k^2 - [k^2 + (k-1)^2]p_1(1-\rho_*^2)]$
$\ge [4D_1k^2\rho_* - [k^2 + (k-1)^2]q_1(1-\rho_*^2)]^2$.

Condition (i) can be simplified to some extent and it amounts to verifying

(i)' $2[(1-\rho_y)(1-\rho_z)-(\rho_*-\rho_{**}^2] \geq (1-\rho_*^2)(1-\rho_z)$.

Likewise, condition (ii) amounts to a similar inequality by replacing ρ_z in the RHS of (i)' by ρ_y.

Next, we replace condition (iii) by a simpler condition, viz.,

(iii)' $[2D_1 - r_1(1-\rho_*^2)][2D_1 - p_1(1-\rho_*^2)] \geq [2D_1\rho_* - q_1(1-\rho_*^2)]^2$.

Routine simplification of the above, in view of the relation $D_1 = p_1r_1 - q_1^2$, leads to the condition

(iii)" $4D_1 - 2(p_1 + r_1 - 2q_1) + (1-\rho_*^2) \geq 0$.

These conditions notwithstanding, we can readily verify the applicability of the conditions on the correlation parameters. For example, when $\rho_* = \rho_{**} = \rho_{00}$ and $\rho_y = \rho_z = \rho_0$, conditions (i) or (ii) further simplify to $\rho_{00} \leq \rho_0 \leq [1+\rho_{00}^2]/2$, while condition (iii)" simplifies to
$\rho_0 \leq (1-\rho_{00})/2$, or, $\rho_0 \geq (1+\rho_{00})/2$. By combining the two conditions, we deduce that the claim holds whenever $\rho_0 \leq (1-\rho_{00})/2$.

Thus, finally, the condition to be satisfied is: $\rho_{00} \leq \rho_0 \leq (1-\rho_{00})/2$.

The above analysis relates to matrix domination which covers all convex optimality criteria such as A- and D-optimality, vide Pukelsheim (1993). We could push the arguments further and derive less stringent conditions on the correlation parameters. Instead of matrix domination, let us consider only D-optimality. In that case, the two expressions to be compared are the determinants of the above matrices.

For Type (I) matrix, it is given by $k^4/(1-\rho_*^2)$ and for the Type (III) matrix [considering only the first part involving $\mathbf{t}'\mathbf{t}$], it is given by $[k^2+(k-1)^2]^2/16D_1$. Replacing $(k-1)^2$ by k^2, the ratio of the above two expressions will exceed unity if $4D_1 > (1-\rho_*^2)$. In the particular case: $\rho_y = \rho_z = \rho_0; \rho_* = \rho_{**} = \rho_{00}$, this condition simplifies to: $\rho_0 < (1+\rho_{00}^2/2)/2$. On the other hand, matrix domination required: $\rho_0 < (1-\rho_{00})/2$ which is more stringent than what is required under determinant domination.

The above analysis suggests that the parametric relations may enable us to make a choice of Type (I) or Type (II) design as a replacement for all types of Type (III) designs.

Remark. Whenever the above conditions are met, we may confine to only these two types of designs and proceed to find an optimum design. Otherwise, it is a matter of comparison among three competitors—one of each type.

5.3 Towards D-Optimality: Comparison of Type (I) and Type (II) Designs

We can now take up a relative comparison of the above two symmetric designs and this we do with reference to the model parameters viz., $\rho_y, \rho_z, \rho^*, \rho^{**}$. The ratio of the above two determinants is given by $R_k = \frac{9k^2[(1-\rho_y)(1-\rho_z)-(\rho^*-\rho^{**})^2]}{(k+1)^2(1-\rho^{*2})}$.

For simplicity, we assume, as before, $\rho_y = \rho_z = \rho_0$; $\rho^* = \rho^{**} = \rho_1$. Then, $R_k = \frac{9k^2(1-\rho_0)^2}{(k+1)^2(1-\rho_1^2)}$. For $k = 1$, $R_1 = \frac{9(1-\rho_0)^2}{4(1-\rho_1^2)}$ and we can derive conditions for this to exceed 1. For example, when $\rho_0 = \rho_1$, we need the condition $\rho_0 < 5/13$. Again, when $\rho_0 = 2\rho_1$, we need the condition $\rho_0 < \frac{9-\sqrt(31)}{10}$. Note also that, trivially, in case of $\rho_0 = \rho_1 = 0$, $R_k = 9k^2/(k+1)^2 > 1$, as expected. For the case of $k = 5$, $R_5 = \frac{225(1-\rho_0)^2}{36(1-\rho_1^2)}$. For $R_5 > 1$, we need the condition $25(1-\rho_0)^2 > 4(1-\rho_1^2)$ which (i) for $\rho_0 = \rho_1$, amounts to $\rho_0 < 21/29$; (ii) for $\rho_0 = 2\rho_1$, amounts to "no restriction" on $\rho_0 > 0$ at all, and (iii) for $\rho_1 = 0$, amounts to $\rho_0 < 3/5$. These findings are generally in accordance with D-optimal designs for high values of ρ.

5.4 Concluding Remarks

Following previous work, we have dealt with linear growth models for two continuous response variables and studied some aspects of optimality for most efficient estimation of the slope parameters involved in the two models. Model variations incorporating quadratic growth for either/both the response variables would be highly interesting to pursue in future. Following earlier studies, optimal prediction problems could be taken up as well. It transpires that optimality study is a highly non-trivial exercise.

Acknowledgements The authors are highly thankful to the anonymous referee for pointing out some shortcomings of the original version of the manuscript.

References

Abt, M., Gaffke, N., Liski, E. P., & Sinha, B. K. (1998). Optimal designs in growth curve models: II: Correlated model for quadratic growth: Optimal designs for slope parameter estimation and growth prediction. *Journal of Statistical Planning and Inference, 67*, 287–296.

Abt, M., Liski, E. P., Mandal, N. K., & Sinha, B. K. (1997). Optimal designs in growth curve models: Part I: Correlated model for linear growth: Optimal designs for slope parameter estimation and growth prediction. *Journal of Statistical Planning and Inference, 64*, 141–150.

Pukelsheim, F. (1993). Optimal design of experiments. New York: Wiley.

Chapter 6
Optimal-Time Harvest of Elephant Foot Yam and Related Theoretical Issues

Ratan Dasgupta

Abstract We assess the optimal harvest time based on growth curve and growth derivative by analysing real data sets on yam stems from farm. We estimate the derivative of stem growth and proliferation rate by a robust technique proposed in (Dasgupta 2013a, Nonuniform rates of convergence to normality for two sample U-statistics in non iid case with applications, in this volume). Large sample properties of the technique are investigated. Common pattern of growth from several observed growth curves is studied. With an application of extended Mahalanobis distance (Dasgupta, 2008, Proceedings of ISI Platinum Jubilee conference, World Scientific, pp 367–382), we compare the harvest scenarios over different years. Distribution of extended Mahalanobis distance in multinormal case is shown to be weighted average of two independent components, viz. an F variable and a limiting Chi-square variable. Limiting distribution is also obtained in general set-up without multinormal assumption. In this context distribution robustness, power robustness and high-dimensional data analysis are also discussed. The method of estimating growth rates may have applications in cancer treatment related studies and for stopping time rules in general. Some modeling issues are discussed. Several classes of theoretical growth curves and their limiting properties are investigated. The procedures developed are applicable while taking decisions of hidden variables based on observable variables.

MSC classifications 2000: Primary 62P10, 62G08; Secondary 65D10, 65D25

R. Dasgupta (✉)
Theoretical Statistics and Mathematics Unit, Indian Statistical Institute, Kolkata, India
e-mail: ratandasgupta@gmail.com

R. Dasgupta (ed.), *Advances in Growth Curve Models*, Springer Proceedings in Mathematics & Statistics 46, DOI 10.1007/978-1-4614-6862-2_6,

6.1 Introduction

Production of Elephant foot Yam is a boon to farmers in barren land. In fertile land of West Bengal the production can be five times the initial seed weight. It can grow in lateritic soil like that of Giridih, Jharkhand although the production may comparatively be less in such types of soil full of gravels. Yam is a staple food for poor; this is also cultivated by tribals in barren land inside forest. Cultivators sometimes leave many un-harvested as emergency food. Un-harvested yam of a season sprouts in the next season resulting in a larger sized yam. Massive growth of Yam-stems above surface is an indication of much deposition of carbohydrates i.e., large underground yam due to higher level of photosynthesis.

Yam-stem has good market value as tasty edible vegetable. Multiple stems may sprout from a single seed-corm. If the stems are not harvested in time it may become hard and then pale with little market price. On the other hand, if stems are harvested at a young stage, then it may reduce the size of underground yam incurring financial loss. Thus, it is important to find out the optimal time for Yam harvest depending on the rate of stem growth so that farmers may benefit from selling both Yam and its stem. If the rate of stem growth becomes insignificant after a time period, farmers may decide to harvest the crop, rather than waiting further for a little more additional increase in yam weight. They may like to save time, sell both yam and stem for enhanced income and proceed for cultivating the next crop.

The growth of stems e.g., height, girth at different locations of stem, may be continuously recorded over time. The weight of the stem can be indirectly assessed by multiplying the volume of the stem that is tapered towards the top of stem, with stem density obtained via separate destructive testing. The plant leaves are edible at tender stage. At latter stages these may be used for compost manure. Flower of Elephant Foot Yam can also be used as edible vegetable having good market price. Total vegetation above surface is termed above-ground biomass and this nonlinear combination of variables may turn out to be a potentially good predictor of underground Yam.

From growth records, a mean response curve for stem growth may be computed by non-parametric lowess regression; see Cleveland (1981). Estimate of growth derivatives may also be derived by a robust technique (Dasgupta 2013a) to decide about the right time for harvest. Combining growth curves of different stems sprouted from a single seed-corm, one may get an overall growth curve and growth rate.

We consider the problem of comparing the yield scenarios over different years. The aim is to get a single index combining two or more associated variables. Both stem and yam are considered in the production scenario in a year. The problem could be solved in an indirect manner by converting the income from each segment in money value and then adding these up. However, it has to be noted that the proposed index should take care of association between variables, be free from inflation and variation of market price for commodities over years. As for example, retail price of potato in West Bengal varied highly from Rs. 3 to 25 in recent years. Thus an index based on selling price may not properly reflect the production scenarios. Mahalanobis distance between two populations takes into account correlation structure of variables, and a relevant index may be based on such considerations.

We develop appropriate analysis when (i) the dispersion matrix of the variables remain same over years and (ii) the matrices vary over years. Invoking the extended definition of Mahalanobis distance $\Delta^2 = \Delta_I^2 + \Delta_{II}^2$ (Dasgupta 2008) for cases where dispersion matrices are distinct we propose a measure $R_d = \Delta_{II}^2/\Delta_I^2$ of dispersion-matrix heterogeneity, relative to mean heterogeneity towards diversity in different populations. This reveals relative importance of each component in extended Mahalanobis distance. Under the multivariate normal set-up, we obtain a limiting null distribution for two independent components in extended Mahalanobis distance. This turns out to be a weighted average of an F variable and a limiting Chi-square variable. The same is investigated for general non-normal case.

The growth model $\{y(t), t \geq 0\}$ developed has implication in study of cancerous tumour in patients. When availability of nutrients is limited in a confined space, the growth rate increases in the beginning; and then it gradually slows down. Depending on the nature of proliferation rate $F(y) = \frac{1}{y}\frac{d}{dt}y(t)$, Gomphertz curve with infinite proliferation rate at origin, or a suitable modification of it may explain the nature of growth in cancerous tumour. Level of anticancer antibody detected via Anti-Malignin Antibody in Serum (AMAS) blood test if found to be greater than 135 micrograms per milliliter (mcg/mL) is a (95%–99%) sure indicator of cancer. Doctor may change/stop medication to patients depending on the rate of change in anticancer antibody level.

High degree of observed association of predictor variables with response variable e.g., Yam-stem/above-ground biomass (anticancer antibody level) for underground Yam (cancerous tumour) may lead to an appropriate model. The problem is non-trivial as it may not always be possible to observe the response variable continuously over time, whereas for predictor variables it may be possible. For example, trans-rectal ultrasound (TRUS) guided prostate biopsy may cause significant discomfort and pain. Recovery progress may be monitored by direct measurement on sparsely selected time points, although indirect assessment is possible in many situations leading to a growth model for the latter to be theoretically inferred on the former (hidden) variable of interest.

Analysis of dispersion and subsequent covariance adjustment with structural parameters, as in growth model due to Rao; Pothoff and Roy, viz., $(U, \mathbf{XB}, \mathbf{\Sigma} \otimes \mathbf{I_n}); \mathbf{B} = \mathbf{\Theta}_0 + \mathbf{\Theta H}$, i.e., $(U - \mathbf{X\Theta}_0, \mathbf{X\Theta H}, \mathbf{\Sigma} \otimes \mathbf{I_n})$ is of little use here, since we have only restricted recordings of response-growth; whereas observable auxiliary variables are continuously recorded over time, providing valuable information; selection of concomitant variables in yields of plants should be made with caution in least squares analysis, (Rao 1974, p 288, 561).

In cancer studies, when it comes to observe the growth of tumors, which may not have a regular geometric shape, the usual three-dimensional coordinates may not convey the entire information. MRI and other functional data analysis models are more commonly used in this respect. A similar situation may arise even for Elephant-foot yam. However, for the cultivated variety "Bidhan-kusum" of Yam in fertile land with homogeneous texture we observed near spherical shapes (with slight suppression near the main stem) of circular cross section in Yam for most of the cases. In many cases tumour development in some particular regions e.g., fibroid tumour and breast tumour may have near spherical shapes. In unfertile lateritic Jharkhand soil full of gravels that may sometime block uninterrupted growth underground, yam may look like a cylinder, or a right sign, an inverted comma, and may even like an abstract art piece of irregular shape occasionally. Yam stems resemble tapered cylinders. Such conical-segment shape of stem can be reconstructed from the measurement of girth taken on various location of a stem along its height. These tapered shaped stems have circular cross sections with decreasing diameter towards the top of a plant.

Stopping time problems may sometimes involve consideration of derivatives. As for example see Xie and Schlick (2000), where an iterative univariate minimizer in each step of a descent method for minimizing a multivariate function is used. The proposed techniques of derivative estimation may have applications in similar studies when suitably adapted to multivariate random vectors.

The paper is arranged as follows. In Sect. 6.2 we explain the technique of estimating growth rate and proliferation rate. Consistency of the proposed technique is investigated. In Sect. 6.3 we examine a production index based on statistical distance between two populations. Of the two components of extended Mahalanobis distance, the second distance-component involving dispersions is related to likelihood ratio test statistic for testing homogeneity of dispersion matrices. Limiting distributions of the distance components are investigated with/without multivariate normal assumption. Estimations of the indices are possible even when the observation vectors are highly dependent over different individuals.

In Sect. 6.4 we analyse growth data of two different stems sprouting from a yam seed-corm. Similar features of growth curves lead to the possibility of further exploring a combined pattern; see Fig. 6.9, where growth of four sprouts from a seed corm is plotted. The common pattern of growth is plotted in Figs. 6.11–6.14. The proliferation rates are shown in Figs. 6.15–6.20.

In Sect. 6.5 we analyse two sets of multivariate longitudinal growth data over the production periods in years 2008 and 2009 from Indian Statistical Institute, Giridih farm. Yam production of second year with sprouted seed-corm seems to be better. In observed data, variation of mean components compared to heterogeneity of dispersion matrices is seen to be prominent in the relative index R_d proposed from extended Mahalanobis distance.

Section 6.6 investigates the common pattern in growth over different sprouts and estimation of growth rate. Several techniques for obtaining an overall growth curve are suggested and implemented with real data. Their performances are assessed in estimating a smooth curve (Figs. 6.11–6.14). Proliferation rates (Figs. 6.15–6.20) are also studied in Sect. 6.6 to check the suitability of Gomphertz model that is commonly used to explain tumour growth. It is observed that proliferation rate remain bounded within a neighbourhood of origin in observed data; the rate is finite at origin for a logistic function having initial stage of growth as approximately exponential; this indicates some modification to Gomphertz curve may be appropriate to model stem growth/above-ground biomass of Yam, to be inferred on yam yield. Identification of suitable combination of variables, possibly non-linear; as a potentially good predictor for underground yam, could suggest a similar growth curve for yam.

Pattern of the observed proliferation rates provide ample indication about the appropriate class of growth curves suitable for a particular situation. Such possibilities are explored in Sect. 6.6 and limiting behaviour of models based on proliferation rates is investigated to identify some well-known growth curves that appear in the limit.

6.2 Estimation of Growth Rate

For estimation of growth rate, there has been an extensive literature, starting from the Robbins–Monro and Kiefer–Wolfowitz stochastic approximations in the early 1950 to modern exploratory high-dimensional data analysis. See the references Clarke et al. (2008), Kiefer and Wolfowitz (1952), Robbins and Monro (1951). In this paper, the technique of estimating growth rate adopted, as proposed in Dasgupta (2013a), is somewhat different. First of all, individual slopes near a point are calculated and then these are suitably weighted & combined for that particular point. These estimates are next smoothed by non-parametric regression over the entire range to estimate growth rate curve. The issue of smoothing is pertinent in respect of fast convergence. The proposed technique of computation adopted via lowess smoothing, a non-parametric regression technique; of the highly non-robust divided differences is shown to be quite satisfactory by different examples in Dasgupta (2013a). As mentioned therein lowess technique plays a significant role in non parametric regression by down weighting, but not totally ignoring outliers; prompting us to evaluate derivative by this method.

Lowess, a local polynomial regression estimator with smooth tricubic kernel and variable bandwidth based on k-th nearest neighbour, employs weighted least square criterion that assigns less weights to distant observations, to have a robust estimate of response curve insensitive to large-residual outliers, by down-weighting these over several iterations. However, lowess does not provide an explicit functional form of response variable with predictor variables.

Denote $(x_i, y_i), i = 1, \cdots, n$ to be the stem growth records y_i at time x_i. Let $\hat{y}$ be lowess estimate of response and $\beta = \beta_n = \max_{1 \leq i < n} |x_i - x_{i+1}| \to 0$, as $n \to \infty$.

Let $y_i = g(x_i) + \epsilon_i$, where g is continuously differentiable and ϵ_i are errors. Convergence results may be derived under distribution assumptions on (X, Y) with realized values (x, y) and response $g(x) = E(Y|X = x)$. For each value of $x = x_i$ compute for the neighbourhood points the following crude estimates of slope: $m_i(j) = (\widehat{y_i} - \widehat{y_j})/(x_i - x_j), j = 1, \cdots, n; j \neq i$. These are assigned weights depending on the normalized distance $d = d_{i,j} = |x_i - x_j|/\beta$. A particular choice could be $w_1 = w_1(d) = w_1(d_{ij}) = w_1(d(x_i - x_j)) \propto d^{-1.5} = O(|x_i - x_j|^{-1.5})$. Weight functions like $w_2(d) \propto e^{-\alpha d}, \alpha > 0$ may also be considered, assigning comparatively lower weight to distant points. Weights $w_1(d_{ij})$, $w_2(d_{ij})$ are standardized with sum of the weights over j as one. From $m_i(j)$ values, $i = 1, \cdots, n$, one may calculate $\widetilde{m_i}$, the weighted average $\sum_j w(d_{ij})m_i(j)$, or other measures of central tendency like median or trimmed mean etc.

Lowess regression to $(x_i, \widetilde{m_i}), i = 1, \cdots, n$ with weight as smooth tricubic function e.g., $u(x) = \frac{70}{81}(1-|x|^3)^3, |x| < 1, u(x) = 0$, otherwise; provide estimated derivatives of the growth curve at different time points. The kernel $u(x)$ is smooth at -1 and 1. For robustness at two stages, one may like to use tricubic weight function based on nearest neighbour in place of w in the first stage itself. Selection of weight function will be discussed in more detail in Sect. 6.4, in relation to some practical examples.

We sketch underlying justification of the technique. Growth estimate from lowess curve is weighted least square (ls) fit of a low degree polynomial p usually of degree ≤ 2, to $(x_i, y_i), i = 1, \cdots, n$; locally at the point $x = x_i$, i.e., $\widehat{y_i} = p_{ls}^{(i)}(x)|_{x=x_i} = p^{(i)}(x_i)$. With smooth weight function the estimates $p(x)$ are smooth and continuous, $|p_{ls}^{(i)}(x) - p_{ls}^{(j)}(x)| \rightarrow 0$, as $|x_i - x_j| \rightarrow 0$, for $x = x_i, x_j \in [a, b]$.

Let lowess estimate satisfy $\widehat{y_i} = g(x_i) + R$ where $R = R_n = o(1)$, $n \rightarrow \infty$. We assume that R is a locally Lipschitz function of order $\alpha > 1$.

As $\beta \rightarrow 0$, there are sufficiently many observations in a small neighbourhood of x_i, where the derivative of response is continuous. Empirical slope estimates in that neighbourhood are close to the derivative, and weighted average of empirical slopes is then a consistent estimate. Note that distant x values from x_i are down-weighted to have negligible contribution to the sum $\widetilde{m_i}$.

In other words, for small grid spacing,

$$(\hat{y}(x_i) - \hat{y}(x_{i+1}))/(x_i - x_{i+1}) = \frac{d}{dx}p(x)|_{x=x_i}(1 + o(1)) \tag{6.1}$$

as $|x_i - x_{i+1}| \rightarrow 0$. Again,

$$(\hat{y}(x_i) - \hat{y}(x_{i+1}))/(x_i - x_{i+1}) = g'(x_i) + o(1), \tag{6.2}$$

as $\widehat{y_i} = g(x_i) + R = g(x_i) + o(1)$, $n \rightarrow \infty$, and R is Lipschitz.
Thus, $m_i(j) = (\widehat{y_i} - \widehat{y_j})/(x_i - x_j) = g'(x_i) + o(1)$ when $|x_i - x_j| \rightarrow 0$.
$\widetilde{m_i} = \sum_j w(d_{ij})m_i(j) = g'(x_i) + o(1)$, as $w(d_{ij}), j \neq i$ is concentrated near x_i.

The $o(1)$ term is negligible for sufficiently large n and small grid spacing.
$\widetilde{m_i}$ are linear combination of $\hat{y}$ values, these in turn are least square estimates and hence linear function of basic observations $y_i, i = 1, \cdots, n$.

Under standard assumptions CLT holds for an average statistic like $\widetilde{m_i}$ that has convergence rate $O(n^{-1/2})$ in Berry–Esseen theorem see e.g., Helmers (1981),

Helmers and Huskova (1984), Bjerve (1977) for error bound in normal approximation with appropriate conditions on weight function in L-statistics.

In lowess regression the parameter f specifies the fraction of sample considered for local regression at a point, observations outside that nearest neighbour are assigned zero weight, much like a trimmed L-statistic, see e.g., Bjerve (1977), Ghosh and Dasgupta (1978).

Under similar conditions and standardization, one may have $O(n^{-1/2})$ as error bound in normal approximation for standardized $\widetilde{m_i}$, an intermediate estimate of growth derivative g'.

Finally, lowess smoothing of $(x_i, \widetilde{m_i}), i = 1, \cdots, n$; with tricubic weight u, is taken as the estimate of g'. As the issue of smoothing is pertinent for fast convergence, proposed smooth lowess estimates of derivatives have sharper convergence rates than usual Berry – Esseen bound.

Higher order derivatives of g may be obtained in a similar manner.

From (6.1), one may like to consider $p'(x_i)$ as an alternate estimate of $g'(x_i)$. However, the proposed estimate based on two stage smoothing is expected to perform better.

Performance of the proposed technique is seen to be good while calculating $g'(x)$ in some examples, see Dasgupta (2013a) for $g(x) = \log x, x > 0$.

Convergence results and CLT for local M-type of regression estimators with variable bandwidth under appropriate assumptions are also obtained by Fan and Jiang (1997), Hall (2010).

One may decide to harvest the yam and its stem if the computed rate of growth for response curve falls below a preassigned value over time.

Estimation of the proliferation rate $F(y) = \frac{1}{y}\frac{d}{dt}y(t) = \frac{d}{dt}\log y(t)$ follows the similar steps.

6.3 Generalized Mahalanobis Distance, Yam Production Index

Both Yam and its stem have contribution to farmers' income. For a valid comparison over different years observations on these two variables have to be taken into account. The above-ground biomass i.e., weight of Yam stems and leaves above ground and their relations to yield of Yam are being studied at ISI Giridih farm. These two variables are closely related, as a high level of biomass is an indicator of high yield of underground Yam production due to extensive photosynthesis in sunlight.

Elephant foot yam flower also has a market value as an edible tasty item, see http://latha468.wordpress.com/kadukum-kariveppum-vattalmulakum/.

In the following analysis one may also incorporate farmers' income from this third component.

For two populations with means $\mu^{(1)}$ and $\mu^{(2)}$ (the mean yield/income of several component variables like weight of yam, stem, flower; considered over years $i = 1, 2$) with *a common dispersion matrix* $\sum$, the Mahalanobis distance squared between the two means or, two populations is defined as,

$$\Delta^2(\mu^{(1)}, \mu^{(2)}) = (\mu^{(1)} - \mu^{(2)})' \sum^{-1} (\mu^{(1)} - \mu^{(2)}) \tag{6.3}$$

Let $x_1^{(1)}, \cdots, x_{n_1}^{(1)}$ be a sample of size n_1, from a population with mean vector and dispersion matrix as $(\mu^{(1)}, \sum)$ and $x_1^{(2)}, \cdots, x_{n_2}^{(2)}$ be a sample of size n_2, from $(\mu^{(2)}, \sum)$. Then, estimate of $\mu^{(1)}$ is $\overline{x}^{(1)} = \sum_{i=1}^{n_1} x_i^{(1)}/n_1$, of $\mu^{(2)}$ is $\overline{x}^{(2)} = \sum_{i=1}^{n_2} x_i^{(2)}/n_2$ and an unbiased estimate S of the common dispersion matrix $\sum$ is given by,

$$(n_1 + n_2 - 2)S = \sum_{i=1}^{n_1} (x_i^{(1)} - \overline{x}^{(1)})(x_i^{(1)} - \overline{x}^{(1)})' + \sum_{i=1}^{n_2} (x_i^{(2)} - \overline{x}^{(2)})(x_i^{(2)} - \overline{x}^{(2)})' \tag{6.4}$$

$$\text{i.e.,}\quad n\ S = (n_1 - 1)S_1 + (n_2 - 1)S_2, \quad n = n_1 + n_2 - 2$$

An estimate of Δ^2 above is provided by sample Mahalanobis distance squared,

$$D^2 = (\overline{x}^{(1)} - \overline{x}^{(2)})' S^{-1} (\overline{x}^{(1)} - \overline{x}^{(2)}) \tag{6.5}$$

The following results of Dasgupta (2008) state the properties of Mahalanobis distance for two highly correlated variables. These are relevant for dimension reduction. For the sake of completeness we restate the results. Here n is the sample size and ρ is population correlation coefficient between two highly correlated variables and these may represent e.g., yield of Yam and its vegetative growth.

Proposition A. *The D^2 statistic is unperturbed when an additional variable with high correlation is included to the set of variables. Under the assumption of linear regression, the error term in the limiting D^2 approximation by Δ^2, deleting the correlated coordinate from Δ^2, is $O_p(n^{-1}(1 - \rho^2)^{-1})$.*

Remark A. The condition of linear regression is satisfied for normal distribution. When $|\rho| = 1$, then both Σ and S are singular and the Mahalanobis distance is not defined. If the absolute value of correlation between any two variables in the set of variables is high $|\rho|$ is near 1, any one of the two correlated variables may be dropped. The growth of n has to be proportional to $(1 - \rho^2)^{-1}$, $|\rho| \to 1$, so that error of approximation is small.

Remark B. Approximation rate of Proposition A may hold "almost surely". By Marcinkiewicz–Zygmund strong law of large numbers (MZSLLN) see e.g., Chow and Teicher (1978), one may obtain $\bar{X}_n - \mu = o(n^{-\gamma/(1+\gamma)})$ a.s., where $X, X_1, X_2, \cdots$ are iid random variables with $EX = \mu$ and $E|X|^{1+\gamma} < \infty$,

$0 < \gamma < 1$. Thus, assuming the existence of fourth moment of coordinate random variables, each element of the matrix S converges to corresponding population value at the rate $o(n^{-\gamma^*})$ a.s., $0 < \gamma^* < 1$, with an application of MZSLLN to x_i^2, y_i^2, $x_i y_i$, etc. The matrix S is then "almost surely" within $o(n^{-\gamma^*})$ neighbourhood of the parameter Σ, i.e., $S = (1 + o(n^{-\gamma^*}))\Sigma$, a.s. Next, recall spectral decomposition of a matrix and use the relationship, $(A + UV')^{-1} = A^{-1} - \frac{(A^{-1}U)(V'A^{-1})}{1+V'A^{-1}U}$, where A is non-singular and U and V are two column vectors, to obtain $S^{-1} = \Sigma^{-1}(1 + o(n^{-\gamma^*}))$, a.s. The error rate in approximation (by deleting the correlated variable in Δ^2) mentioned in Proposition A is then $|\Delta^2 - D^2| = o(n^{-\gamma^*}(1 - \rho^2)^{-1})$ a.s., $0 < \gamma^* < 1$.

Remark C. In agricultural productions the yields of nearby plots/pits may sometimes be dependent due to leakage of nutrients and treatments within adjacent plots, yields from far distant plots may be taken to be independent. Proposition A, Remarks A and B remain valid even if the random variables (over different individuals) are nonstationary and strong-mixing with polynomial decay, as MZSLLN holds for such variables, see Dasgupta (2013b).

In view of Proposition A and Remark A one may concentrate on the single variable for comparison if the correlation coefficient is found to be high from repeated experiments, e,g., if $\rho^2 \approx 0.9$. The error of approximation in Mahalanobis distance Δ^2, D^2 based on yearly records of yield by a single coordinate is of approximate order $n^{-1}(1 - \rho^2)^{-1}$.

One may estimate ρ by sample correlation, then test for high value of ρ to decide about deleting the highly dependent variable from calculation of Mahalanobis distance.

When correlation is moderate between two variables, one may not be able to delete any component and an yield index should take into account of all the variables.

When the dispersion matrix of the variables remain same in different scenarios, a quality index based on D^2 statistic is proposed by Dasgupta (2008). One may have a similar measure for agricultural production. Consider a multivariate population with mean μ and dispersion matrix Σ. The Mahalanobis distance of the point μ from origin 0 (considered as the base point or ground level with no production) is $\mu'\ \Sigma^{-1}\mu = \Delta_0^2$, $\mu' = (\mu_1, \ldots, \mu_p)$. Let μ_i, the i th coordinate of μ, denote the mean of i th characteristic x_i of an agricultural product; higher the value of characteristic, better is the production. As before, the components in Δ_0^2 may be replaced by estimates from sample.

One may interpret the distance Δ_0 as an yield index of the agricultural production, with respect to the base level 0, *and proceed for yield comparison. The index is invariant under change of scale of measurements.*

Further note that the choice of origin as "base level" is natural in this case, as zero weight of Yam, stem and yam-flower mean no production.

Proposition 2. *Under the assumption of a common dispersion matrix, the index Δ_0^2, combining more than one variable can be estimated from yield production data and may be compared over years to find the particular year farmer had best production. Larger the value of the index better is the production situation.*

Now assume that the dispersion matrices are different in two different production scenario. Let F_1, F_2 be two distribution function with densities $f_1(x)$ and $f_2(x)$ with respect to some measure ν. Bhattacharyya affinity between these two populations is defined as,

$$d = d_{f_1,f_2} = \int \{f_1(x) f_2(x)\}^{1/2} d\nu \tag{6.6}$$

see Bhattacharyya (1943, 1946). Consideration of the function d goes back to Hellinger (1909), and sometimes this is also refereed as Hellinger affinity. Calculation of Bhattacharya affinity for two multinormal distribution leads to an analogue of Mahalanobis distance (Dasgupta 2008).

The extended definition of Δ^2 to the case $\Sigma_1 \neq \Sigma_2$, for *two multivariate normal densities* $\phi_1 = N_p(\mu^{(1)}, \Sigma_1)$ and $\phi_2 = N_p(\mu^{(2)}, \Sigma_2)$, reduces to,

$$\Delta^2_{\phi_1,\phi_2} = (\mu^{(1)} - \mu^{(2)})' \Sigma^{-1} (\mu^{(1)} - \mu^{(2)}) + 4 \log \frac{|\Sigma|}{(|\Sigma_1||\Sigma_2|)^{1/2}}; \ \Sigma = (\Sigma_1 + \Sigma_2)/2. \tag{6.7}$$

Mahalanobis distance in general case is related to Hellinger distance, a measure whose general properties are extensively studied. As a result this inherits all the nice properties of the latter distance. It has to be observed that the above expression (6.7) reduces to usual Mahalanobis distance squared when the dispersion matrices are equal. The above can be estimated from sample by,

$$D^2_{\phi_1,\phi_2} = (\overline{x}^{(1)} - \overline{x}^{(2)})' \bar{S}^{-1} (\overline{x}^{(1)} - \overline{x}^{(2)}) + 4 \log \frac{|\bar{S}|}{(|S_1||S_2|)^{1/2}} \tag{6.8}$$

$$\text{where } \bar{S} = \hat{\Sigma} = (\hat{\Sigma}_1 + \hat{\Sigma}_2)/2, \\ = (S_1 + S_2)/2.$$

Multivariate CLT holds under mild assumptions and by SLLN D^2 converges to Δ^2, thus extending Mahalanobis distance (6.7), (6.8) in a general set-up without multinormal assumption, where dispersions in two populations may be different.

Index (6.8) may then be interpreted as distance-squared between two production scenarios. Distance from origin is obtained by dropping one of the coordinates $\mu^{(2)}, \overline{x}^{(2)}$, etc. in the above, and the resulting expression $\Delta^2|_{\mu^{(i)}=0}, D^2|_{\overline{x}^{(i)}=0}, i = 1, 2$; can be interpreted as a production index for the second and first years, respectively. Here Σ is the mean of Σ_1 and Σ_2. The geometric mean of $|\Sigma_1|$ and $|\Sigma_2|$ appears in the expression in denominator.

Distribution of Extended Mahalanobis Distance and High Dimensional Data

A straightforward generalization may be made in a situation where we have to compare the production scenario over k years with varying dispersion matrices $\Sigma_1, \cdots, \Sigma_k$. The production index in i th population and sample are then

$$\Delta_i^2 = (\mu^{(i)})'\Sigma^{-1}(\mu^{(i)}) + 4 \log\frac{|\Sigma|}{(|\Sigma_1|\cdots|\Sigma_k|)^{1/k}} = \Delta_{i,I}^2 + \Delta_{i,II}^2 \quad (6.9)$$

$$D_i^2 = (\overline{x}^{(i)})'\bar{S}^{-1}(\overline{x}^{(i)}) + 4 \log\frac{|\bar{S}|}{(|S_1|\cdots|S_k|)^{1/k}} = D_{i,I}^2 + D_{i,II}^2, \text{ say;} \quad (6.10)$$

$$\text{where } \Sigma = (\Sigma_1 + \cdots + \Sigma_k)/k, \; \bar{S} = \hat{\Sigma} = (\hat{\Sigma}_1 + \cdots + \hat{\Sigma}_k)/k, \\ = (S_1 + \cdots + S_k)/k.$$

The quantities Δ_i^2 may be estimated by D_i^2 in the sample, $i = 1, \cdots, k$.

It is interesting to note that for equal sample sizes, the second component in Δ_i^2 is related to the likelihood ratio test (LRT) statistic for the hypothesis of equality of dispersion matrices, $H_o : \Sigma_1 = \cdots = \Sigma_k$. Extended Mahalanobis distance depends on diversity of dispersion matrices partially. The ratio of two components in extended Mahalanobis distance $R_d = \Delta_{i,II}^2/\Delta_{i,I}^2$ represents a measure of dispersion-heterogeneity, relative to mean heterogeneity towards diversity in different populations.

It is true that conventional use of Mahalanobis (studentized) D^2 statistics rests heavily on the assumption that the underlying distributions are all multivariate normal. In such applications, this stringent assumption may be rarely justified. As such, researchers have investigated ellipsoidal symmetric (ES) multivariate distributions for which Mahalanobis distance plays a basic role in maintaining affine equivariance (invariance) structures. Even for such ES distributions, the exact sampling distribution of the Mahalanobis distance is not precisely known. In multinormal case we shall shortly derive exact sampling distribution of extended Mahalanobis distance based on pooled dispersion. Also for general case it is possible to obtain a limiting null distribution for components in extended Mahalanobis distance, as the exponent of the limiting multivariate normal distribution is a Chi-square distribution. To see this write the terms in (6.8) for general case as

$$D^2 = (\overline{x}^{(1)} - \overline{x}^{(2)})'\bar{S}^{-1}(\overline{x}^{(1)} - \overline{x}^{(2)}) + 4 \log\frac{|\bar{S}|}{(|S_1||S_2|)^{1/2}} = D_I^2 + D_{II}^2, \text{ say.} \quad (6.11)$$

Now, by SLLN under the assumption of existence of the two dispersion matrices one may write

$$\bar{S} = \hat{\Sigma} = (\hat{\Sigma}_1 + \hat{\Sigma}_2)/2 = (S_1 + S_2)/2 \to (\Sigma_1 + \Sigma_2)/2 = \Sigma, \quad \bar{S}^{-1} \to \Sigma^{-1}, \quad (6.12)$$

almost surely. Assume that $n_1/n_2 \rightarrow 1$. The first component of (6.11) is approximately a Chi-square variable with an application of multivariate central limit theorem

$$\overline{x}^{(1)} - \overline{x}^{(2)} \rightarrow^L N_p\left(\mu^{(1)} - \mu^{(2)}, \frac{\Sigma_1}{n_1} + \frac{\Sigma_2}{n_2}\right) \tag{6.13}$$

$$D_I^2 = (\overline{x}^{(1)} - \overline{x}^{(2)})'\bar{S}^{-1}(\overline{x}^{(1)} - \overline{x}^{(2)}) \tag{6.14}$$

$$= (\overline{x}^{(1)} - \overline{x}^{(2)})'\Sigma^{-1}(\overline{x}^{(1)} - \overline{x}^{(2)})(1 + o_p(1))$$

$$nD_I^2 = (1 + o_p(1))\sum_{i=1}^{p} \xi_i^2, \tag{6.15}$$

where $(\xi_1, \cdots, \xi_p)$ is asymptotically $N_p(0, I_p)$. The characteristic function of $\sum_{i=1}^{p} \xi_i^2$ converges to that of a Chi-square variable.

Asymptotic distribution of the second component $D_{II}^2 = 4\log\frac{|\bar{S}|}{(|S_1||S_2|)^{1/2}}$ is considered by Bai et al. (2009) even for *non-Gaussian* case. After suitable standardization based on sample sizes n_1, n_2 and dimension p, the statistic D_{II}^2 follows normal law in the limit.

It is also possible to obtain a more precise form of the distribution of $D^2 = (\overline{x}^{(1)} - \overline{x}^{(2)})'\bar{S}^{-1}(\overline{x}^{(1)} - \overline{x}^{(2)}) + 4\log\frac{|\bar{S}|}{(|S_1||S_2|)^{1/2}}$ given in (6.11) for *multivariate normal* set-up. Under the assumption of same multivariate normal distribution in two population, distribution of proposed index D_i^2 in (6.9) is a weighted convolution of F and a limiting Chi-square variable, the variables being independent. Noncentrality parameters appear along with the above in non-null case when two multivariate distributions have different parameters. Similar observations hold for the estimated ratio $\hat{R}_d$.

These result may be obtained with an application of standard results on Wishart distribution as follows.

Let $W^{(i)} = W_p^{(i)} = W_p^{(i)}(k_i, \Sigma), i = 1, 2$ be two independent p-dimensional central Wishart variables with parameter Σ and degrees of freedom k_i. Then $\frac{|W^{(i)}|}{|W^{(1)}+W^{(2)}|}, i = 1, 2$ are independent of $W^{(1)}+W^{(2)}$. Thus $\frac{|W^{(1)}||W^{(2)}|}{|W^{(1)}+W^{(2)}|}$ is independent of $W^{(1)} + W^{(2)}$. Since the mean vector and the matrix $\bar{S} = (S_1 + S_2)/2$ are independent in multivariate normal set-up, it follows that the two components D_I^2 and D_{II}^2 in (6.11) are independently distributed.

In multinormal set-up the two components in (6.11) are independently distributed, first component is a Hotelling's generalized T^2 variable having an F distribution and the second component is related to the likelihood ratio test (LRT) statistics for testing equality of dispersion in multivariate normal set-up, having a limiting Chi-square distribution with $\frac{1}{2}p(p+1)$ degrees of freedom.

Approximate values of the critical points for the distribution of extended Mahalanobis distance (6.11) may easily be formed via simulation from these two independent distributions.

Additional covariates such as weather characteristics, fertilizer, labour expertise and cost, irritation level in yam (in presence of calcium oxalate crystals), market demand and texture of cooked yam, may be taken into account, as in high dimensional data analysis. The number p of important variables we identified in farm experiment on yam is about 10, which is moderate.

Dimension reduction is possible in a set of highly correlated variables, e.g., see Proposition A.

As the dimension p increases the d.f. of limiting Chi-square distribution increases like $O_e(p^2)$. This problem of high dimension may be solved by considering limiting normal distribution of the modified test statistic after suitable standardisation.

The pseudo-likelihood ratio test for high dimensions, as proposed in Bai et al. (2009) performs well even in small or moderate dimensions p.

Assumption on normality of variables is a stringent assumption. However, in (6.14)–(6.15) CLT may be used. Berry–Esseen bound provides a rate of convergence to normality as $O(n^{-1/2})$. The value of $n = n_1 + n_2$ considered here is around 200. CLT has applications in a broad class of real life situations, thereby allowing computation of limiting distribution of the first component nD_I^2 in general case.

Proposed statistic D_I^2 is robust in the sense that one need not be confined to Gaussian distributions, as we require only multivariate CLT that holds under mild assumption of finite dispersion matrices. Standardized D_{II}^2 has a limit distribution in general, vide (Bai et al. 2009).

Power Robustness

To address the issue of power-robustness in the context of subtle departures from multinormal to elliptically symmetric to general multivariate distributions, note that the proposed measure is derived from Hellinger distance. $d_H(f_1, f_2) = 1 - d(f_1, f_2) = 1 - \int \{f_1(x) f_2(x)\}^{1/2} d\nu$ is continuous in its arguments, in the sense that those subtle departures of f_1, f_2 from multinormal, say in terms of normal mixtures with main part plus a second part with negligible mixing proportion of the form $\alpha\phi_1 + (1-\alpha)\phi_2, \alpha \uparrow 1$ results in negligible change in Hellinger distance and that in turn causes little perturbation of order $(1 + o(1))$ on the statistic based on Hellinger distance, with a little change in probability of the critical region under alternative, preserving power-robustness.

To see this consider $f_1 = f(\theta), f_2 = f(\mu)$ two multinormal densities and $f_2^* = \alpha f_2 + (1-\alpha) f(\mu + \delta\mu)$, a multinormal mixture; $||\delta\mu||$ is small and $\alpha \uparrow 1$.

Next write $f_2^* = f(\mu) + (1-\alpha) D f(\mu) \delta\mu (1 + o(1))$, where $Df(\mu)$ is the vector of first derivative of f at μ and $o(1)$ term goes to zero as $||\delta\mu|| \to 0$.

Then $d(f_1, f_2^*) = \int \{f_1(x) f_2^*(x)\}^{1/2} d\nu = d(f(\theta), f(\mu))(1 + O((1 - \alpha)||\delta\mu||))$.

Since $d(f(\theta), f(\mu))$ is related to the extended Mahalabobis distance squared, see (1.6) of Dasgupta (2013c) and Dasgupta (2008), r.h.s. of the above expression indicates that perturbation of statistic of interest is negligible, being of order

$O((1-\alpha)||\delta\mu||)$; thus ensuring power robustness for subtle departure from multinormality.

The second components in the expressions of Δ_i^2 and D_i^2 in (6.9), (6.10) do not involve i, ranking of the scenarios may then be done by the first components, viz., $\Delta_{i,I}^2 = (\mu^{(i)})'\Sigma^{-1}(\mu^{(i)})$ and its estimate $D_{i,I}^2 = (\overline{x}^{(i)})'\bar{S}^{-1}(\overline{x}^{(i)})$. Calculation of the first components, however, requires the information on $\Sigma_1, \cdots, \Sigma_k$ via $S_1, \cdots, S_k$. We have the following.

Proposition 3. *When the dispersion matrices vary over scenarios, the one which has the largest value of* $D_{i,I}^2 = (\overline{x}^{(i)})'\bar{S}^{-1}(\overline{x}^{(i)})$, *among all years of production is the best production scenario.*

In the above proposition, comparisons are made with base level 0 i.e., nil production.

6.4 Analysis of Yam-Stem Growth

Data for analysis in this and subsequent sections are from experiments on growth curve estimation based on Elephant foot yam cultivation conducted at Indian Statistical Institute's Giridih farm at Jharkhand.

Consideration of the cost function is implicit in the fact that the farmers would not like to keep crop in farmland unless there is some gain in growth e.g., if the growth rate is below a level they would like to harvest the crop. This part is taken into account while estimating the growth rate and checking when the rate falls below a level. In Fig. 6.1 we show the growth of main stem that sprouted from a seed corm of weight 350 g with moderately rough seed skin texture. The lowess estimate of growth curve is also shown. In Figs. 6.1–6.8, lowess estimates are shown by points.

Figure 6.2 shows the growth of second stem connected by lines along with the smooth lowess estimate of growth curve shown by points. Averaged values of estimated growth rate $\widetilde{m_i} = \sum_j w(|x_i - x_j|)m_i(j)$, with polynomially decaying weights $w(d) \propto d^{-1.5}$ for different time point x_i are shown in Fig. 6.3, the figure also shows lowess estimate (with $f = 1/3$, regulating the proportion of data set used by lowess smoothing) of growth rate. Figure 6.4 shows the same with exponentially decaying weights $w(d) \propto e^{-.1d}$. The weights are standardized, the sum of weights being one. Estimated growth rate via lowess curve does not depend much on the choice of weight function used in $\widetilde{m_i}$, the weighted mean over all j. Lowess estimates are close to original curves.

Similar figures for the main stem of the same corm are shown in Dasgupta (2013a), those are of same type as that for secondary stem shown here in Figs. 6.2–6.4. For main stem, in Figs. 6.5, 6.6 we plot the growth rate with weight functions $w(d) \propto d^{-1.5}$ and $w(d) \propto e^{-.1d}$ for $f = 1/3$ by lowess regression. These figures are counterparts of Figs. 6.3, 6.4.

It will be appropriate now to discuss about the choice of the weight function and whether one should consider median or some other robust estimate instead of mean

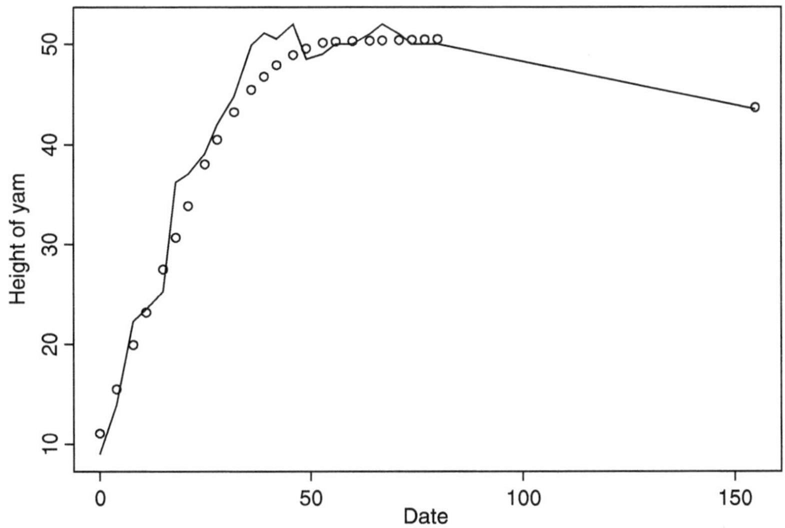

Fig. 6.1 Lowess fit for height curve of Yam plant 9, stem 1

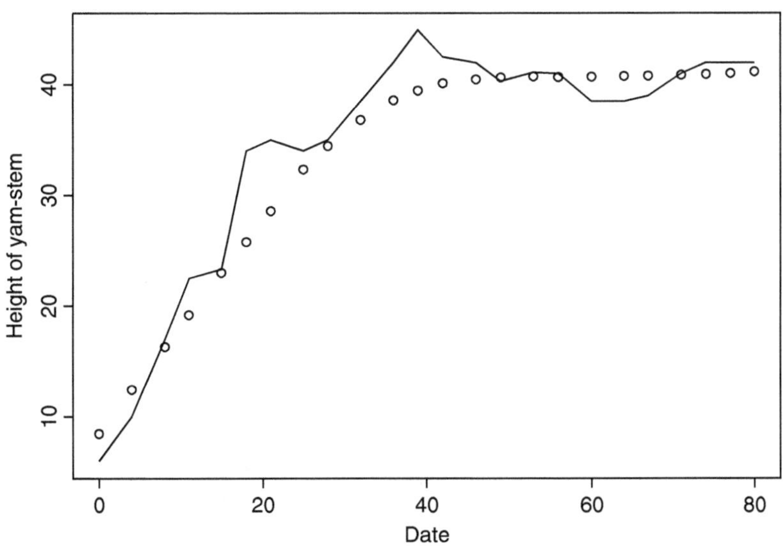

Fig. 6.2 Lowess fit for height curve of Yam plant 9, stem 2

while combining $m_i(j)$ values. We observed that using too much localized estimate like considering median instead of mean, along with selecting sharp exponential decrease of weights e.g., $w(d) \propto e^{-d}$ do not produce smooth growth rate curve. On the other hand a trimmed mean of successive three observations viz., average of unique median plus values immediately above and below it/median as average

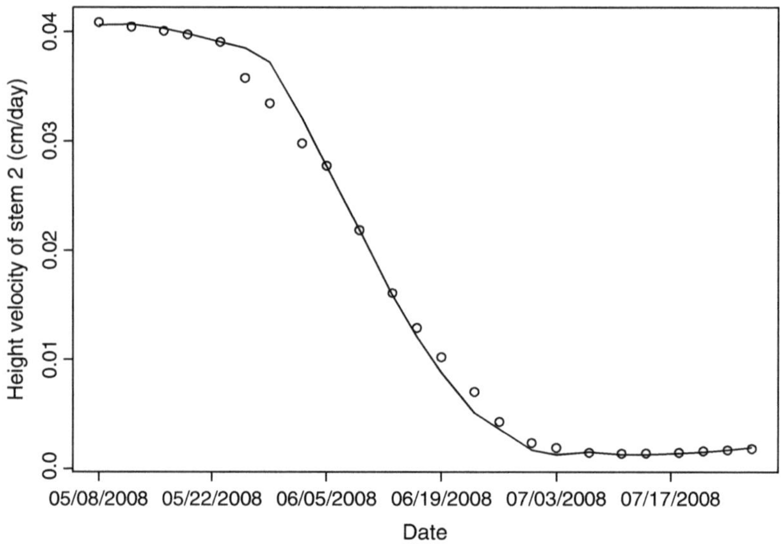

Fig. 6.3 Height velocity curve with local averaging through $x^\wedge(-1.5)$

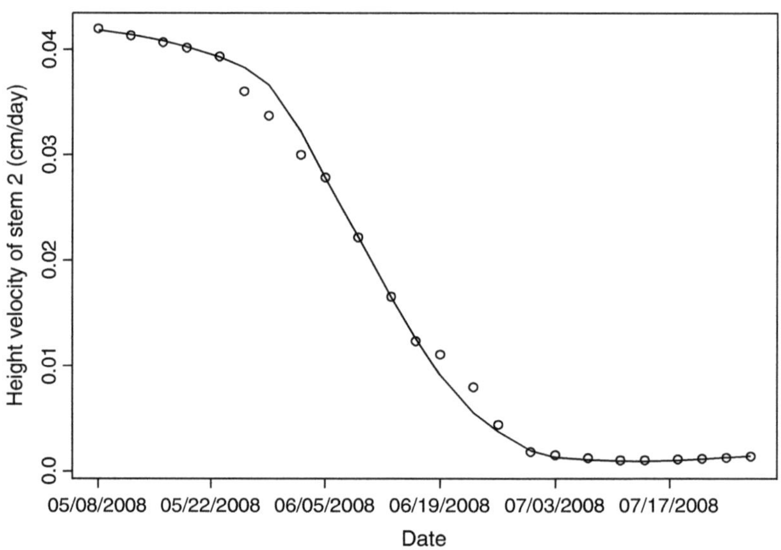

Fig. 6.4 Height velocity curve with local averaging through $\exp(-.1x)$

of two order statistic in the middle when there is no unique median, with lower exponential decay of weights e.g., $w(d) \propto e^{-.01d}$ with $f = 1/3$ are more appropriate for lowess regression that provides smooth growth rate of yam-stem. Figures 6.7, 6.8 represent two such growth-rate curves for stem 2 and stem 1, respectively, based on trimmed mean.

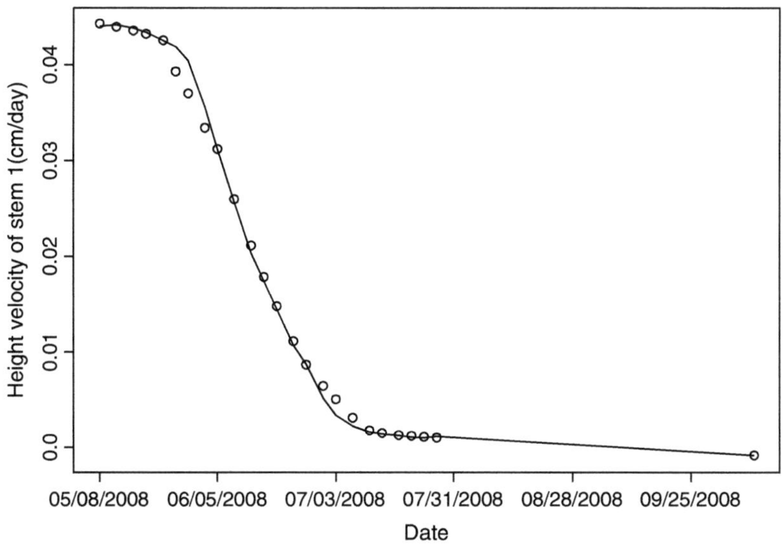

Fig. 6.5 Height velocity curve with local averaging through $x^\wedge(-1.5)$

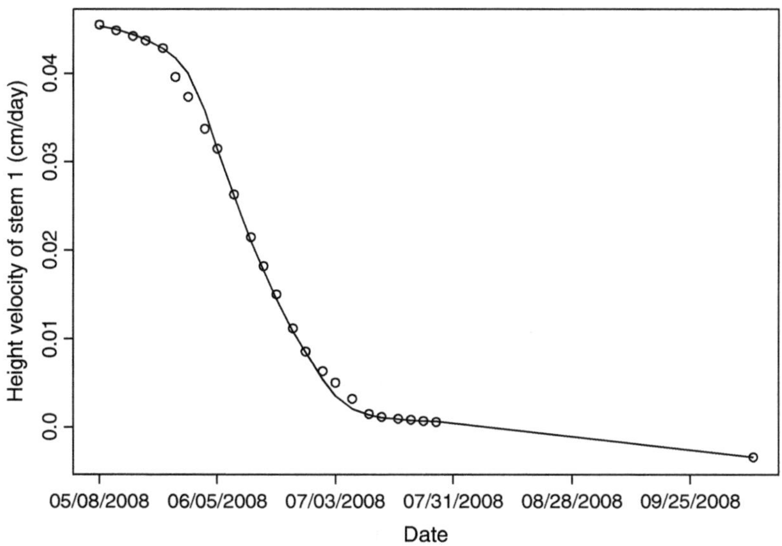

Fig. 6.6 Height velocity curve with local averaging through $\exp(-.1x)$

In Fig. 6.9 we plot the growth curve of all the four stems viz., main stem (stem 1), and three other auxiliary stems. Among these stems, the main stem and stem 2 showed similar growths over a long period of time; other two stems were short lived.

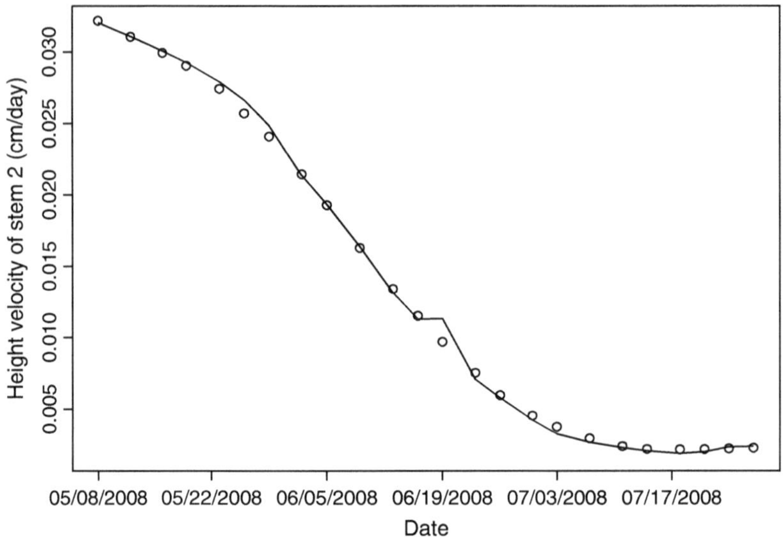

Fig. 6.7 Height velocity curve with trimmed mean through $\exp(-.01x)$

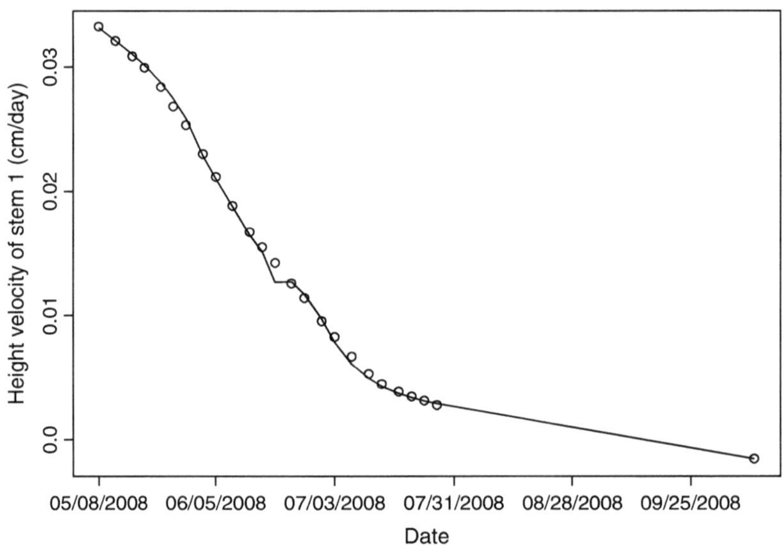

Fig. 6.8 Height velocity curve with trimmed mean through $\exp(-.01x)$

We observe that the figures reveal similar patterns for two stems, thus we explore the possibility of combining the features to obtain a precise estimate of growth rate in Sect. 6.6. The starting points of stem 3 and stem 4 may also be shifted to origin from where the other two stems started, to get additional information about the common pattern of the growth curve for this seed-corm.

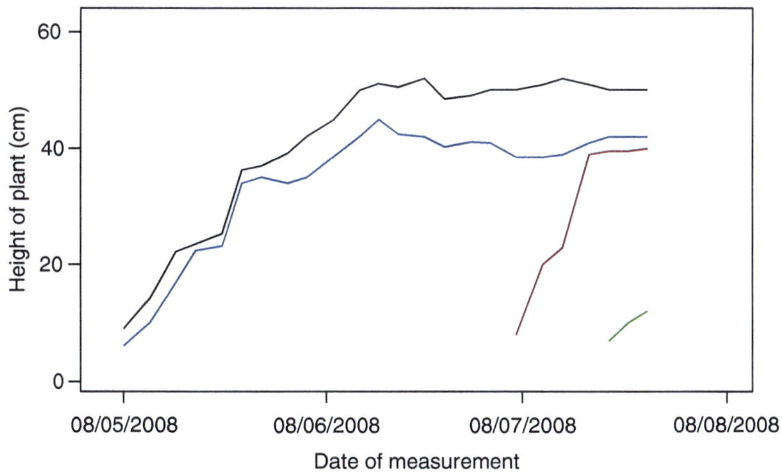

Fig. 6.9 Growth of four stems of plant number 9

6.5 Comparison of Yield Scenarios Over Two Production Periods

We analyse yam data for 2 years taking into account several variables of interest. For each seed-corm, yam weight at final harvest (x_1), maximum height of stem attained (x_2), and number of sprouts (x_3) above 10 cm of height are considered.

Farmers are mainly interested in the first variable. However, other two variables relate to vegetative growth above surface and are responsible for underground yam deposition; the latter variables are also of interest from economical point of view as yam-stems have market value. In growth experiments conducted at Indian Statistical Institute's Giridih farm out of 100 plantations in year 2008, three seed-corms did not germinate. In the repeated experiment of next year 2009, the number of sprouts per seed-corm was higher compared to that for the year 2008, as the seed-corms were already in sprouting stage when plantations were made a little bit late in time. That year 18 seed-corms did not germinate, 19 yield observations were nil, one seed-corm germinated and the stem-height crossed the threshold mark of 10 cm for consideration in analysis, but it died prematurely resulting in nil yield for yam.

Yam weight being a continuous variable, high mass concentration at a single point is unusual, this is incompatible with the assumption of approximate normal distribution (with possibility of distinct dispersion matrices) under which the analogue of Mahalanobis distance is derived in Dasgupta (2008), appealing to the criterion of "Bhattacharya-affinity" of two densities. We compare the scenarios via Mahalanobis distance criterion deleting the null vectors, i.e., observations with $(x_1, x_2, x_3) = 0$. Another way of comparison is to bootstrap the large deviation probabilities (Dasgupta 2010, 2013a).

Although in the experiment, pits of nongerminating seed-corm remained empty till the end of experiment, a farmer may either replace these by new seed-corms or utilize the vacant space for intercropping with some other vegetables such as chilli, brinjal, etc.

The estimated mean vector and dispersion matrix of $x = (x_1, x_2, x_3)$ are as follows.

For the year 2008, $\overline{x}^{(1)} = (1.5637629, 62.73608, 2.288660)$;

$$S_1 = \begin{pmatrix} 0.60670110 & 6.2782555 & -0.07193084 \\ 6.27825552 & 119.3014970 & -0.50739905 \\ -0.07193084 & -0.5073991 & 0.87414089 \end{pmatrix}$$

For the year 2009, $\overline{x}^{(2)} = (3.111481, 74.27284, 6.243902)$;

$$S_2 = \begin{pmatrix} 2.993733 & 29.98738 & 2.231375 \\ 29.987376 & 498.54175 & 34.102238 \\ 2.231375 & 34.10224 & 11.516049 \end{pmatrix}$$

Recall that for comparison over 2 years $k = 2$, and distance squared from origin

$$D_i^2 = (\overline{x}^{(i)})'\bar{S}^{-1}(\overline{x}^{(i)}) + 4 \log \frac{|\bar{S}|}{(|S_1||S_2|)^{1/2}}, \ D_1^2 = 20.09078, \ D_2^2 = 22.14735.$$

The second year is of higher production scenario, as this one is more "Mahalanobis-distant" from origin than the first year. This finding is in consistent with the fact that the second mean vector is higher than the first mean vector in each coordinate. Estimated Mahalanobis distance squared between two populations is 4.537006, of which the contribution from the logarithmic term involving heterogeneity of dispersion matrix is $4 \log \frac{|\bar{S}|}{(|S_1||S_2|)^{1/2}} = 1.112433$.

From two components in the distance-squared Δ^2, a measure of relative heterogeneity (from $\mu_o = 0$) due to dispersions w.r.t. mean vectors of several multivariate populations is $R_d(i) = \Delta_{2i}^2/\Delta_{1i}^2$, with the convention that $0/0 = 1$. For a selected ideal value of mean $\mu_o \neq 0$, one may consider $(\mu - \mu_o)$ in place of μ in the above definition. Small value of R_d is an indication of less contribution from dispersion-heterogeneity, relative to mean towards diversity in populations reflected in extended Mahalanobis distance.

As an estimate of the above one may consider $\hat{R}_d(i) = D_{2i}^2/D_{1i}^2$.

Estimated value of these in the present case are as follows.

For the year 2008, $\hat{R}_d(1) = 1.112433/(20.09078 - 1.112433) = 0.05861591$, and for the year 2009, $\hat{R}_d(2) = 1.112433/(22.14735 - 1.112433) = 0.05288507$.

Contribution to heterogeneity of scenarios comes mainly from mean part rather than dispersion.

Figure 6.10 shows coefficient of variation of all the three variables (x_1, x_2, x_3), this exhibit decreasing trend in the second year, and so does the line connecting the

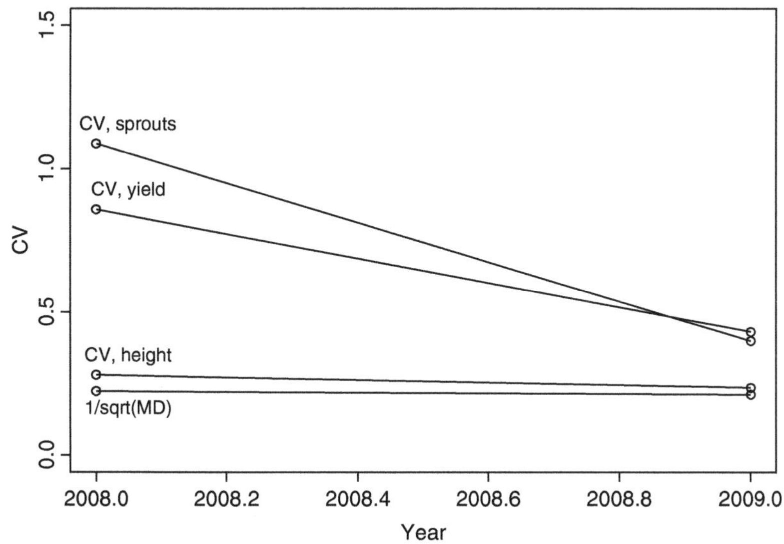

Fig. 6.10 Comparison of production scenarios over 2 years

values of $1/D$. Decrease in coefficient of variation for (x_1, x_2, x_3) from the year 2008 to 2009 is as follows; from 0.8580084 to 0.4312164 for x_1, from 0.2801604 to 0.2366433 for x_2, and from 1.0875337 to 0.3986281 for x_3.

6.6 Some Standard Growth Models

The observed growth of Yam-stems to some extent resembles Gomphertz curve

$$y(t) = a\exp(b\exp(ct)); a > 0, b < 0, c < 0. \tag{6.16}$$

The curve and its modifications have been successfully used to explain the growth of cancerous tumours. In a confined space where the availability of nutrients is limited, growth rate is high in the beginning and then it slows down due to competition for nutrients.

Selecting different form of proliferation rate $F(y) = \frac{d}{dt}\log y(t)$, one may obtain logistic, generalized logistic, Gomphertz, and other type of growth curves. For logistic curve the proliferation rate $F(y)$ is finite, whereas for Gomphertz curve this is unbounded.

An appropriate model may be selected from the observed pattern of proliferation rate $F(y) = \frac{1}{y}\frac{d}{dt}y(t) = \frac{d}{dt}\log y(t)$.

For cancerous tumours it may be difficult to directly and continuously record the stages of growth. However, indirect assessment is possible via Anti-Malignin Antibody in Serum (AMAS) blood test. Anticancer antibody with threshold value 135 micrograms per milliliter (mcg/mL) is seen to be present amongst 95%–99%

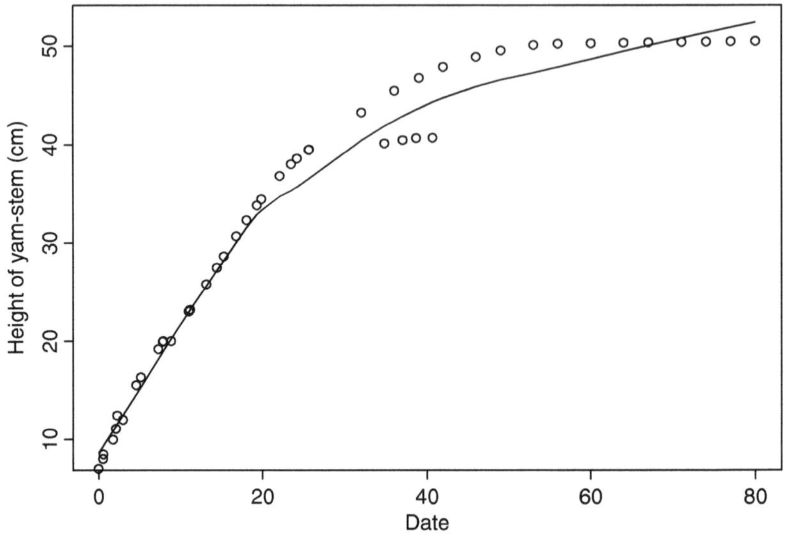

Fig. 6.11 Lowess (on AM of x values) for combined height of plant 9

patients at early stage of cancer and during treatment, except in terminal cases. Thus, one may monitor the stage of cancerous tumour and response to treatment via the AMAS level recorded frequently over time, rather than intrusive and often painful direct observations. Doctor may like to change medication if the rate of *decrease* of antibody level is not satisfactory. Presences of hormones are important indicative agents in sex-related cancers such as cancer of the breast, uterus, prostate, ovary, and testis, and of thyroid cancer and bone cancer. Some visible symptoms such as unexplained weight loss, fatigue and red pinpoints in skin are common in child cancer. These variables may continuously be observed over time, rather than direct observation on tumour. A similar analysis of Yam-like growth in a confined space with limited nutrients thus seems possible.

We may combine the four stem-growth curves of plant 9 as shown in Fig. 6.9, to obtain an overall picture of growth pattern. For studying the common features in stems from start time zero, the time of individual sprouting are all considered to be zero. Imagine that all the four curves are shifted to origin by translation, without disturbing the patterns. The observed values of y are ordered and the largest value of y, up to which at least one of the curves is monotone (nondecreasing) is found. Then for each observed y, the x values i.e., quantiles are obtained by linear interpolation from the stem-growth curves if that value of y falls within the admissible (monotonic) range of a particular curve after translation; otherwise that curve is rejected from consideration for that y value. Next the average of these admissible quantiles (x-values) is plotted against the y value in (x, y) plane. A lowess curve fitted to this scatter plot is shown in Fig. 6.11 as a continuous curve, to be interpreted as overall growth curve of plant with seed-corm no. 9. In Figs. 6.11–6.20, lowess estimates are shown as continuous broken lines.

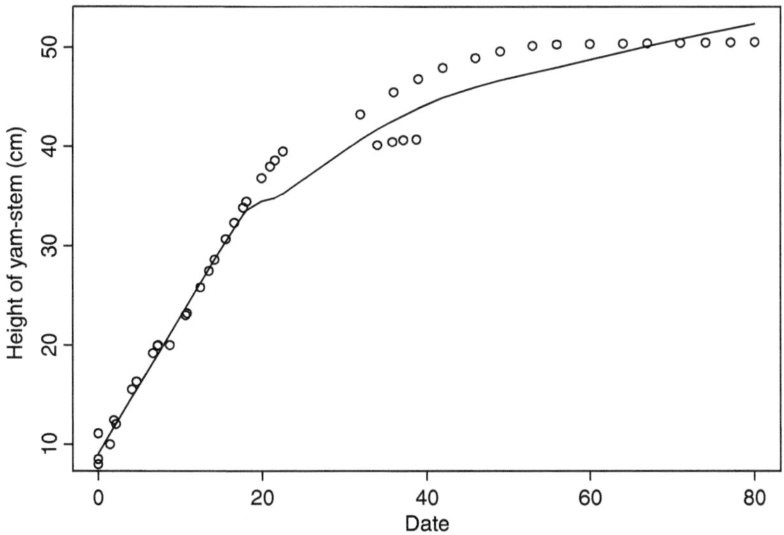

Fig. 6.12 Lowess (on GM of x values) for combined height of plant 9

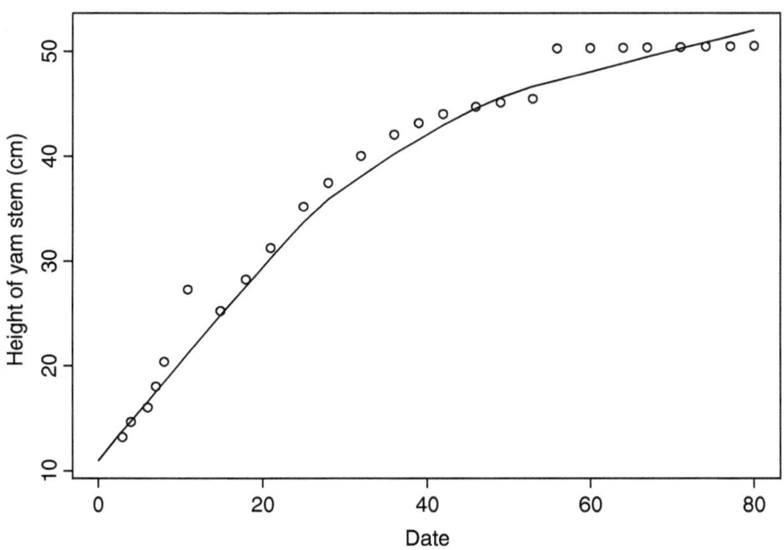

Fig. 6.13 Lowess (on AM of y values) for combined height of plant 9

Figure 6.12 shows the same characteristics like Fig. 6.11, where we consider Geometric Mean (GM) of the quantiles instead of Arithmetic Mean (AM). Under a multiplicative model GM is proper measure of central tendency. The lowess curve shows slight indication of bi-phasic growth in Fig. 6.12.

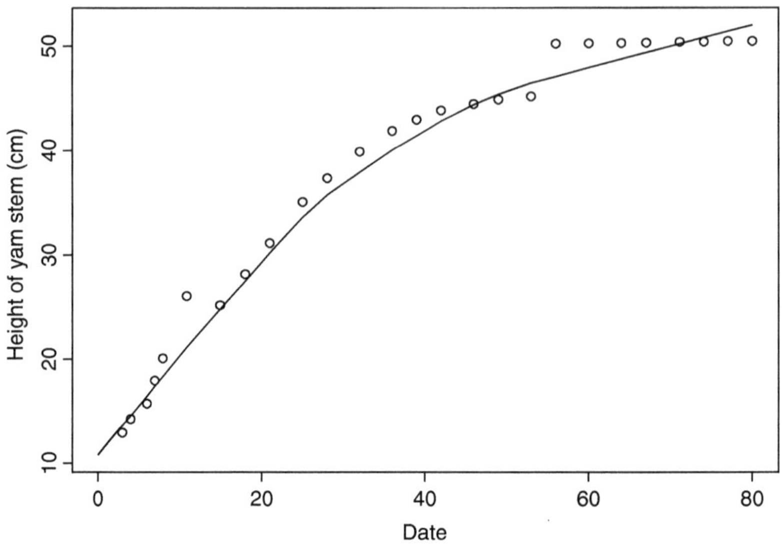

Fig. 6.14 Lowess (on GM of y values) for combined height of plant 9

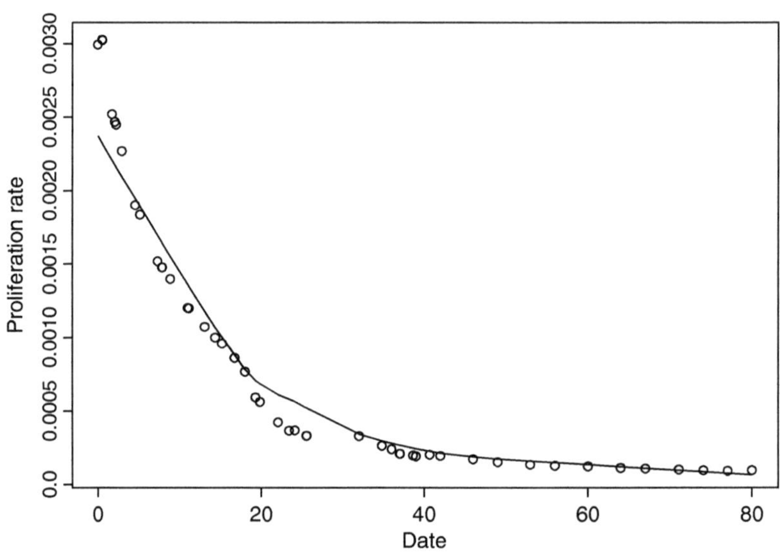

Fig. 6.15 Proliferation rate of plant 9 with weight ~d^(−1.5) Ref. Fig. 6.11

In Fig. 6.13, for each observed value of x, we consider the average of corresponding y values obtained from different curves to get the scatter plot. The lowess curve is shown as a continuous broken line. Figure 6.14 shows the same, where instead of AM the GM of y values is considered.

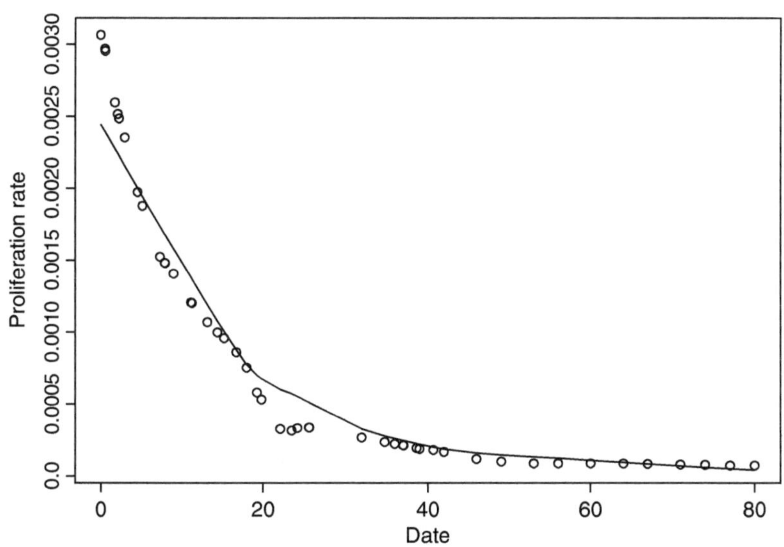

Fig. 6.16 Proliferation rate of plant 9 with weight∼e^(−d), Ref. Fig. 6.11

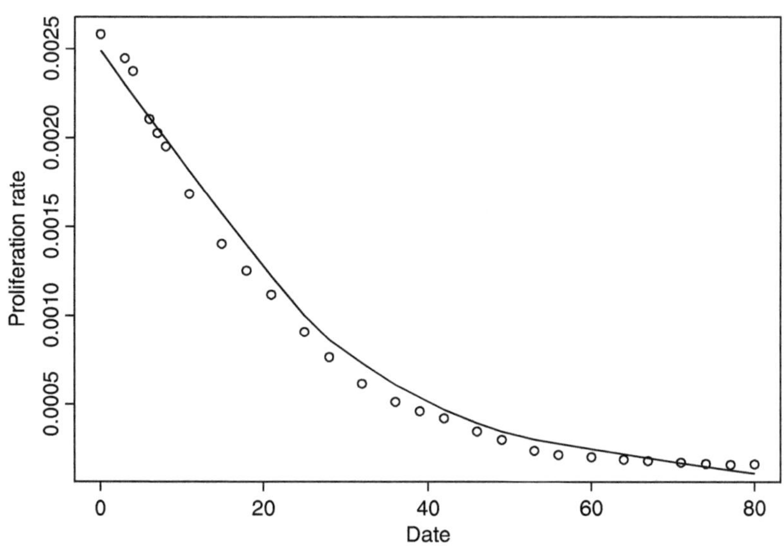

Fig. 6.17 Proliferation rate of plant 9 with weight∼d^(−1.5) Ref. Fig. 6.13

Although the curves show more or less similar features, curves computed from AM seem to be a little bit smoother than those from GM.

Figure 6.15 plots the point estimate of proliferation rate $\frac{d}{dt} \log y(t)$ obtained by the technique of computing derivative explained in Sect. 6.2, taking the (lowess) estimates given in Fig. 6.11 as input values. A lowess curve is fitted to the point

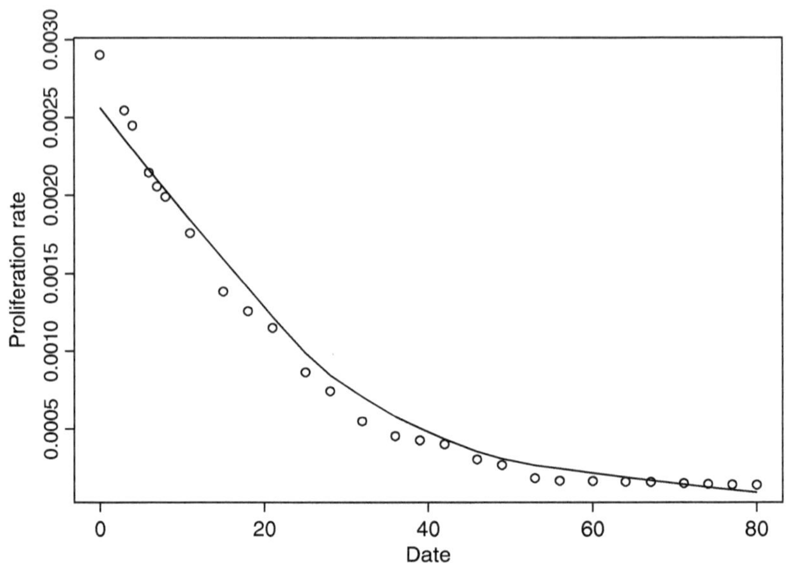

Fig. 6.18 Proliferation rate of plant 9 with weight∼e^(−d), Ref. Fig. 6.13

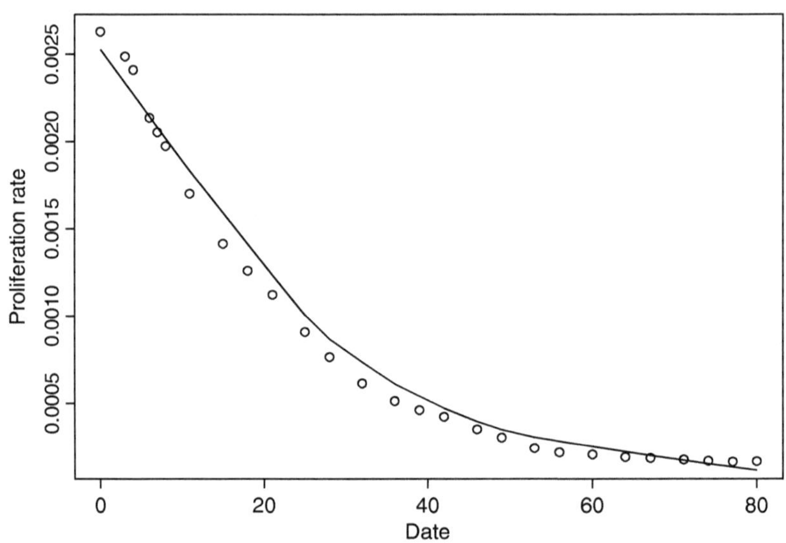

Fig. 6.19 Proliferation rate of plant 9 with weight∼d^(−1.5) Ref. Fig. 6.14

estimates of Fig. 6.15 with weight function $w_1 = w(d) \propto d^{-1.5}$, this is shown as a continuous broken line in Fig. 6.15. Figure 6.16 plots proliferation rate with input from Fig. 6.11, and lowess estimate with weight function $w_2 = w(d) \propto e^{-d}$.

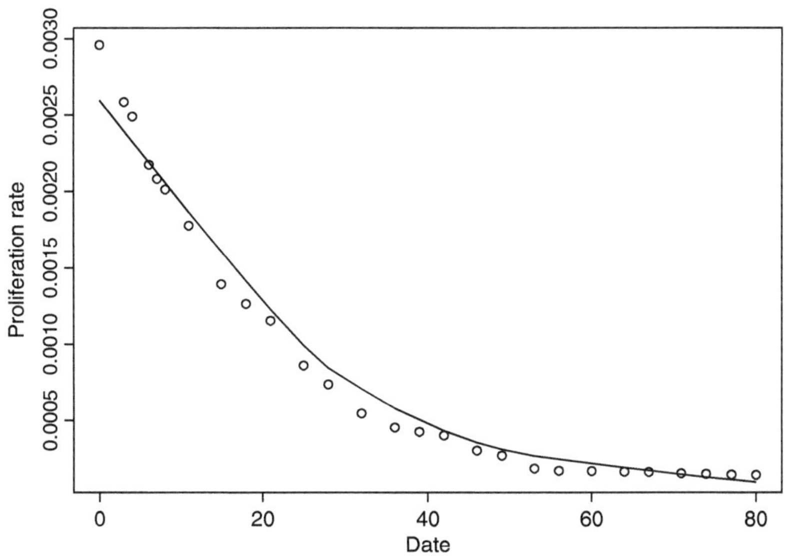

Fig. 6.20 Proliferation rate of plant 9 with weight∼e^(−*d*), Ref. Fig. 6.14

Figure 6.17 refers to proliferation rate with input from Fig. 6.13, and corresponding lowess estimate with weight function w_1. Figure 6.18 is similar to Fig. 6.17 with a different choice of weight function w_2.

Among the two curves computed from GM, Fig. 6.14 is relatively smoother than Fig. 6.12. In Figs. 6.19 and 6.20 we show the proliferation rate from the smooth growth curve in Fig. 6.14, based on weight functions w_1 and w_2, respectively.

The figures suggest that the proliferation rate is about 0.003 near origin. Gomphertz curve has limiting proliferation rate unbounded at origin. For $\alpha > 0$, $\nu > 0$, Gomphertz curve may be obtained as limiting form ($\nu \to \infty$) of generalized logistic function with proliferation rate of the type

$$y'(t)/y(t) = \alpha\nu\left[1 - \left\{\frac{y(t)}{k}\right\}^{1/\nu}\right] \approx -\alpha\log(y(t)/k) = \alpha\log(k/y(t)) \quad (6.17)$$

as $\nu \to \infty$, where $k(>0)$ is the maximum attainable size .

With finite ν the rate in (6.17) is finite at origin; however, the limiting form involves $\log y(t)$, and proliferation rate is infinite at origin with $y(0) = 0$.

The type of growth curve that is appropriate for a particular situation may be found from the lowess curve fitted on the basic points of proliferation rate, obtained as derivative of growth observations as suggested in Sect. 6.2 and implemented above. One may consider proliferation decaying in a slower rate than that for generalized logistic function, e.g., consider

$$y'(t)/y(t) = \alpha\nu\left[1 - \left\{\log\left(1 + (e-1)\frac{y(t)}{k}\right)\right\}^{1/\nu}\right]$$
$$\approx -\alpha \log\left\{\log\left(1 + (e-1)\frac{y(t)}{k}\right)\right\} \tag{6.18}$$

as $\nu \to \infty$. A faster decay of proliferation rate may also be considered, e.g., for $\delta > 0$, consider

$$y'(t)/y(t) = \alpha\nu[1 - e^{\{(y(t)/k)^{\delta}-1\}/\nu}] \approx \alpha\left[1 - \left\{\frac{y(t)}{k}\right\}^{\delta}\right] \tag{6.19}$$

as $\nu \to \infty$. It is interesting to observe that in the limit $\nu \to \infty$, one obtains generalized logistic function with polynomially decaying proliferation rate in the r.h.s. of (6.19).

Once an appropriate class of growth curve is identified for a particular situation, the effect of nutrients, soil and seed types, environment, etc. may be reflected in the parameters of the fitted growth curve in more general situations and comparison of growth scenarios are possible in terms of relevant parameters.

Acknowledgements Thanks are due to the reviewers for constructive comments that improved the presentation. Professor Debasis Sengupta helped in computer programming.

References

Bhattacharyya, A. (1943). On a measure of divergence between two statistical populations defined by their probability distributions. *Bulletin of the Calcutta Math. Society, 35*, 99–109.

Bhattacharyya, A. (1946). On a measure of divergence between two multinomial populations. *Sankhyā, A. 7*, 401–406.

Bai, Z., Jiang, D., Yao, J., & Zheng, S. (2009). Corrections to LRT on large dimensional covariance matrix by RMT, http://arxiv.org/pdf/0902.0552.pdf.

Bjerve, S. (1977). Error bounds for linear combinations of order statistics. *Annals of Statistics*, 357–369.

Chow, Y.S., & Teicher, H. (1978). *Probability theory: independence, interchangebility, martingales*. New York: Springer.

Clarke, R., Ressom, H.W., Wang, A., Xuan, J., Minetta, C., Liu, M.C., Gehan, E.A., & Wang, Y. (2008). The properties of high-dimensional data spaces: implications for exploring gene and protein expression data. *Nature Reviews Cancer, 8*, 37–49.

Cleveland, W.S. (1981). LOWESS: A program for smoothing scatterplots by robust locally weighted regression. *The American Statistician, 35*, 54.

Dasgupta, R. (2008). Quality index and Mahalanobis D^2 statistics. Advances in Multivariate statistical methods. In *Proc. of ISI Platinum Jubilee Conference* (pp. 367–382.). World Scientific.

Dasgupta, R. (2010). Bootstrap of deviation probabilities with applications. *Journal of Multivariate Analysis, 101*(9), 2137–2148.

Dasgupta, R. (2013a). Nonuniform rates of convergence to normality for two sample U-statistics in non iid case with applications. To appear in this volume as chapter 4.

Dasgupta, R. (2013b). Moment bounds for strong-mixing processes with applications. In *Proc. of ISI Platinum Jubilee Conference*. World Scientific. In press.

Dasgupta, R. (2013c). Yam Growth Experiment and Above-Ground Biomass as Possible Predictor. To appear in this volume as chapter 1.

Fan, J., & Jiang, J. (1997). Variable bandwidth and one step local M-estimator. Preprint, (http://scholar.google.co.in/scholar_url?hl=en&q=http://www.researchgate.net/publication/2302878_Variable_bandwidth_and_One-step_Local_M-Estimator/file/79e415124f0043a8af.pdf&sa=X&scisig=AAGBfm1d_QYEFeL1VeJRcmleQJ4n9GlrMQ&oi=scholarr&ei=YLqUUfKzHc3_rAfo2oHABA&sqi=2&ved=0CCoQgAMoADAA).

Ghosh, M., & Dasgupta, R. (1978). On some nonuniform rates of convergence to normality. *Sankhya A, 40*, 347–368.

Hall, B. (2010). *Nonparametric estimation of derivatives with applications*. Doctoral Dissertation, University of Kentucky.

Hellinger, E.D. (1909). Neue Begrundung der Theorie quadratischen Formen von unendlichen vielen Veranderlichen. *Journal fur Reine und Angewandte Mathematik, 136*, 210–271.

Helmers, R. (1981). A Berry-Esseen theorem for linear combinations of order statistics. *The Annals of Probability*, 342–347.

Helmers, R., & Huskova, M. (1984). A Berry-Esseen bounds for L-statistics with unbounded weight function. In *Proc. 3rd Prague Symp. on Asymptotic Statist.*

Kiefer, J., & Wolfowitz, J. (1952). Stochastic estimation of the maximum of a regression function. *Annals of Mathematical Statistics, 23*(3), 462–466.

Rao, C.R. (1974). *Linear statistical inference*, 2nd edn. New Delhi: Wiley Eastern.

Robbins, H., & Monro, S. (1951). A stochastic approximation method. *Annals of Mathematical Statistics, 22*(3), 400–407.

Xie, D., & Schlick, T. (2000). A more lenient stopping rule for line search algorithms. *Optimization Methods and Software*, 1–18.

Chapter 7
Evolution of Scour and Velocity Fluctuations Due to Turbulence Around Cylinders

H. Maity, R. Dasgupta, and B.S. Mazumder

Abstract The study is aimed at investigating the turbulence characteristics in scour geometry developed near a circular cylinder placed over the sand bed transverse to the flow. An obstacle of length 10 cm, placed on a sand bed develops a crescent-shaped scour mark on the bed. The scour is caused by generation of vortex developed on the upstream side of the obstacle. The turbulent flow field within the scour mark was measured using an acoustic Doppler velocimeter (ADV). We estimate the joint probability density function of fluctuating velocity components (u', w') applying the cumulant-discard method to the Gram–Charlier series at different locations over the scour mark. The scour marks named as current crescents preserved in geological record are traditionally used as indicators of palaeo-current direction. We further study the evolution of scour width till a state of equilibrium is attained. The scour-width growth curve is estimated by lowess nonparametric regression and smoothing spline techniques. Scour geometry is an indicator of velocity of past waterflow, preservation of fossils in ancient riverbed, etc. With an application of a robust nonparametric method (Dasgupta, 2013, Non uniform Rates of Convergence to Normality for Two sample U-statistics in Non IID Case with Applications, appearing in this volume as chapter 4) we estimate the first and higher order derivatives of growth curve in the present context and interpret the results.

H. Maity • B.S. Mazumder
Fluvial Mechanics Laboratory, Physics and Applied Mathematics Unit,
Indian Statistical Institute, Kolkata - 700 108, India
e-mail: maitym_r@isical.ac.in; bijoy@isical.ac.in

R. Dasgupta (✉)
Theoretical Statistics and Mathematics Unit, Indian Statistical Institute, Kolkata - 700 108, India
e-mail: ratandasgupta@gmail.com; rdgupta@isical.ac.in

R. Dasgupta (ed.), *Advances in Growth Curve Models*, Springer Proceedings in Mathematics & Statistics 46, DOI 10.1007/978-1-4614-6862-2_7,

7.1 Introduction

Scour mark around any object placed in the sediment bed usually develops due to the interaction of the local flow field with the sediment bed and the object. Investigations had been made to quantify the mean flow across the scour marks generated by different types of obstacles placed on sand bed (non-cohesive). The obstacle marks named as current crescent preserved in geological record are traditionally used as an indicator of palaeo-current direction (Sengupta 1966; Melville and Raudkivi 1977; Allen 1982; Karcz 1968), which are available in cross-bedded sediments depending on the orientation and plunge of the long axes of the pebbles. Sengupta et al. (2005) conducted experiments in a laboratory flume to generate the crescent scour mark using obstacles placed on sand bed. The formation of scour mark at the upstream of cylinder showed like a crescent scour structure, which is morphologically akin to the current crescent preserved in geological record (Sengupta 1966, 2007). Catano-Lopera and Garcia (2007) investigated experimentally the geometry of scour hole and the flow structure around short cylinders under the action of wave alone and combined wave and current. Their experimental evidence indicates both width and length of the scour hole generated due to the short cylinder primarily depend on the Keulegan–Carpenter number and the cylinder aspect ratio. Recently, Mazumder et al. (2011) investigated the mean flow and turbulence statistics in equilibrium scour marks developed near the static short circular cylinder placed over the sand bed transverse to the flow. They showed that the dimensionless scour-width increases with increase of cylinder Reynolds number, for a fixed sediment Froude number; and proposed a relation between the scour-width, the sediment Froude number, and the cylinder diameter. They also reported the process of evolution of scour mark around the object using digital photo camera. However, our knowledge on the evolution of scour width with time associated with statistical distributions is deficient (Fig. 7.1).

This investigation is aimed at studying the evolution of scour mark generated around the static short circular cylinders placed over the sediment bed transverse to the flow and to estimate the joint probability density function of fluctuating velocity components (u', w') over the equilibrium scour geometry. We study the growth of scour-width upto a state of equilibrium and model scour width growth curve by lowess nonparametric regression and smoothing spline techniques. Scour geometry is related to velocity of past waterflow and its direction, preservation of fossils in ancient riverbed, etc. With an application of a robust nonparametric method proposed in Dasgupta (2013), we estimate the first and higher order derivatives of the growth curve and discuss their applications.

7.2 Experimental Setup

Experiments were conducted in a re-circulating "closed circuit" laboratory flume (Mazumder et al. 2005) specially designed at the Fluvial Mechanics Laboratory (FML) of Physics and Earth Science Division, Indian Statistical Institute, Calcutta.

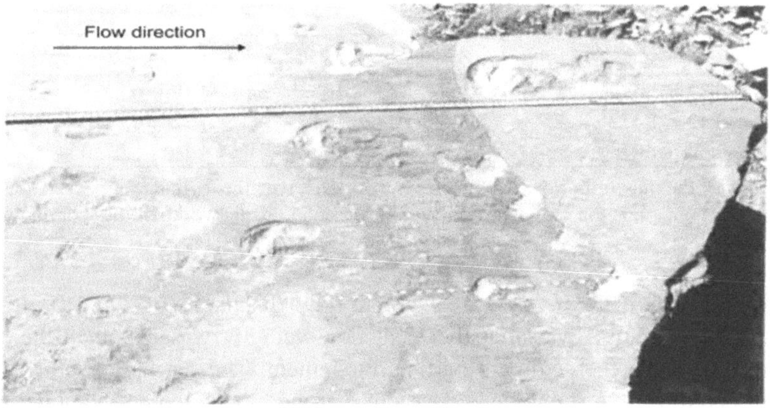

Fig. 7.1 Bedding plane showing current crescents. Proterozoic Kaimur formation, Maihar, M. P.

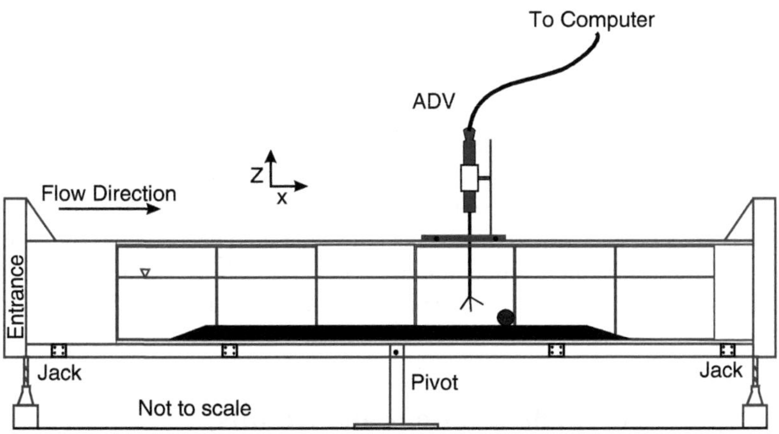

Fig. 7.2 Schematic diagram of the experimental set-up

Both the experimental and the re-circulating channels of the flume have identical dimensions of 10m length, 0.50m width, and 0.50m height. For details of test channel, experiments, experimental conditions, and the results associated with the mean flows and turbulence characteristics around the scour geometry, the paper by Mazumder et al. (2011) may be refereed.

A sand bed of thickness $h' = 4$ cm and 5m long covering the entire width (50 cm) of the flume was laid at the bottom. The median particle diameter d_{50} of the sand was 0.25 mm and the standard geometric deviation $\sigma_g = 0.685$. The specific gravity of sediments used for the experiments was 2.65. A series of experiments was conducted over the sediment bed of known grain-size distribution using three different circular cylinders of diameters $D_c = 3.2$, 4.2, and 6 cm of fixed length $L = 10$ cm placed at the center line of the flume. For each experiment, single cylinder was placed at the center line over the sand bed transverse to the flow at the measuring station 6m downstream of the channel inlet (Fig. 7.2). Flow depth

Table 7.1 Experimental values of flow parameters

D_c(cm)	a_r	$Q \times 10^{-2}(\mathrm{m}^3/s)$	u_m(cm/s)	$Fr(=\frac{u_m}{\sqrt{gh}})$	F_s	$Re(=\frac{u_m h}{\nu})$	w_s(cm)
3.2	0.32	1.472	26.63	0.167	4.19	69,238	3.5
4.2	0.42	1.472	26.20	0.164	4.12	68,120	4.5
6.0	0.60	1.472	26.45	0.166	4.16	68,770	6.5

where u_m is the maximum fluid velocity, $a_r(= D_c Ł)$ is the cylinder aspect ratio, $F_s = u_m/\sqrt{(\gamma_s - 1)gd_{50}}$, is the sediment Froude number, and w_s is the width of the scour hole

was kept constant at $H_w = 0.30$ m. A flow discharge of $Q = 0.015\ \mathrm{m}^3/s$ was chosen in such a way that the local flow velocity was less than the critical velocity to initiate the sediment particle movement. The discharge setup was left undisturbed to form a scour-shaped structure around the cylinder and consequently to attain perfect equilibrium conditions in the scour mark. The hydraulic slope of the flume was negligible and it was an order of 0.0001. Once the equilibrium is attained, the vertical velocity profiles from upstream to downstream along the flume centerline were measured using a SonTek 5cm down-looking three-dimensional Micro-acoustic Doppler velocimeter (ADV) for 3 min at a sampling rate of 40Hz to ensure full characterization of the turbulence phenomena. The velocity data were collected along the scour marks with the lowest point in each profile being 0.30 cm above the flume bed and the highest point being 18 cm for each profile for the flow Reynolds number $Re(= u_m h/\nu$ approx 6.87×10^4 (where u_m is the maximum fluid velocity, $h = H_w - h'$ is the depth of water above the sand bed, and ν is the kinematic viscosity of fluid) and the Froude number $Fr(= u_m/\sqrt{gh}) \approx 0.165$. The values of the flow parameters used for the experiments are provided in Table 7.1.

7.3 Experimental Observations and Results

7.3.1 Evolution of Scour Marks Around Cylinders

A series of photographs (Fig. 7.6a1–a12) was taken about 15–20 min interval during the experiment starting from the initial stage of flat bed condition to the equilibrium to envisage the time-dependent evolution of scour mark around a short cylinder diameter of 3.2 cm under the low flow discharge. It is observed from the figures that the initial scour takes place mostly at the upstream of the cylinder; and consequently the deposition takes place at the downstream of the cylinder. During the process of evolution the scour mark in the upstream side parallel to the cylinder was not symmetric, which may be due to the nonlinearity of turbulence and wakes; but eventually with time (after about 2 h) the scour mark tends to become symmetric in size and shape with deepening about the center line of the

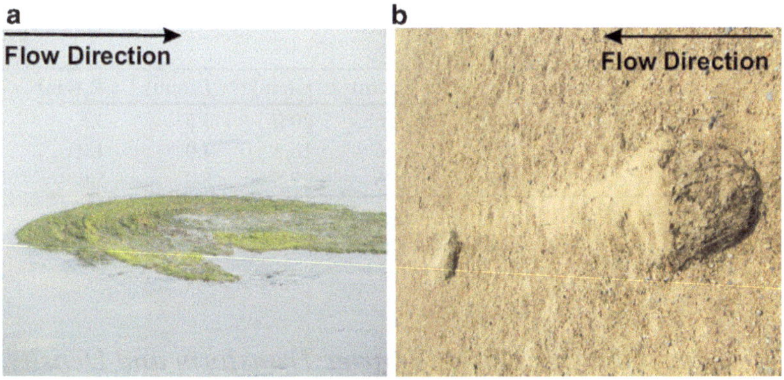

Fig. 7.3 (**a**) and (**b**) Recent "current-crescents" on Subarnarekha River bed

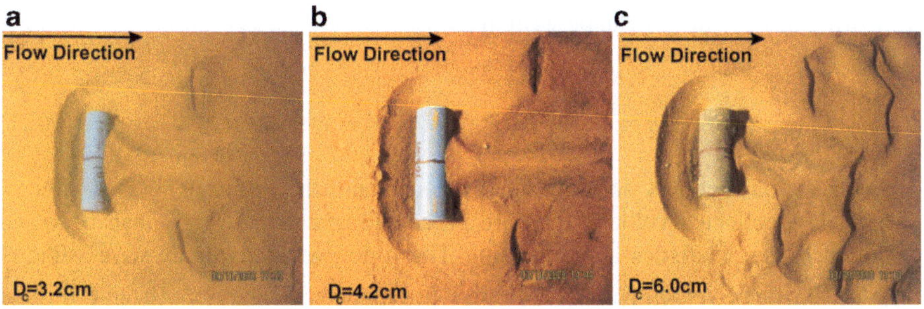

Fig. 7.4 The equilibrium scour holes developed at the upstream of three different cylinders of diameters (D_c = 3.2, 4.2, and 6.0 cm) 6.0 cm) for $Q = 0.015\ \mathrm{m}^3/s$ (after Mazumder et al. 2011)

cylinder (Fig. 7.6a1–a12), which shows the transitional phenomena of the process. In fact, these vortices are the responsible ingredients to transport sediment and scour around the cylinder. In a similar way, two more experiments using two different short cylinders of diameters D = 4.2 and 6.0 cm of identical length with the same flow Reynolds number or discharge were performed; and the photographs of equilibrium scour marks developed in the upstream sides of all three cylinders were taken for analysis (Fig. 7.4). The diameter of cylinder leads to increase the width of scour mark. The equilibrium conditions occur at different times (t_e = 178, 195 and 300 min) for different diameters of the cylinders. Observed values of depth and width of scour marks, length of the ridge, width of left-end and right-end scour marks generated due to three different cylinder diameters are tabulated in Table 7.2. Detailed explanations are given in paper by Mazumder et al. (2011).

Table 7.2 Values of observed parameters for the Reynolds number $Re = 6.76 \times 10^4$

D_c(cm)	w_s(cm)	t_e(min)	s_h(cm)	r_e(cm)	L_c(cm)	R_c(cm)
3.2	3.5	178	2.56	20.0	3.5	3.5
4.2	4.5	195	2.76	16.5	4.0	4.0
6.0	6.5	300	2.86	7.5	5.0	5.0

where t_e is the time for equilibrium scour-width, s_h is the equilibrium scour depth, r_e is the equilibrium ridge length, L_c is the left-end scour width, and R_c is the right-end scour width

7.3.2 *Gram–Charlier Series, Fourier Transform and Density Estimation Via Moments*

For the instantaneous velocity components (u, v, w) in the (x, y, z)-directions, the following three relations can be written as

$$u = \bar{u} + u', \; v = \bar{v} + v', w = \bar{w} + w' \tag{7.1}$$

where over bar denotes time-averaged velocity and the prime denotes its fluctuations. The collected velocity data are processed to calculate the mean flow and turbulence characteristics at each point. The time averaged stream-wise velocity $\bar{u}$, vertical mean velocity $\bar{w}$, stream-wise turbulence intensity $\sqrt{\overline{u'^2}}$, and vertical turbulence intensity $\sqrt{\overline{w'^2}}$ are defined as

$$\bar{u} = \frac{1}{n}\sum_{i=1}^{n} u_i \tag{7.2}$$

$$\bar{w} = \frac{1}{n}\sum_{i=1}^{n} w_i \tag{7.3}$$

$$\sqrt{\overline{u'^2}} = \sqrt{\frac{1}{n}\sum_{i=1}^{n} (u_i - \bar{u})^2} \tag{7.4}$$

$$\sqrt{\overline{w'^2}} = \sqrt{\frac{1}{n}\sum_{i=1}^{n} (w_i - \bar{w})^2} \tag{7.5}$$

where n is the total number of velocity observations at each point.

In order to estimate the joint density function $p(\hat{u}, \hat{w})$ using the cumulant discard method, we summarize here briefly the method and related theory following Nakagawa and Nezu (1977) and Raupach (1981). We normalize the velocity fluctuations u' and w' by standard deviation (r.m.s) in each direction so that $\hat{u} = u'/\sqrt{\overline{u'^2}}$, and $\hat{w} = w'/\sqrt{\overline{w'^2}}$. Denote the joint probability density function of $\hat{u}$ and $\hat{w}$ by $p(\hat{u}, \hat{w})$, its characteristic function by $\chi(\alpha, \beta)$, moment of $\overline{\hat{u}^s \hat{w}^t}$ by m_{st} and corresponding cumulant by q_{st}, the joint probability density function $p(\hat{u}, \hat{w})$ can be related to

$$\chi(\alpha, \beta) = \iint_{-\infty}^{\infty} e^{i(\hat{u}\alpha + \hat{w}\beta)}\, p(\hat{u}, \hat{w})\, d\hat{u}\, d\hat{w} \tag{7.6}$$

where $\chi(\alpha, \beta)$ is the Fourier transform of $p(\hat{u}, \hat{w})$, and α and β are its arguments. Here m_{st} and q_{st} correspond, respectively, to the coefficients in Taylor series expansions of $\chi(\alpha, \beta)$ and $\ln \chi(\alpha, \beta)$; and hence the relationships between the moments m_{st} and the cumulants q_{st} are successively obtained (Nakagawa and Nezu 1977). Using inverse transform of (7.6) in which the terms of $\chi(\alpha, \beta)$ less than fourth order are taken into account, $p(\hat{u}, \hat{w})$ can be written as

$$p(\hat{u}, \hat{w}) = \frac{1}{2\pi} \iint_{-\infty}^{\infty} e^{-i(\hat{u}\alpha + \hat{w}\beta)}\, \chi(\alpha, \beta)\, d\alpha\, d\beta \tag{7.7}$$

Since $p(\hat{u}, \hat{w})$ be the joint density function of $\hat{u}$ and $\hat{w}$ it has zero mean and unit variance. Assuming all the moments of $\hat{u}$ and $\hat{w}$ exist and some other conditions (Chambers 1967), the joint density function $p(\hat{u}, \hat{w})$ can be expanded as a series of derivatives of the standard bivariate normal density function $\phi(\hat{u}, \hat{w})$ (Mardia 1970).

$$p(\hat{u}, \hat{w}) = \phi(\hat{u}, \hat{w})[1 + \sum_{s+t=3}^{4} \frac{q_{st}}{s!t!} H_{st}(\hat{u}, \hat{w})] \tag{7.8}$$

where

$$\phi(\hat{u}, \hat{w}) = \frac{1}{2\pi(1 - R)^{0.5}} e^{-\frac{\hat{u}^2 + 2R\hat{u}\hat{w} + \hat{w}^2}{2(1 - R^2)}} \tag{7.9}$$

and $H_{st}(\hat{u}, \hat{w})$ is Hermite polynomial of order $(s+t)$ in two variables. Equation (7.8) represents a joint probability density distribution of the Gram–Charlier type in bivariate case. Approximation of neglecting quintuple terms is closely related to the "quasi-normal" relation, which has been often assumed in the statistical theory of homogeneous turbulence. There is some doubt on the validity of the cumulant discard approximation for intense turbulence.

The joint probability density function $p(\hat{u}, \hat{w})$ given by Eqs. (7.8) and (7.9) using Gram–Charlier type is shown in Fig. 7.5 at the bed level (zero level) at the location D over the scour hole generated by the cylinder diameter $D_c = 3.2$ cm. This probability density function has been plotted for corresponding time interval given in Fig. 7.6. It is clearly observed from the figures that initially the joint probability density function shows normal distribution, while as time increases it deviates from the normality, but gradually it recovers to some extent after a certain time.

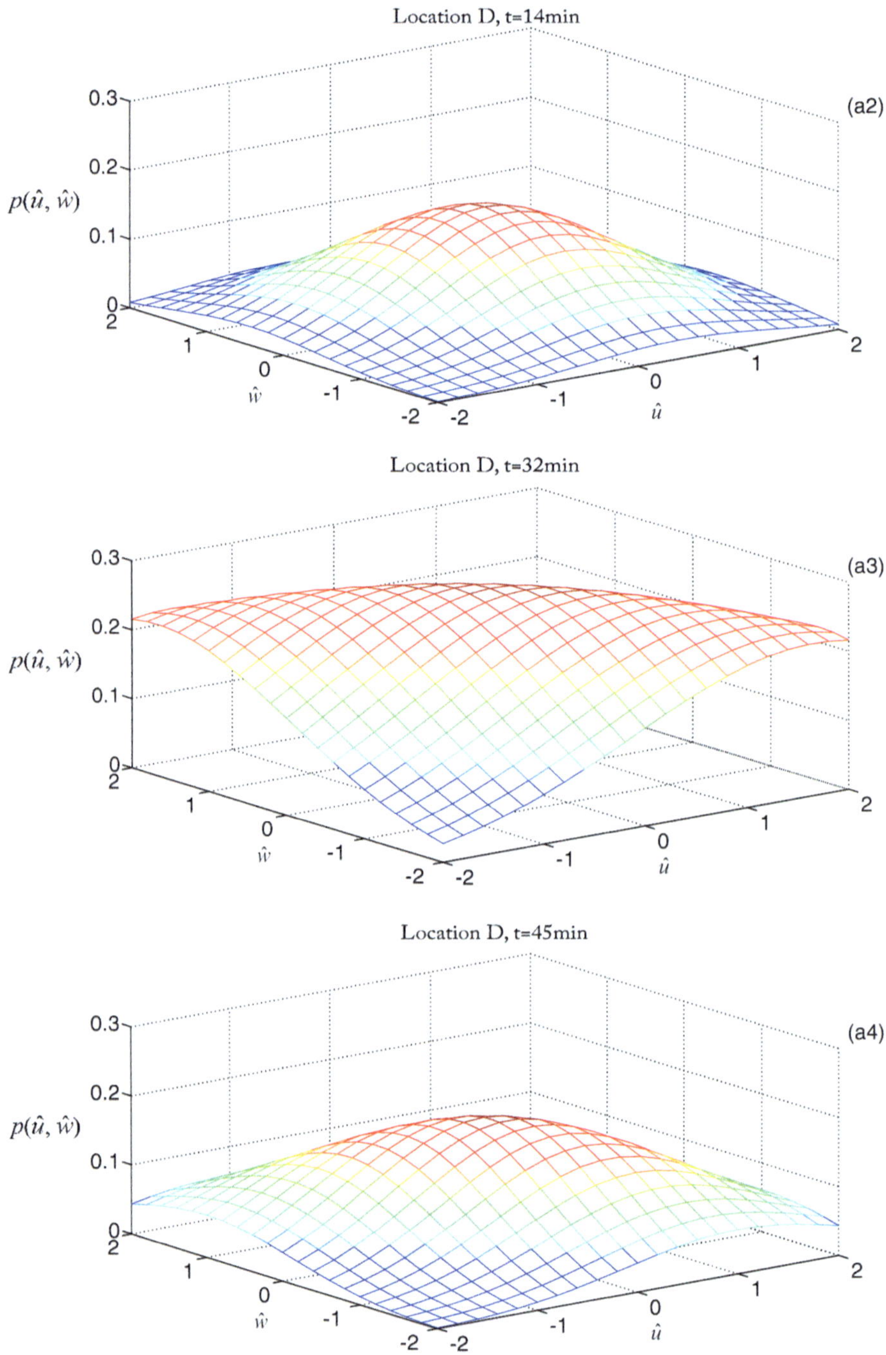

Fig. 7.5 Joint probability density function $p(\hat{u}, \hat{w})$ of fluctuating velocity components (u', w') at the location D shown in Fig. 7.7

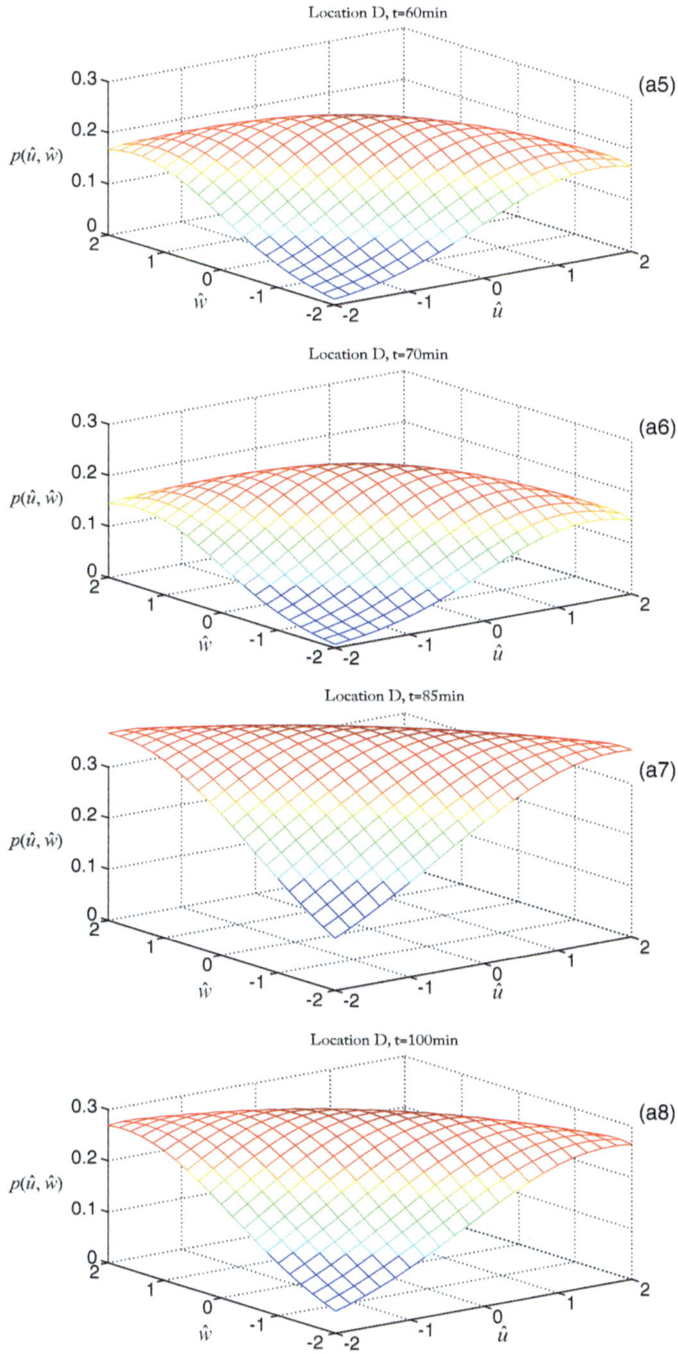

Fig. 7.5 (continued)

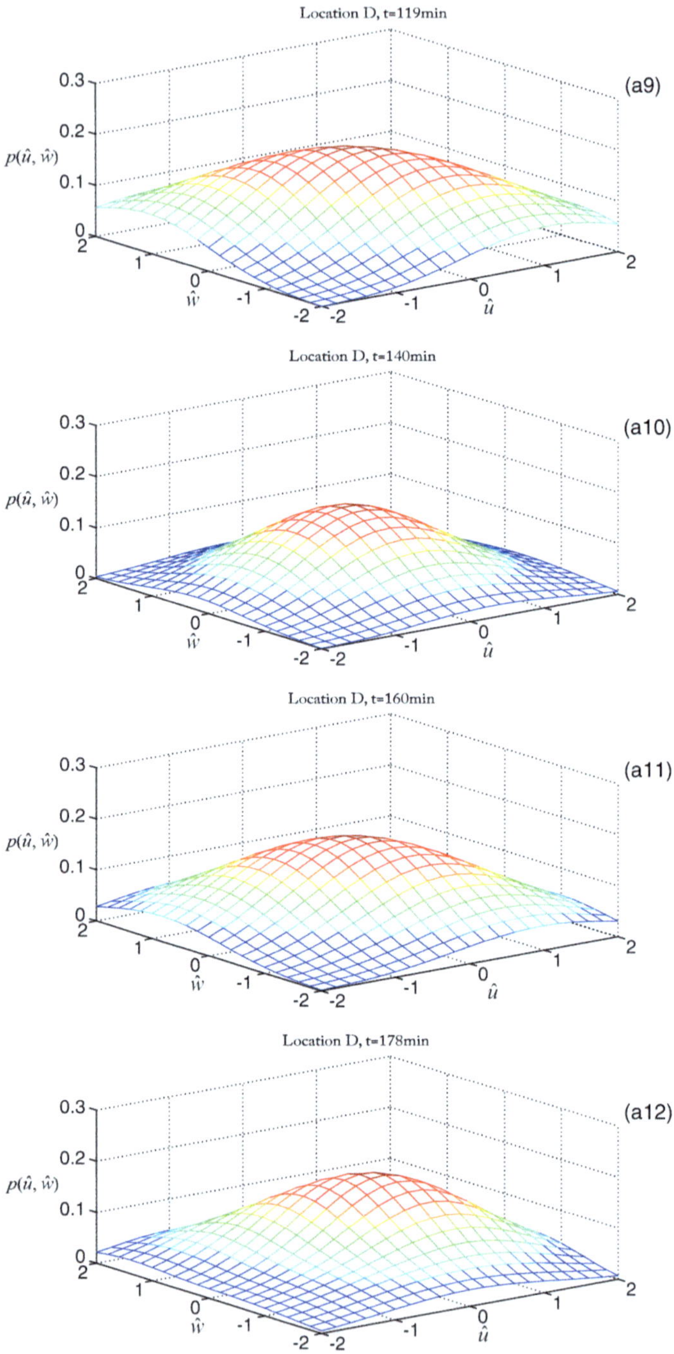

Fig. 7.5 (continued)

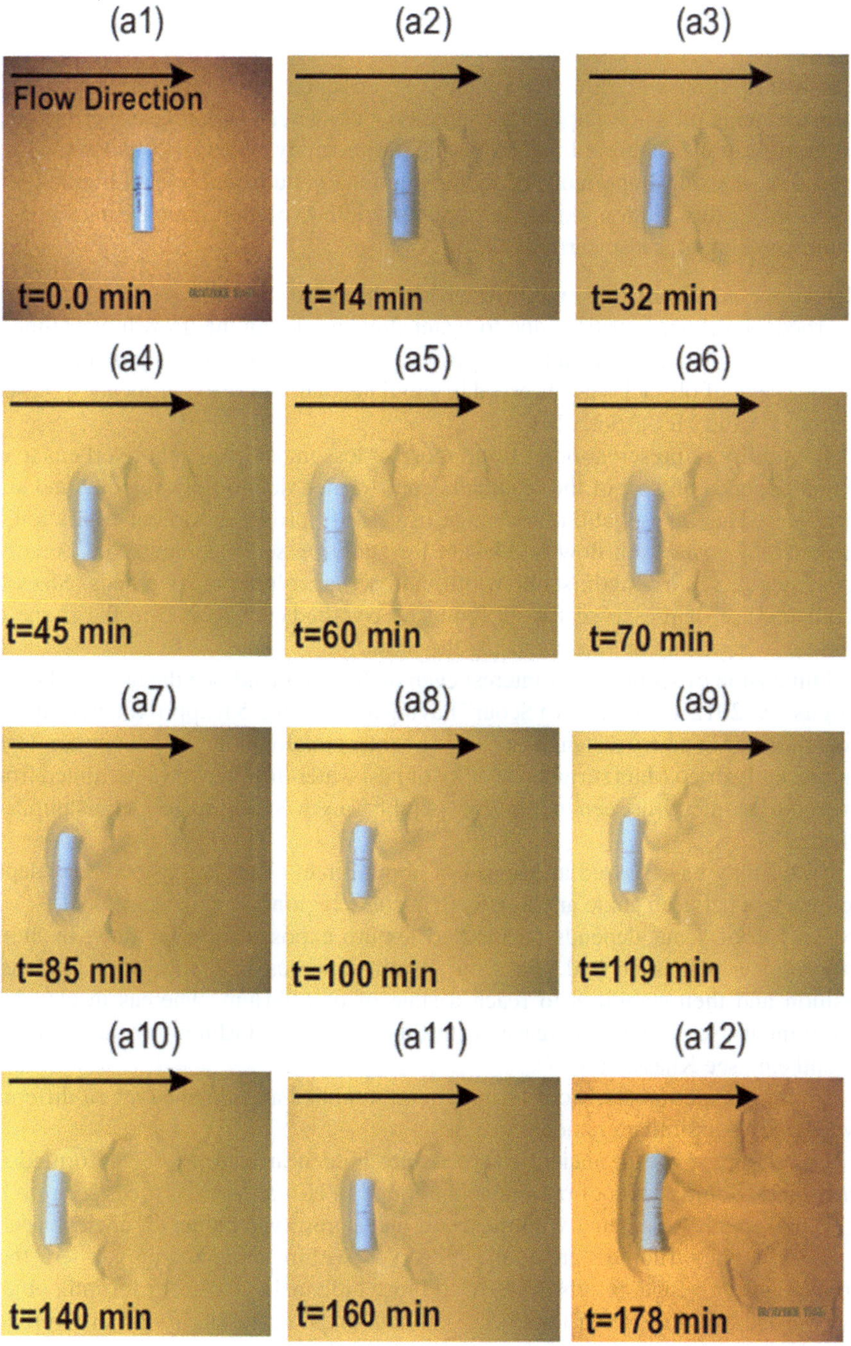

Fig. 7.6 Evolution of scour hole over time around cylinder of diameter (D_c = 3.2 cm) for discharge $Q = 0.015\ \mathrm{m}^3/s$ (after Mazumder et al. 2011)

7.3.3 Growth of Scour Width and State of Equilibrium

Consider a flow with a fixed velocity in an open channel flume. Numerical measurements on some important characteristics of scour are taken over the time till equilibrium is achieved and subsequent data analysis helps to understand the evolution of scour. Importance of such evolutionary study and related literatures is discussed in this section. Analysis of scour-width evolution data arising out of a flume experiment is also carried out.

Relevance of Scour Growth Model and Related Studies. As the scour on riverbeds develops over time due to water flow, modeling the growth over time is relevant to understand the underlying process. Scour formation is highly dependent on the shape of obstacles as observed in simulated laboratory experiments, e.g., see, Thompson and McCarrick (2010).

Possibility of preservation of fossil more or less intact in ancient riverbeds to an extent depends on size of the resultant scour formed due to flow velocity and size of object. This investigation on change in velocity profile and scour formation in presence of obstacle in flow is relevant for study on scour geometry i.e., relative scour depth, scour length, scour width, and scour volume on riverbeds. Sizes of scour in case of preserved fossils found on riverbeds are related to the shape of ancient living being or object causing the scour.

Dimension of scours are of interest even on Martian land, see the image taken on August 19, 2012, of "Goulburn Scour" having a width of 2.5 ft approximately, due to flow of water in a very distant past causing sandy conglomerate, a sedimentary layer on presently dried Mars surface; velocity of past water flow may be determined from the scour geometry, as seen in http://www.jpl.nasa.gov/spaceimages/details.php?id=PIA16187.

Water may have moved at a speed of about 3 feet (1 m) per second, at a depth somewhere between ankle and hip deep, in some regions.

Growth of scour depends on the bed texture exposed to water flow. In flume experiments with sand bed, the scour depth increases with time in a nonlinear fashion and then stabilizes to reach a state of equilibrium, whereas in clay bed experiments, scour depth increases with time in a linear fashion to reach an upper bound e.g., see Khassaf (2007).

In flume experiments, scour formation on sand bed and clay bed are of different types due to dissimilar erosion processes.

Fossils preserved in ancient riverbeds are hidden around the scour formed by the object. Some fossils of phragmacone found in Himalayan riverbeds seem to be well preserved. Associated with these are other preserved beings of ancient years, e.g., sometimes trilobites are found preserved within phragmocones, where they crawled in for refuge, fossils of host and those hide inside the scour forming object in riverbed are discovered together. Size and shape of fossils found in riverbeds may be related to the width of the scour formed by the obstacle of interest that imprinted its shape in river flow. An almost complete fossil skeleton of a 47 feet long sauropod from the early Jurassic period (about 160 million years back)

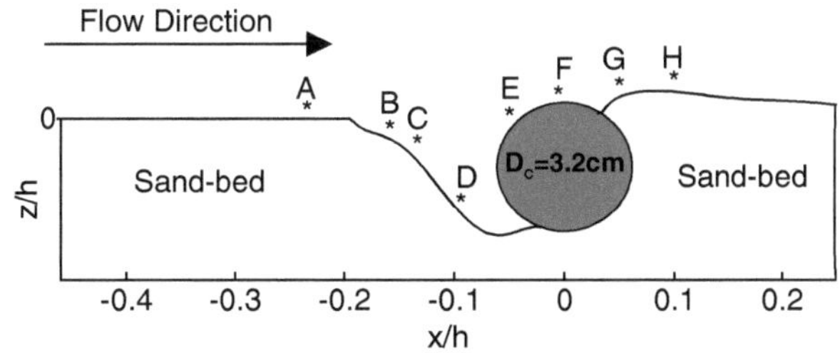

Fig. 7.7 Schematic diagram of equilibrium scour mark developed at the upstream of cylinder of diameter $D_c = 3.2$ cm. Here A, B, C, D, E, F, G, and H are the locations of measurements

was discovered by Indian Statistical Institute geologists during a 1958 exploration in the Pranhita–Godavari valley. Subsequently, two near complete reptilian fossil specimens from the basal siltstone were found by some local villagers from a quarry around 1970. River Godavari, its tributaries like Manjra river, Mula river and surrounding basin located in south India, are well known for a number of discovered fossils, see e.g., http://biosciencediscovery.com/attachments/File/baber.pdf, http://palaeontologicalsociety.in/vol41/v1.pdf

In this context one may study the growth of scour over time in experimental flume till a state of equilibrium, if and when attained in order to infer about such phenomenon in riverbed. Scour characteristics, those are directly associated with the scour width, as for example area of scour, or volume of scour may have a linear relationship in logarithmic scale with scour width. Studying growth of scour width also sheds light on development of these characteristics over time.

Studies conducted with the purpose of predicting scour have resulted in various empirical equations those are based on laboratory results and field data. These differ from each other with respect to the factors considered in constructing the scour model. See e.g., Khwairakpam et al. (2012).

Empirical equations with Froude number and relative flow depth (inflow depth per pier diameter) are considered therein to model entire scour geometry at equilibrium, in a flume experiment on scour hole characteristics around a single vertical pier in Clearwater conditions (Fig. 7.7).

Analysis of Scour Growth. In the present flume experiment with unidirectional flow, scour is formed around the cylindrical obstacle placed on flume bed, axis of cylinder is placed perpendicular to flow direction. Scour width increases with the flow velocity for a particular cylinder diameter. With a fixed moderate velocity, width increases with time and reaches a steady state after a certain time. For a fixed velocity, width of scour increases with cylinder diameter. Here we analyse scour width due to a short cylinder of diameter 3.2 cm of fixed length $L = 10$ cm under the low flow discharge $Q = 0.015\ \mathrm{m}^3/s$.

One may examine flume experimental data via nonparametric regression techniques to see the evolution of scour over time till an equilibrium is achieved. Such nonparametric modeling is robust with respect to presence of outliers in the collected data and may correctly capture the inherent features present in a data set.

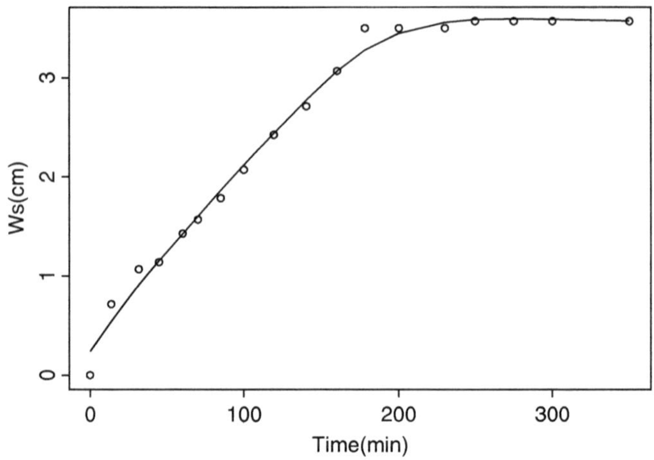

Fig. 7.8 Growth curve (spline) of scour width

To estimate the growth curve of the response variable scour width $\mu(t)$ from the set of data $(t_i, y_i); i = 1, \cdots, m$; stabilizing after a long period, we may assume the model $y_i = \mu(t_i) + \epsilon_i$, where ϵ_i are zero mean, identically distributed random variables with common variance σ_ϵ^2. A cubic smoothing spline $s(t)$ that minimizes $J(\lambda) = \sum_{i=1}^{m}\{y_i - s(t_i)\}^2 + \lambda \int\{s''(t)\}^2 dt$ with λ as "constant of roughness penalty" is chosen. The procedure is a balance between ordinary least square and smoothing of data set. λ may be taken as .001, so as to preserve the main features of the diagram joining the points by straight line. The resultant growth curve is shown in Fig. 7.8, after an extra point is added at 350 min with same y value as that for 300 min, to check for equilibrium towards the end of the curve. Estimated growth curve is close to most of the observed data points.

Another nonparametric regression technique insensitive to outliers is locally weighted scatterplot smoothing (lowess). The technique attempts to fit a weighted linear least squares regression over the span of data assigning little weight to distant observations. A smoothing parameter f regulates the ruggedness of data points. The lowess estimates with $f = 0.61$ joined by lines is shown in Fig. 7.9. The graph maintains the main features of data set intact and seems to be insensitive towards some distant points near the end.

One may compute derivative of the growth function i.e., velocity function of the scour width, in order to find the time of maximum velocity of growth and to check whether equilibrium is really achieved towards the end of data collection.

With an application of a robust nonparametric technique proposed in Dasgupta (2013), we estimate the derivative of a function at a point based on slope estimates $m_i(j)$ at x_i on the basis of neighbor data points $(x_j, y_j), j \neq i$. Let $\hat{y}$ be lowess estimate of y at x for the observed point (x, y).

Consider $m_i(j) = (\widehat{y_i} - \widehat{y_j})/(x_i - x_j), j \neq i$; the slope of the line joining the points $\{(x_i, \widehat{y_i}), (x_j, \widehat{y_j})\}$. A robust estimate of derivative of the function at x_i with

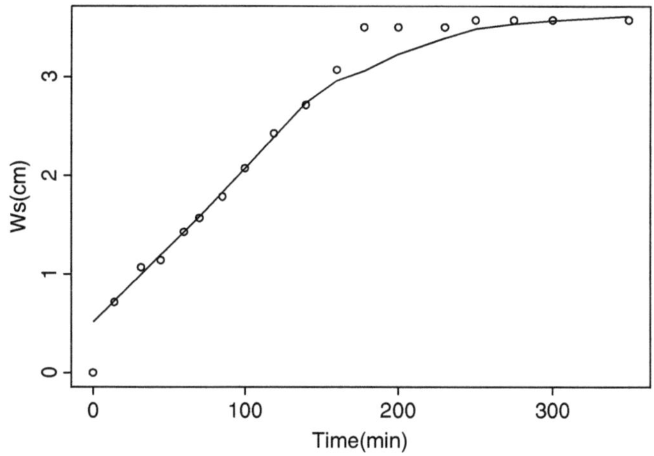

Fig. 7.9 Growth curve (lowess) of scour width

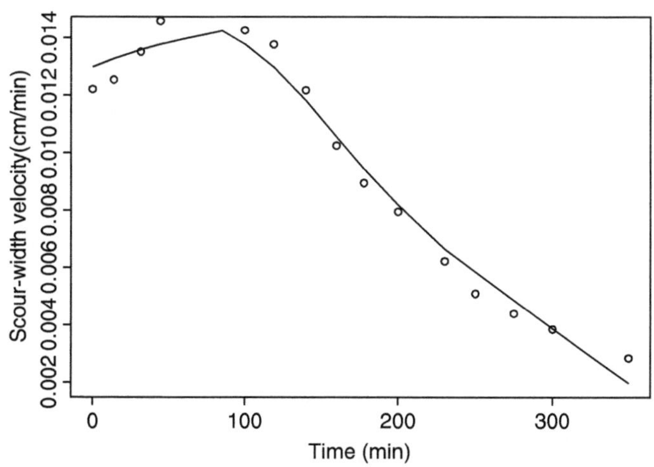

Fig. 7.10 Scour width velocity with trimmed mean, wt. exp $(-.01\ x)$

$f = .61$, (say) is lowess estimate obtained from $m_i(j)$ suitably averaged over j by a normalized weight function $w(d)$ of polynomial/ exponential decay based on distance $d = |x_i - x_j|$, $w(d)$ typically assigns more weight to nearby points.

In Fig. 7.10 we consider $w(d) \propto e^{-.01d}$ and plot the trimmed mean of successive three observations viz., average of unique median plus values immediately above and below it, of the quantities $w(|x_i - x_j|)m_i(j)$, for a fixed i over different j, $j \neq i$.

These trimmed mean values are shown by points and the corresponding lowess estimates ($f = .61$) of growth rate based on these trimmed mean values are joined by line in the same figure, viz. Fig. 7.10.

From the figure it is seen that the peak velocity is around 85 min.

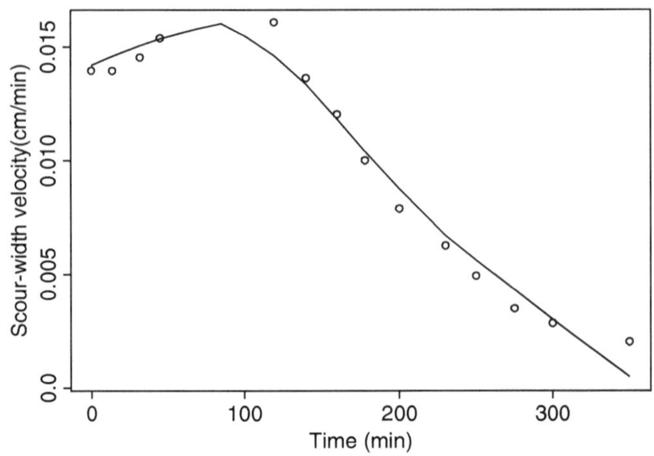

Fig. 7.11 Scour width velocity with trimmed mean, wt. exp $(-.01\ x)$

The picture remains almost same when calculated velocity is based on smooth spline estimates of Fig. 7.9, where observations y_i are closer to spline estimates. The resultant growth rates, with similar calculations as that of Fig. 7.10, are shown in Fig. 7.11.

Study of second derivative of growth curve is of interest from curvature viewpoint and may be obtained in a similar fashion. Recall the procedure of calculating first derivative, based on the inputs of smooth spline growth estimates as shown in Fig. 7.9. We repeat the same procedure with input points as in Fig. 7.11. Differentiating the first derivative (shown in Fig. 7.11) again, we obtain the second derivative or, instantaneous acceleration of growth, see Fig. 7.12.

The function in Fig. 7.12 lies below the line $y = 0$. With negative second derivative, the growth function curves downwards. Magnitude of curvature/acceleration is seen to be maximum at 230 min.

It appears from Figs. 7.10 to 7.12 that first and second derivatives of scour width growth are small and approaching zero, but still away from zero towards the end of experiment indicating that equilibrium is yet to be achieved. Experimental data collection should probably have continued for some more time.

We may further calculate the jerk (or jolt) of the scour width growth curve. The rate of change of acceleration named jolt has analogue in the sensation of varying thrust on a passenger exerted from seatrest in a moving car with varying rate of acceleration.

Proceeding in a similar manner it is possible to obtain the third derivative (jerk) of growth curve taking the inputs from Fig. 7.12 and differentiating the graph again. The basic points and the smooth spline estimates (joined by line) of third derivative are shown in Fig. 7.13. The highest magnitude of jolt is observed at 60 min from the start of experiment.

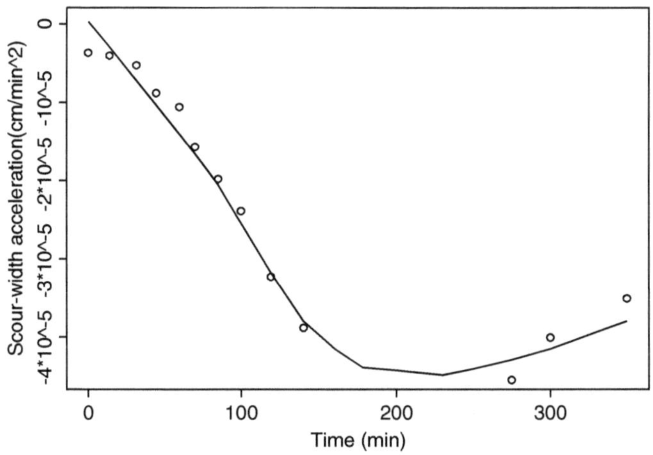

Fig. 7.12 Scour-width second derivative with trimmed mean, wt. exp $(-.01\ x)$

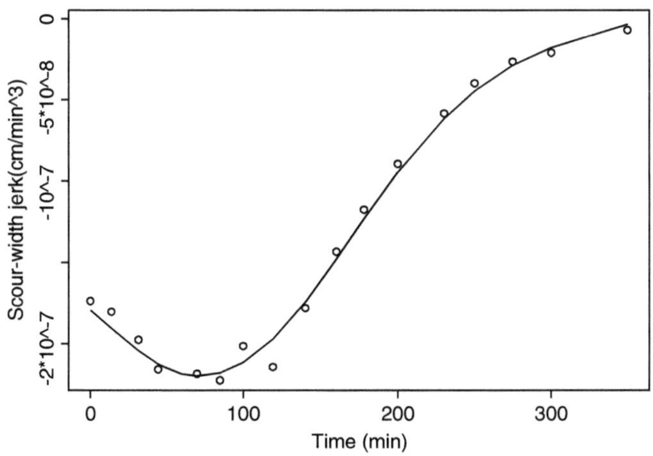

Fig. 7.13 Scour-width third derivative with trimmed mean, wt. exp $(-.01\ x)$

7.4 Conclusions

This work was focused on the study of scouring processes around short cylinders of fixed length with different diameters placed over the sand bed transverse to the flow. The main objective of this work was to understand of the scouring processes with time. This study would be helpful for engineers to modify their structural designs or using some protective measures for meeting the problems faced due to scouring. Also studying the processes involved in scour formation would help to address many sediment transport problems. It may be mentioned here that the current crescents preserved in geological record are traditionally used as indicators of palaeo-current direction, e.g., see Fig. 7.3. Scour geometry is related to preservation of fossils in ancient riverbeds. Studies on growth curves presented here are relevant for assessment of foundation scour in a bridge that is created by turbulence around a footing, design for Camshaft-gear in automobile engines, trajectory planning while integrating sensors in robotic environments, etc.

References

Allen, J.R.L. (1982). *Sedimentary structures their character and physical basis* (vol. 1, 593 p.). Amsterdam: Elsevier.

Catano-Lopera, Y., & Garcia, M.H. (2007). Burial of short cylinders induced by scour under combined waves and currents. *Journal of Waterway Port-coastal and Ocean Engineering-ASCE, 132*(6), 439-449.

Chambers, J.M. (1967). On methods of asymptotic approximation for multivarite distributions. *Biometrika, 54*, 367–384.

Dasgupta, R., (2013). Non uniform Rates of Convergence to Normality for Two sample U-statistics in Non IID Case with Applications. Appearing in this volume as chapter 4.

Karcz, I. (1968). Fluvial obstacle marks from the wadis of the Negev (Southern Israel). *Journal of Sedimentary Petrology, 38*, 1000–1012.

Khassaf, I.S. (2007). Effect of cohesive and noncohesive soils on equilibrium scour depth. *Tikrit Journal of Eng. Sciences*, 73–85.

Khwairakpam, P., Sinha Ray, S., Das, S., Das, R., & Mazumdar, A. (2012). Scour hole characteristics around a vertical pier under Clearwater scour conditions. *ARPN Journal of Engineering and Applied Sciences*, 649–654.

Mardia, K.V. (1970). *Families of bivariate distributions*. London: Griffin.

Mazumder, B.S., Ray, R.N., & Dalal, D.C. (2005). Size distributions of suspended particles in open-channel flow over sediment beds. *Environmetrics, 16*, 149–165.

Mazumder, B.S., Maity, H., & Chadda, T. (2011). Turbulent flow field over fluvial obstacle marks generated in a laboratory flume. *International Journal of Sediment Research, 26*(1), 62–77.

Melville, B.W., & Raudkivi, A.J. (1977). Flow characteristics in Local scour at bridge piers. *Journal of Hydraulic Research, 15*(4), 373–380.

Nakagawa, H., & Nezu, I. (1977). Prediction of the contributions to the Reynolds stress from bursting events in open-channel flows. *Journal of Fluid Mechanics, 80*(1), 99–128.

Raupach, M.R. (1981). Conditional statistics of Reynolds stress in rough-wall and smooth-wall turbulent boundary layers. *Journal of Fluid Mechanics, 108*, 363–382.

Sengupta, S. (1966). Studies on orientation and imbrication of pebbles with respect to cross-stratification. *Journal of Sedimentary Petrology, 36*(2), 362–369.

Sengupta, S., Das, S.S., & Gupta, A.S. (2005). Current crescent as indicator of flow velocity. In S.N. Bora (Ed.) *Some Aspects of Environmental Fluid Mechanics*. Proceedings ICEFM '05 Guwahati (pp. 76–77).

Sengupta, S. (2007). *Introduction to sedimentology*. New Delhi: CBS Publications and Distributors.

Thompson, D.M., & McCarrick, C.R. (2010). A flume experiment on the effect of constriction shape. *Hydrology and Earth System Sciences*, 1321–1330.

Chapter 8
South Pole Ozone Profile and Lower Tolerance Limit

Ratan Dasgupta

Abstract Ozone layer is protecting this green planet from ultraviolet radiation. Ozone in high altitude region is constantly destroyed by chemical reaction involving chlorine and fluorine from human produced chlorofluorocarbons (CFCs), and thus creating an ozone-hole. We study the ozone profile in polar region, model this in terms of Gaussian process, fit a smooth growth curve and predict the lower tolerance limit of ozone concentration that is seen to hold up to the year 2012. The problem is reanalyzed in nonparametric setup and bootstrap resampling technique.

MSC classification 2000: Primary 62P12, 62G15; Secondary 62G32

8.1 Introduction: Ultraviolet Radiation and Ozone Layer

The sun emits radiation over a broad range of wavelengths to which the human eye responds in the region from approximately 400 to 700 nm. Wavelengths from 320 to 400 nm are known as UV-A, that from 280 to 320 nm are called UV-B, and from 200 to 280 nm are known as UV-C; X-rays and gamma ray radiations are in the region 0–200 nm.

Our concern is mainly on UV-B as the atmosphere absorbs virtually all UV-C. Radiation UV-B is partially absorbed by the ozone layer, a thin band in the stratosphere, protecting the earth from its harmful effects.

Depletion of the ozone layer allows the UV radiation to reach earth, causing skin cancer, eye cataract, reduced plant yields, and damage to ocean ecological system.

R. Dasgupta (✉)
Theoretical Statistics and Mathematics Unit, Indian Statistical Institute, Kolkata, India
e-mail: ratandasgupta@gmail.com

R. Dasgupta (ed.), *Advances in Growth Curve Models*, Springer Proceedings in Mathematics & Statistics 46, DOI 10.1007/978-1-4614-6862-2_8,

Several pollutants attack the ozone layer. Chief among these is the class of chemicals known as chlorofluorocarbons (CFCs) that contain both chlorine and fluorine. The CFCs are used as the propellants in gaseous suspension of fine solid or liquid particles; in refrigeration technology as solvents and foam producing agents. The CFCs do not break up in the lower atmosphere, known as the troposphere. Instead these slowly migrate to stratosphere, the altitude region 10–50 km above sea level. There these react with other chemicals under the influence of ultraviolet radiation and release chlorine. Chlorine acts as a catalyst to destroy ozone in the stratosphere. Other pollutants, including nitrous oxide from fertilizers and the pesticide methyl bromide, also attack atmospheric ozone. The balloon borne ozone instruments launched at regular intervals since the year 1986 at the Amundsen-Scott South Pole Station show ozone depletion in the September to October period, following sunrise, when ozone in the 6–14 mile altitude region is almost totally destroyed by chemical reaction of CFCs where stratospheric clouds are formed, thus creating an ozone hole, i.e., a polar region with vertical ozone profile of values below 220 Dobson units (DU).

Air parcels move on isentropic surfaces, i.e., surfaces of equal potential temperature, rather than pressure surfaces. Cold polar air is trapped by the very strong winds of polar night jet, thus forming polar vortex; measured in the units of million square km. During the winter/spring period, when the polar vortex is strongest, warm air outside of the vortex cannot enter inside the vortex of cold polar air. About 17 km above sea level in the south polar region and in-between the 70 and 50 millibar (mb) pressure surfaces, ozone is in greatest abundance in the vertical profile. At that altitude, the peak of the Antarctic polar vortex was about 35 million square km, during September end and early October 2006. Further above, about 28 km above sea level in the south polar region, and in-between the 30 and 20 mb pressure surfaces, the peak area of the polar vortex was about 42 million square km during August 2006. Temperature below −78 °C forms Polar Stratospheric Clouds (PSC) with nitric acid trihydrate and ice within polar vortex. These particles grow in size and number to create cloud-like features. Heterogeneous photochemical destruction of ozone takes place within these PSCs. As the area of low temperatures becomes larger, there is a greater possibility of PSCs forming which cause accumulation of reactive chlorine gases. This destroys ozone, once sunlight returns to the polar region.

Warm air from the mid latitudes cannot mix with the cold polar air and the polar air gets colder due to loss of heat by radiation. The depleted ozone in the vortex is not replenished with ozone rich air from outside the vortex. Cold air remains trapped until mid to late spring when the polar vortex gradually weakens and eventually breaks down. After this, thorough mixing occurs and ozone amounts are replenished. If phasing out of all ozone-depleting substances is continued under the Montreal Protocol, it may take 50–100 years to fully recover the ozone level.

After major volcanic eruptions, particles of sulfuric acid form in the stratosphere. These particles spread around the world and cause ozone depletion up to 1–2 years till the sulfur particles fall from the stratosphere.

The UV radiation of 180–240 nm helps ozone formation and radiation of 200–320 nm causes ozone breakdown. The rate of breakdown can be greatly accelerated by catalysts such as chlorine and nitric oxide.

Vortex at the other pole, viz., arctic polar vortex is much weaker than the Antarctic. The arctic temperatures are several degrees higher, and PSCs are much less common. There PSCs tend to break up earlier in the spring. Thus there is ozone depletion, although no "ozone hole" in the northern hemisphere till now.

At the ground level, chemical reactions between volatile organic compounds, nitrogen oxides, and sunlight produce ozone molecules. This, however, is not a remedy for the damage caused by ozone hole, as the area of the latter is much larger compared to the pockets of ozone in ground level. Ground level ozone causes breathing difficulties and adversely affects plant-photosynthesis. Greenhouse effect of ground level ozone allows the higher atmosphere to cool. Thus more stratospheric clouds are formed with worse effect on ozone hole. Hopefully, this year (2012) the size of the ozone hole was the second smallest in the last 20 years.

The paper is arranged as follows. In Sect. 8.2, we describe measurement of ozone concentration in polar region. Ozone concentrations of past years are also studied. In this context, we study a β-content tolerance limit, i.e., a bound that contains at least $100\beta\%$ of the future observations above it, by modeling the observations as sum total of mean-response $\mu(t)$ varying over time t and a remainder $\epsilon(t)$ modeled by a correlated Gaussian process. We establish an almost sure convergence of empirical distribution of observations, recorded from a weakly correlated nonstationary Gaussian process, to the limiting stationary distribution; the recorded observations need not be equispaced in time. The result stated in Theorem 1 may be of independent interest. In Theorem 2 of Sect. 8.3, a tolerance limit is obtained based on *minimum of correlated normal random variables*; see also Dasgupta and Bhaumik (1995) where the maximum of iid normal random variables are considered. Using Theorem 2, we compute a lower tolerance limit of total ozone for a future year based on ozone-data of the years 1986–2006, in Sect. 8.4. Efficacy of the computed lower tolerance limit is seen to hold by minimum of total ozone level observed in next several years during 2007–2012. Alternative solutions are investigated by bootstrap and analyzing directly the 21 recorded yearly minimum ozone levels. Proof of the results are provided in Appendix.

8.2 Measurement and Analysis of Ozone Concentration

South pole ozone profiles are measured by balloon-borne electrochemical concentration cell (ECC) ozonesondes. An ozonesonde consists of a small piston pump that bubbles ambient air into a cell containing potassium iodide, unbuffered solution. The reaction of ozone and iodide produces a small electrical current in the cell, which is proportional to the amount of ozone. The ozonesonde is also interfaced with a radiosonde, which measures air temperature, pressure, relative humidity and transmits all of the data back to a ground receiving station. Total column ozone is calculated by integrating the ozone partial pressure profile up to the balloon burst altitude and adding a residual amount, based on climatologically ozone tables, to account for ozone above the balloon burst altitude. Data on such experiments are available from internet, e.g., NOAA website.

Usually circular shaped in August and September, the Antarctic polar vortex tends to elongate in October, stretching towards inhabited areas of South America. By November, the polar vortex begins to weaken and ozone rich air begins to mix with the air in the "ozone hole" region. The "ozone hole" is usually gone by late November/early December.

The minimum ozone profiles, i.e., the ozone pressure (in mPa) vs. altitude (in km) along with yearly minimum total ozone from 1986 up to the year 2005 are shown in Fig. 8.1. From this chart we see that minimum typically occurs in early October. However, in 2003, 106 DU was measured on September 26, which is about 13 days earlier than normal. The minimum ozone level for the year 2006 is 93 DU recorded on October 9.

In mid-August, about 6 weeks before the minimum occurs, the ozone profile over south pole generally has a well-defined peak in ozone concentration, occurring at 18 km. In the year 2001, the peak in ozone concentration occurred on 25 June with total ozone 273 DU. Then total ozone drops rapidly at a rate 3–5 DU per day. Nearly all of the ozone destruction centers around the ozone peak in a layer from 14 to 22 km. The record minimum 89 DU was observed on October 12, 1993. Since that year the 14–22 km layer has consistently shown near complete ozone destruction.

The severe ozone hole for the year 2001 occurred on 28 September with total ozone 100 DU, see Fig. 8.2. In each plot in that figure the third curve that starts from the extreme right is the temperature (in °C) vs. altitude (in km) curve.

The 52 observations on total ozone spread over the year 2001 are shown in Table 8.1. To estimate the response curve $\mu(t)$ from the set of data $(t_i, y_i); i = 1, \cdots, m; (m = 52$ here), we may assume the model $y_i = \mu(t_i) + \epsilon_i$ where ϵ_i are zero mean, identically distributed random variables with common variance σ_ϵ^2. A cubic smoothing spline $s(t)$, which minimizes the criterion $J(\lambda) = \sum_{i=1}^{m}\{y_i - s(t_i)\}^2 + \lambda \int \{s''(t)\}^2 dt$ is used. The "constant of roughness penalty" λ is taken as .002, so as to preserve the main features of the diagram joining the points by straight line. See Figs. 8.3–8.6. The components of residual ϵ, vide Fig. 8.7, are measurement errors, sudden climate changes, variation in time, location and altitude of balloon burst, etc. Thus the residuals ϵ arising from several independent causes may be assumed to be normally distributed with mean zero.

The goodness of fit test for normality described above extends to continuous time Gaussian process. The form of the response curves μ over different years exhibits a similar pattern and by separating this systematic part from the observations we obtain the random variable of interest viz. the residuals $\epsilon(t)$, where the time variable t spans over the days of all the successive years. The residuals need not be independent. However, there is a possibility of residuals $\epsilon(t)$ achieving a stationary weak limit as $t \to \infty$. Indeed, in the most ideal situation when there is no ozone

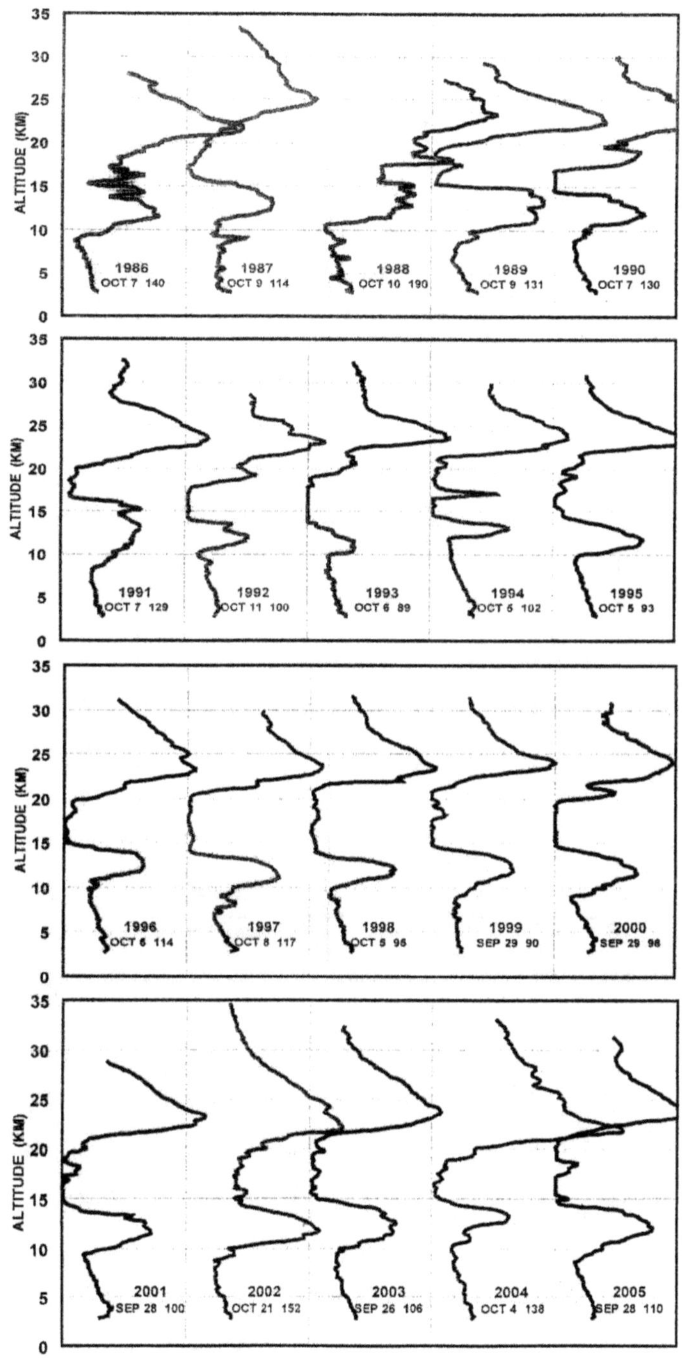

Fig. 8.1 Minimum ozone profiles over the years 1986–2005

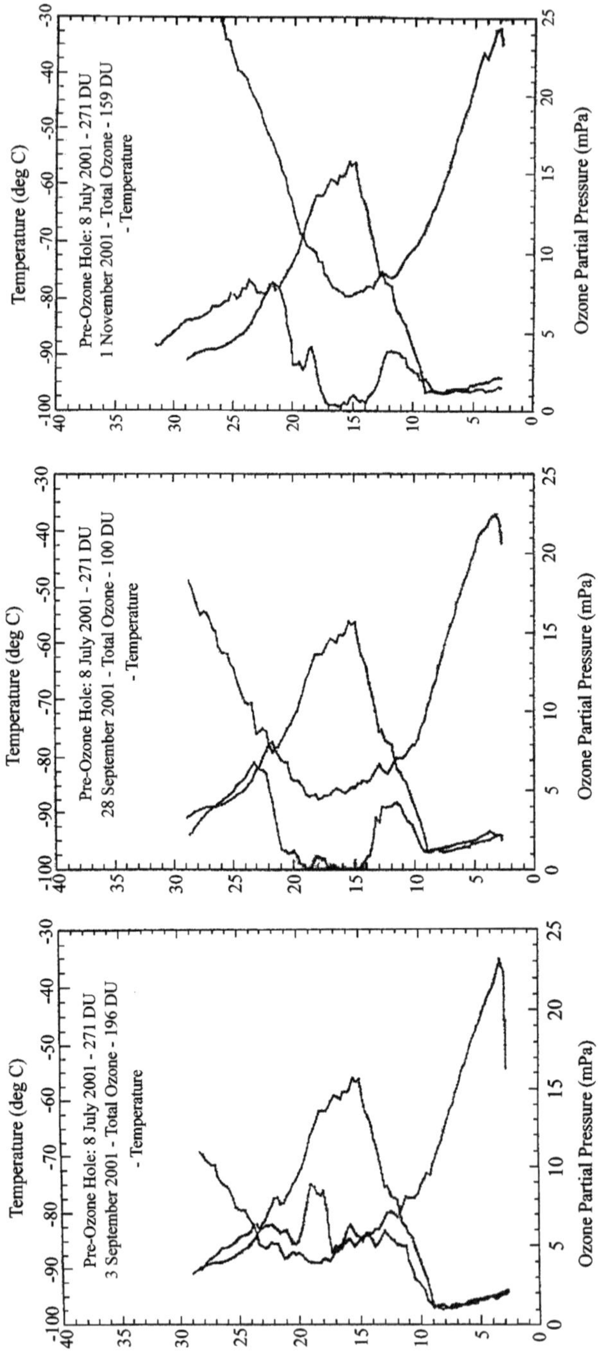

Fig. 8.2 Formation of ozone hole in 2001

Table 8.1 Total ozone over the year 2001

Date	Day no. (t_i)	Ozone (DU)	Residual (ϵ)	$\epsilon/\hat{\sigma}_\epsilon$
1.1.01	1	282	1.7472	0.1623
7.1.01	7	270	−2.1466	−0.1994
14.1.01	14	262	−1.4526	−0.1350
21.1.01	21	260	4.6530	0.4323
28.1.01	28	254	6.7436	0.6265
4.2.01	35	228	13.3455	−1.2399
11.2.01	42	241	−0.8736	−0.0812
18.2.02	49	240	−7.6571	−0.7114
4.3.01	63	278	22.5801	2.0978
11.3.01	70	251	7.0895	0.6587
18.3.01	77	224	−1.4473	−0.1345
25.3.01	84	183	−27.9701	−2.5986
8.4.01	98	227	14.4721	1.3445
15.4.01	105	211	−5.8190	−0.5406
22.4.01	112	232	11.4432	1.0631
29.4.01	119	216	−7.3637	−0.6841
6.5.01	126	208	−20.5960	−1.9135
13.5.01	133	264	28.8132	2.6769
20.5.01	140	238	2.9118	0.2705
27.5.01	147	222	−9.4305	−0.8761
10.6.01	161	224	−12.1891	−1.1324
25.6.01	176	*273	12.0903	1.1233
8.7.01	189	271	3.5907	0.3336
16.7.01	197	260	−4.2343	−0.3934
23.7.01	204	253	−8.2115	−0.7629
4.8.01	216	257	1.5560	0.1446
8.8.01	220	264	13.4029	1.2452
12.8.01	224	242	−1.3461	−0.1251
19.8.01	231	224	−3.2521	−0.3021
24.8.01	236	197	−18.2548	−1.6960
28.8.01	240	205	−0.8485	−0.0788
1.9.01	244	220	24.8779	2.3113
3.9.01	246	186	−2.7222	−0.2529
8.9.01	251	179	8.9651	0.8329
13.9.01	256	151	1.7262	0.1604
15.9.01	258	140	−1.1737	−0.1090
17.9.01	260	122	−11.5917	−1.0770
21.9.01	264	119	−1.7905	−0.1663
23.9.01	266	117	1.2506	0.1162
26.9.01	269	105	−5.0553	−0.4697
28.9.01	271	*100	−7.5853	−0.7047
2.10.01	275	116	10.3225	0.9590
4.10.01	277	101	−5.0164	−0.4661
6.10.01	279	105	2.1707	−0.2071

(continued)

Table 8.1 (continued)

Date	Day no. (t_i)	Ozone (DU)	Residual (ϵ)	$\epsilon/\hat{\sigma}_\epsilon$
8.10.01	281	111	1.9026	0.1768
11.10.01	284	109	−4.2778	−0.3974
14.10.01	287	116	−2.7950	−0.2597
17.10.01	290	130	4.7533	0.4416
20.10.01	293	131	−1.1340	−0.1054
25.10.01	298	161	17.4035	1.6169
28.10.01	301	139	−11.0267	−1.0244
1.11.01	305	159	0.4826	0.0448

The observations marked by a * represent the maximum and minimum of the observations. Here, $\hat{\mu}_\epsilon = 1.913 \times 10^{-15}, \hat{\sigma}_\epsilon^2 = 115.856,\ \hat{\sigma}_\epsilon = 10.7636$.
As already stated, one may model the residuals ϵs as normal random variables with mean zero; and the χ^2 test based on 52 residuals in the present case supports this assumption

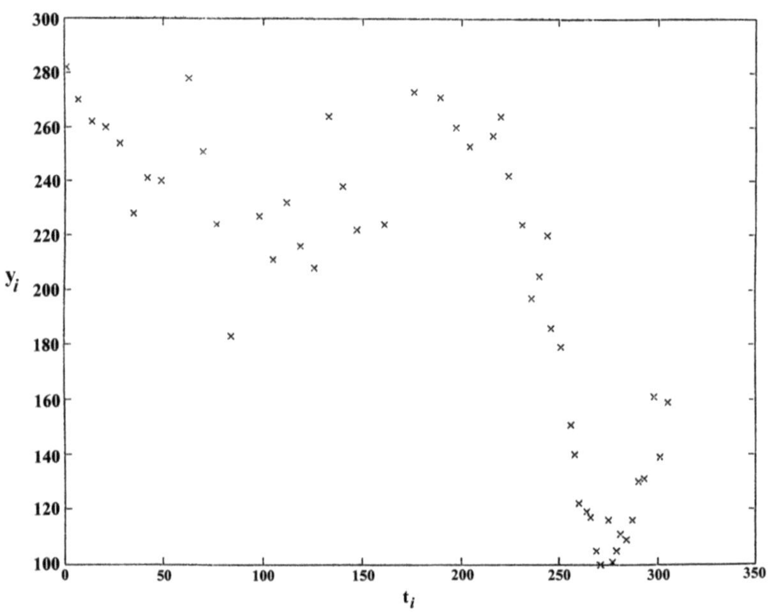

Fig. 8.3 Scatter diagram of ozone for the year 2001

destruction due to complete ban on human produced harmful chemicals, the mean part may be a constant over time and the ozone observations may attain a stationary distribution in the limit as $t \to \infty$.

It is known that the empirical distribution of a weakly correlated (nonstationary) Gaussian process is a consistent estimate of the limiting stationary distribution, e.g., see A1 of Dasgupta (2006). A convergence result, asserting that the empirical

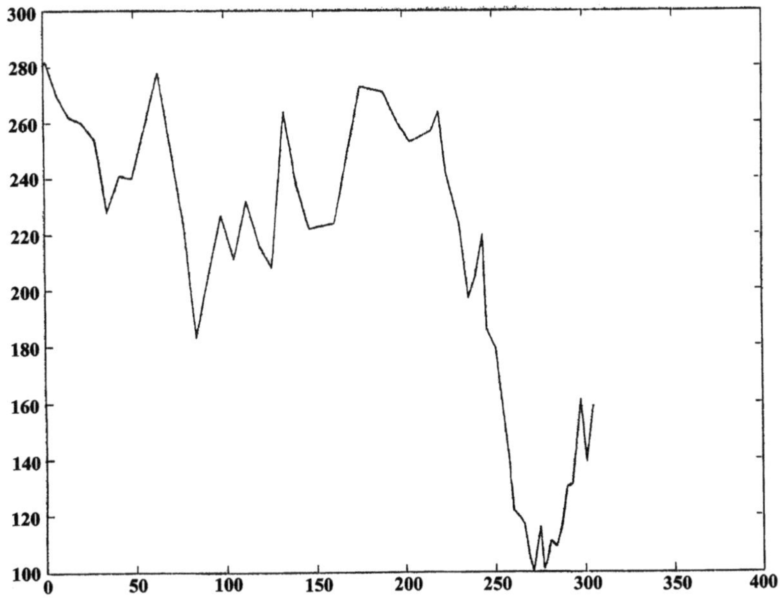

Fig. 8.4 Diagram joining the points by *straight line* for the year 2001

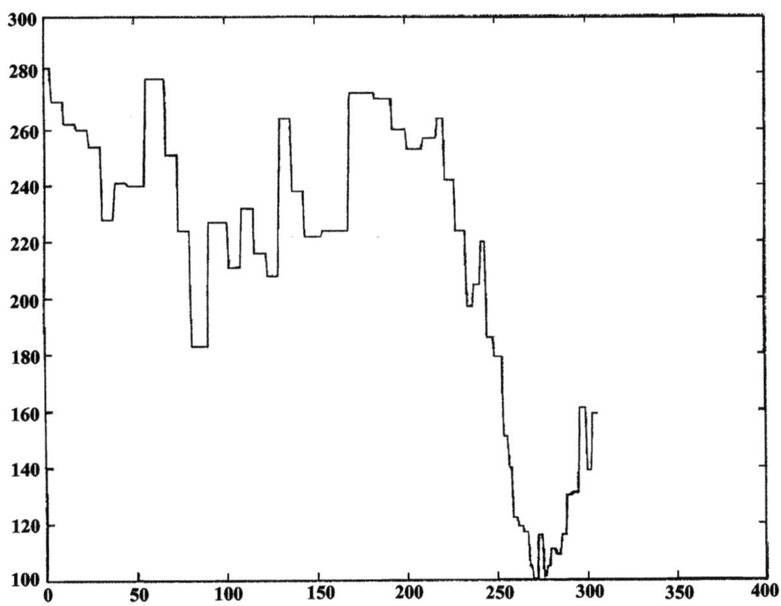

Fig. 8.5 Smoothing the points by nearest neighborhood method for the year 2001

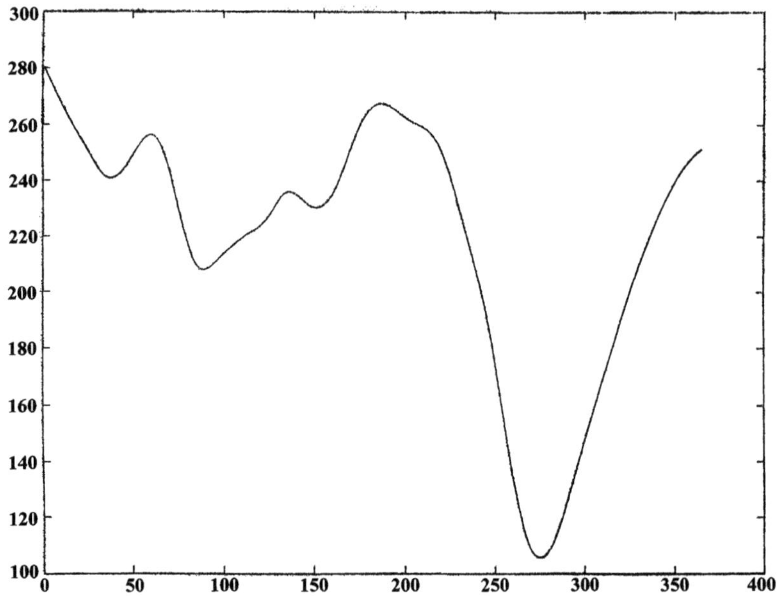

Fig. 8.6 Estimation of mean response for 2001 with roughness penalty .002

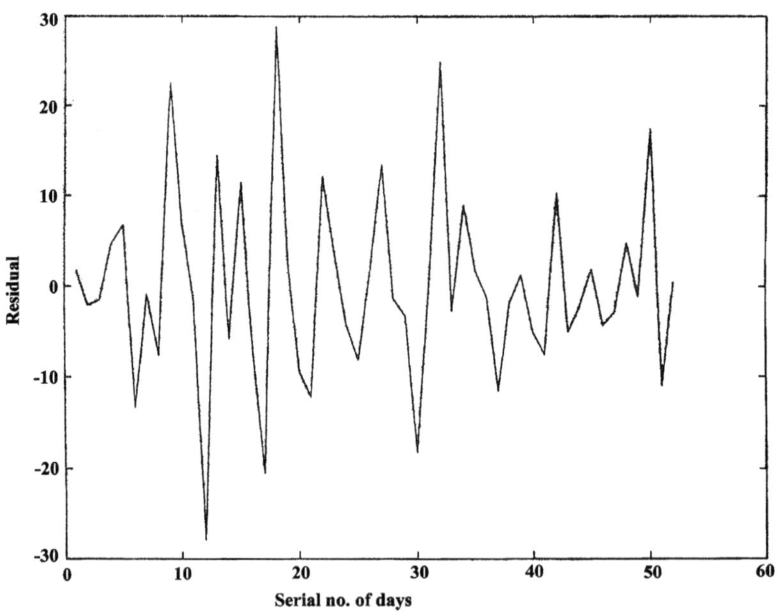

Fig. 8.7 Residuals of the year 2001 joined by *straight line*

Table 8.2 Normal fit for residuals

Class interval of (ϵ / $\hat{\sigma}_\epsilon$)	O_i	e_i
$(-\infty, -1.28]$	3	5.2
$(-1.28, -0.84]$	5	5.2
$(-0.84, -0.52]$	5	5.2
$(-0.52, -0.25]$	7	5.2
$(-0.25, 0.00]$	10	5.2
$(0.00, 0.25]$	6	5.2
$(0.25, 0.52]$	4	5.2
$(0.52, 0.84]$	3	5.2
$(0.84, 1.28]$	4	5.2
$(1.28, \infty)$	5	5.2
Total	52	52

The test statistic, $\chi^2 = \Sigma O_i^2/e_i - m = 7.62$ with 8 degrees of freedom; $\chi^2_{05.8} = 15.51$. p value of significance is $p = 0.47$.

distribution of a nonstationary Gaussian process is a consistent estimate of the limiting stationary distribution, where the recorded observations are not equispaced in time; holds provided the correlation function of the process is polynomially decaying. Specifically, we prove the following in Appendix A.1.

Theorem 1. *Consider a Gaussian process $X(t)$, $0 \le t \le T$ with mean $m(t)$ and covariance kernel $\sigma(t,u) = \sigma(t)\sigma(u)\rho(t,u)$, where $m(t) \to 0$, $\sigma(t) \to \sigma$; $t \to \infty$. Assume $X(t)$ has the weak limit denoted by $X(\infty)$ and the correlation function $|\rho(t,u)| < K|t-u|^{-\beta}$, $K > 0$, $\beta > 0$. Consider the empirical distribution function of the process based on the observations at time points $t_1, t_2, \cdots, t_n$ which are not necessarily equispaced. Let the time interval $[0, T)$ of recording the observations be subdivided into k subintervals and the length of each subinterval and the number of observations in each subinterval increase to ∞. Also let the time gap between two consecutive observations within each subinterval be homogeneous and the number n^* of "isolated" observations which do not fall in any one of the homogeneous subintervals, be negligible compared to n, i.e., $n^* = o(n)$. Then the empirical distribution function of the recorded observations from the process is a strongly consistent estimate for distribution function of the limiting variable $X(\infty)$, as $n \to \infty$.*

Thus the above χ^2 goodness of fit test for (limiting stationary) normal distribution for a nonstationary Gaussian process $\epsilon(t)$ based on the realizations ϵ_i, $i = 1, \cdots, n$; ($n = 52$ here) which are non equispaced in time, is valid as the empirical distribution of ϵ_i, $i = 1, \cdots, n$; converges to the normal distribution, provided the correlation function is polynomially decaying.

Note that in one of the class intervals of Table 8.2, the observed frequency is as high as 10. An approximate normal test of observed frequency 10 against expected frequency 5.2 for the fifth class interval of Table 8.2, under the model, is provided by $\tau = 2.22$, to be compared with tabulated $\tau_{.05} = 1.96$, $\tau_{.01} = 2.58$. However

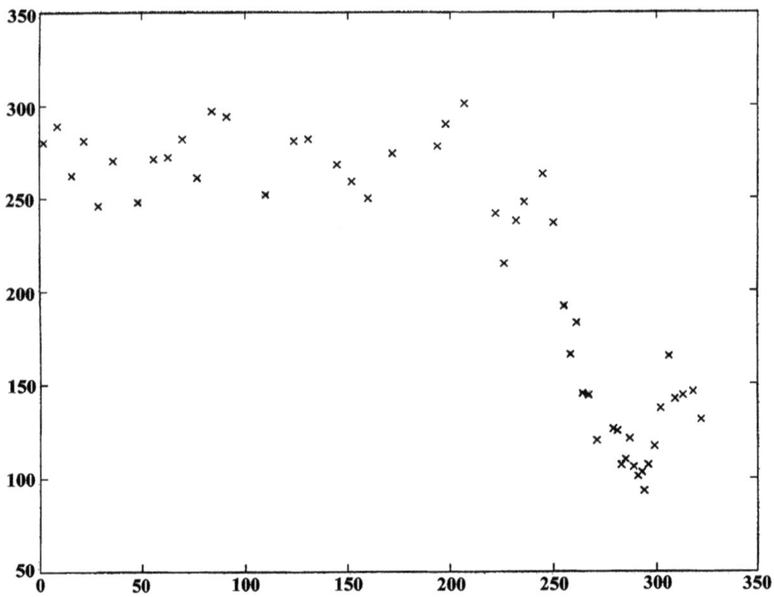

Fig. 8.8 Scatter diagram of ozone for the year 2006

this test is conservative. A more appropriate test would be based on extreme value theory of correlated normal random variables. We state a large sample result in this case; see, e.g., Galambos (1987).

Theorem A. Let $Z_n(r)$ be the maximum of a Gaussian stationary sequence $X_1, X_2, \cdots, X_n$ with zero expectation, unit variance, and correlations $r_m = EX_1X_{1+m}$. Let $a_n = \frac{1}{b_n} - \frac{1}{2}b_n(\log\log n + \log 4\pi),\ b_n = (2\log n)^{-1/2}$.

Then $(Z_n(r) - a_n)/b_n \xrightarrow{d} H_{3,0}(x) = \exp(-e^{-x}), -\infty < x < \infty$ provided, $r_m \log m \to 0$, as $m \to \infty$.

Thus, defining $W_n(r) = \min_{1\leq i\leq n} X_i$, one has $(W_n(r) - c_n)/d_n \xrightarrow{d} L(x) = 1 - e^{-e^x}, x \in (-\infty, \infty)$, with $c_n = -a_n, d_n = b_n$.

Hence the number $\tau = 2.22$, mentioned before may be compared with an approximate $H_{3,0}(x)$ random variable with $a_n = 1.362$, $b_n = .466$, where $n = 10$. Here, for the multinomial cell frequency $r = -\frac{1}{n}(1-\frac{1}{n}) \simeq 0$. p value of significance for $\tau = 2.22$ with respect to $H_{3,0}$ is $p = 0.147$.

Another scatter plot of 52 observations on total ozone spread over the year 2006 is shown in Fig. 8.8. The estimated response curve with cubic smoothing spline and roughness penalty $\lambda = .001,\ .002,$ and $.005$ are shown in Figs. 8.9–8.11, respectively. A smooth growth curve fitted on yearly minimum ozone level for 21 years from 1986 to 2006 is shown in Fig. 8.12. This suggests a possibility of higher value of minimum ozone in the year 2007 compared to the year 2006.

It was indeed so as the minimum ozone value corresponding to the year 2007 is 125 DU.

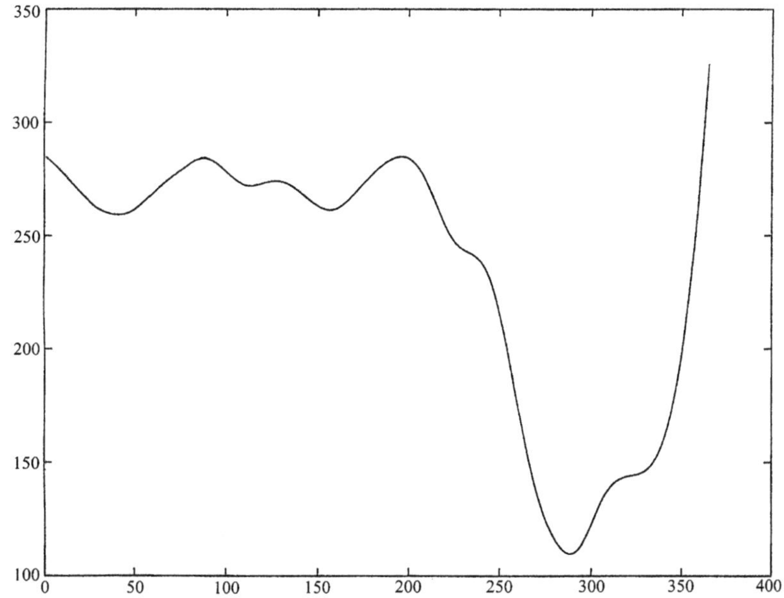

Fig. 8.9 Estimation of mean response for 2006 with roughness penalty .001

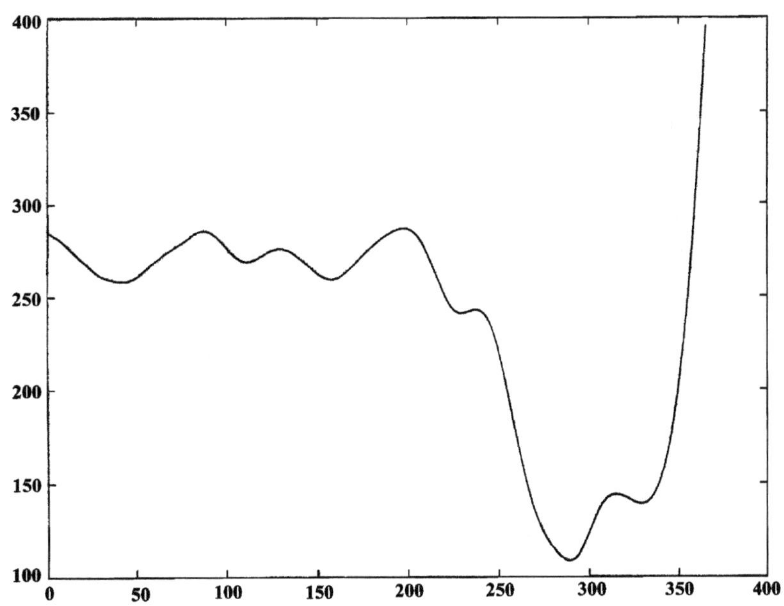

Fig. 8.10 Estimation of mean response for 2006 with roughness penalty .002

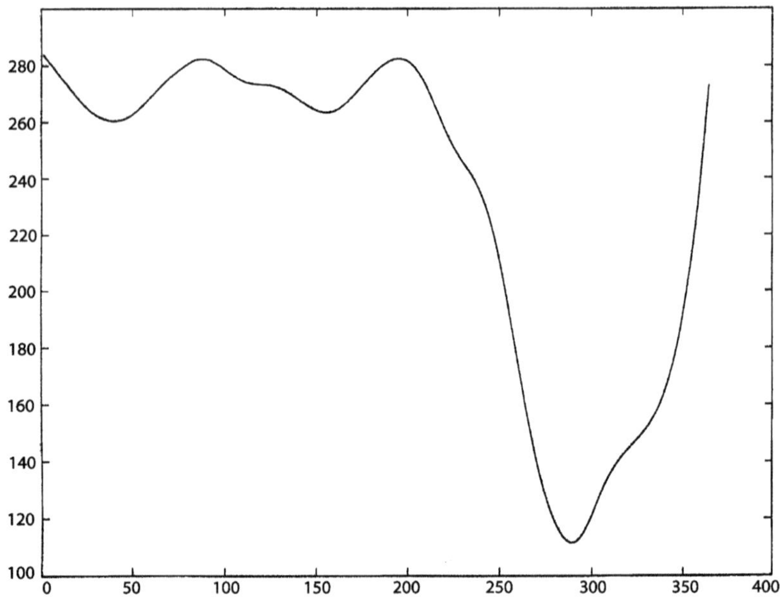

Fig. 8.11 Estimation of mean response for 2006 with roughness penalty .005

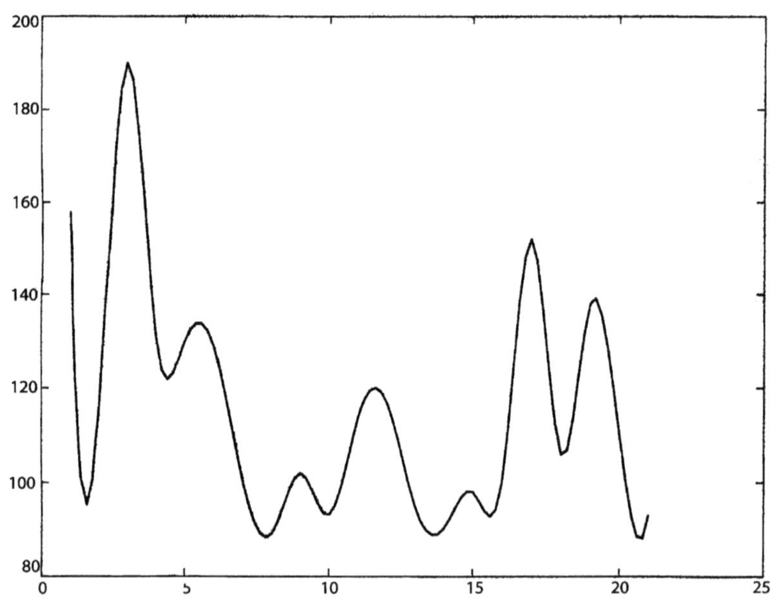

Fig. 8.12 Fitted smooth curve over yearly minimum ozone for 1986–2006

8.3 Lower β-Content Tolerance Limit of Minimum Ozone

A β content tolerance region S contains at least $100\beta\%$ of the future minimum with a high probability γ.

As our aim is to safeguard the minimum level of ozone concentration, we would like to construct a lower bound using the minimum level of ozone concentration of the past years, such that at least $100\beta\%$ of the future minimum observations would be above that bound with a high probability γ.

We observe that the minimum ozone concentration at the south pole over the years 1986–2006 occurred within the time period September 26–October 12, the record minimum being 89 DU on October 12, 1989.

Let $\{X_i \ : \ i \geq 1\}$ be weakly correlated $N(0,1)$ random variables with logarithmically decaying correlation function and $W_n = \min(X_1, \cdots, X_n)$ then $(W_n - c_n)/d_n \xrightarrow{D} W$, with $P(W \leq x) = 1 - \exp(-e^x)$, $-\infty < x < \infty$ and $c_n = -a_n = -(2\log n)^{1/2} + \frac{1}{2}(\log\log n + \log 4\pi)/(2\log n)^{1/2}$, $d_n = b_n = (2\log n)^{-1/2}$, vide Theorem A.

In our application the scaled residuals take the role of normal variables $\{X_i : i \geq 1\}$. The χ^2 goodness of fit vide Table 8.2 may be supplemented by checking convergence of distribution function mentioned in Theorem 1. Equivalently, the observed quantiles should converge to the corresponding theoretical normal quantiles. The quantile–quantile plot for 52 scaled residuals $\epsilon/\hat{\sigma}_\epsilon$ of Table 8.1 is given in Fig. 8.13. The observed data is seen to cluster around the line $y = x$, as expected. The best fitted line to the data has intercept -0.0526 ($\simeq$0) and slope 0.9734 ($\simeq$1), with R^2 of linear regression as 0.9645. For the given data points, ratio of sum of

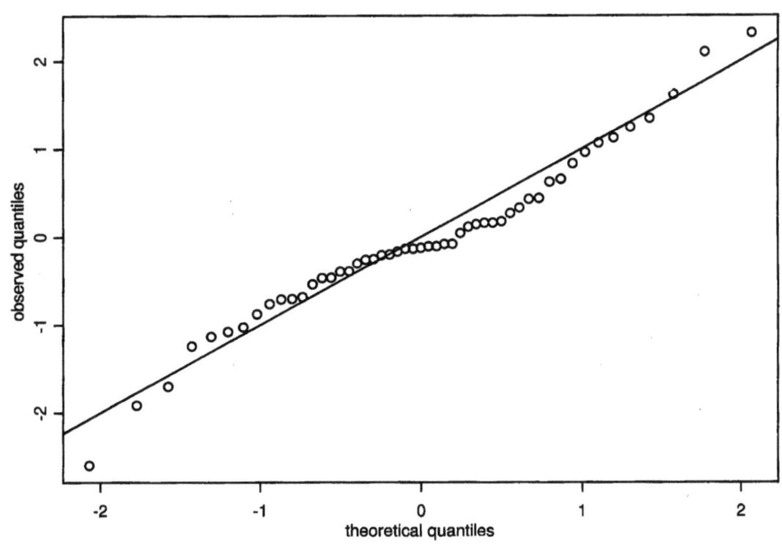

Fig. 8.13 Normal plot for standardized residuals

squares of distances from the best fitted line, to the sum of squares of distances from the theoretical line $y = x$ is 0.9003. This indicates the estimations via graphical methods may be quite efficient.

Over the years 1986–2006, the minimum level of ozone occurred during the period September 26 (corresponding to the year 2003) to October 12 (corresponding to the year 1993) and the number of days in that span is $n = 17$. Consider $Y_1, \cdots, Y_n$ the total ozone level on those 17 days. Since the response curve $\mu(t)$ is continuous and may be taken to be a constant μ within a short range of time, denoting $w_n^* = \min(Y_1, \cdots, Y_n)$ and observing that the residual ϵ_is may be taken to be weakly correlated $N(0, \sigma^2)$ random variables, one may write

$$(w_n^* - \mu)/\sigma \sim W_n \xrightarrow{D} c_n + d_n W, \text{ i.e., } w_n^* \xrightarrow{D} \mu + \sigma(c_n + d_n W) = \mu^* + \sigma^* W,$$

where $\mu^* = \mu + c_n\sigma, \ \sigma^* = d_n\sigma$.

Next, consider the record minimum of ozone level $Z_{(1)}$ over k years, i.e., $Z_{(1)} = \min(Z_1, \cdots, Z_k)$. Here $Z_{(1)} = 89$ DU with $k = 21$ years from 1986 to 2006. Let $\delta(\mu, \sigma) = \{\mu + \sigma \log(-\log \beta)\}/\ [\mu + \sigma \log\{-\log(1 - (1 - \gamma)^{1/k})\}]$.

We have the following result.

Theorem 2. *Let the response curve $\mu(t)$ of total ozone Y spread over the years be continuous and the residuals ϵs be weakly correlated $N(0, \sigma^2)$ random variables with logarithmically decaying correlation function. Then $Z_{(1)}\delta(\hat{\mu}^*, \hat{\sigma}^*)$ is an approximate β content lower tolerance limit for minimum ozone level Z for a future year, i.e.,*

$$P_{Z_{(1)}}[P_Y\{Z \geq Z_{(1)}\delta(\hat{\mu}^*, \hat{\sigma}^*)\} \geq \beta] = \gamma + o(1), \ \ as \ \ k \to \infty,$$

where $\hat{\mu}^*$ and $\hat{\sigma}^*$ are consistent estimates of μ^* and σ^* respectively.

Proof of the theorem is given in Appendix A.2.

In the present case $k = 21, Z_{(1)} = 89, \ \hat{\sigma} = 10.7636, \ \hat{\mu}$ may be taken average of μ over the period September 26–October 12. From the response curve of 2001 data $\hat{\mu} = 105.86$, estimated as average for these days, see Appendix A.3. Recommended values of β and γ are $\beta = .9, \ \gamma = .95$.

8.4 Lower Tolerance Limit Based on Data of the Years 1986–2006

We take $n = 17$, the number of days in the time period September 26–October 12. Then,

$$c_n = -(2 \log n)^{1/2} + \frac{1}{2}(\log \log n + \log 4\pi)/(2 \ \log \ n)^{1/2} = -1.63$$

$$d_n = (2 \log n)^{-1/2} = 0.42$$

These are centering and scaling factors of extreme value distribution based on $N(0, 1)$ random variables.

Regarding the choice of $n = 17$, note that values of c_n and d_n are logarithmic and lower order function of n and thus insensitive to slight variation of n; e.g., $d_{30} = 0.385$, $c_{30} = -1.6$. Hence this technique of computing the lower bound is robust with regard to slight variation of the particular date when actual minimum of yearly total ozone is attained.

Next, the centering and scaling factor of extreme value distribution based on minimum ozone of a single year are as follows:

$$\hat{\mu}^* = \hat{\mu} + c_n\hat{\sigma} = 88.315,$$
$$\hat{\sigma}^* = d_n\hat{\sigma} = 4.521.$$

Also, $\delta(\hat{\mu}^*, \hat{\sigma}^*) = .8541$, which is a shrinking factor of record minimum to obtain the lower tolerance limit for a future minimum. We obtain $Z_{(1)}\delta(\hat{\mu}^*, \hat{\sigma}^*) = 76.02$.

The above may be interpreted as 90% of the future minimum of yearly total ozone is going to be $\geq$76.02 DU with a high probability nearly .95.

One may like to use an estimate of σ based only on the data of the time period September 26–October 12, when minimum ozone concentration occurred over the years 1986–2006. Then, $\hat{\sigma} = 5.622$ from the data of the year 2001. In that case

$$\hat{\mu}^* = \hat{\mu} + c_n\hat{\sigma} = 96.696,$$
$$\hat{\sigma}^* = d_n\hat{\sigma} = 2.3612.$$

Also, $\delta(\hat{\mu}^*, \hat{\sigma}^*) = .9291$. We obtain $Z_{(1)}\delta(\hat{\mu}^*, \hat{\sigma}^*) = 82.69$.

One may fit the extreme value distribution $L_{\mu,\sigma}(x) = P(w \leq x) = 1 - \exp(-e^{\frac{x-\mu}{\sigma}})$ to the yearly minimum of ozone level for 21 years during 1986–2006, by probability plot, see Fig. 8.14. The estimated value of μ and σ from the plot is $\hat{\mu} = 121.4818$, $\hat{\sigma} = 19.1939$. The value of R^2 for linear regression is quite high; $R^2 = .8344$ indicating a good fit. Estimate of $P(Y \geq 82.69)$ is $L_{\hat{\mu},\hat{\sigma}}(82.69) = .8319$. This is slightly lower than the specified value of $\beta = .9$.

We may also bootstrap the empirical distribution of the minimum ozone over years to obtain another solution of the problem. The sample median and the sample interquartile range may be taken as nonparametric estimates of the corresponding location and scale parameters of the distribution. Equating the empirical distribution with $L_{\mu,\sigma}(x)$ at three quartiles $\hat{x}_{1/4}, \hat{x}_{med}, \hat{x}_{3/4}$, we may obtain the following nonparametric estimates of the location and scale parameters μ and σ. These estimates are slight variations of the usual nonparametric estimates:

$$\hat{\sigma} = [\hat{x}_{3/4} - \hat{x}_{1/4}]/[\log\log 4 - \log\log(4/3)]; \quad \hat{\mu} = \hat{x}_{med} - \hat{\sigma}\log\log 2 \qquad (8.1)$$

The estimates based only on $\hat{x}_{med}, \hat{x}_{3/4}$ are as follows:

$$\hat{\sigma} = [\hat{x}_{3/4} - \hat{x}_{med}]/\log 2\ ; \quad \hat{\mu} = \hat{x}_{med} - \hat{\sigma}\log\log 2 \qquad (8.2)$$

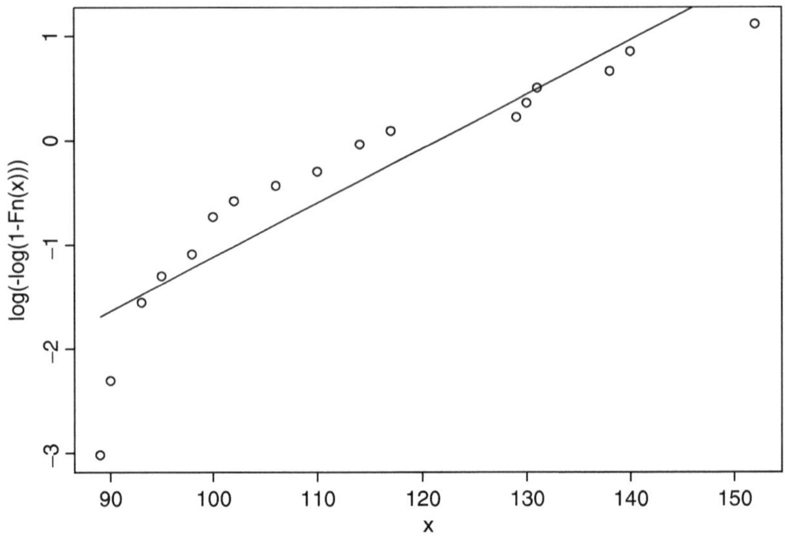

Fig. 8.14 Extreme value fit of ozone level for 21 years

The distribution $L_{\mu,\sigma}(x)$ being negatively skewed, the third sample quartile is more stable than the first. Thus the estimates given in (8.2) are expected to perform better.

Bootstrapping the distribution of the minimum ozone over 21 years are done as follows. Draw a simple random sample x_i^*, $1 \leq i \leq 21$ of size 21 with replacement from the empirical distribution of the minimum ozone over 21 years and obtain standardized pseudo minimum values $\frac{x_i^* - \hat{\mu}}{\hat{\sigma}}$ from (8.1)/(8.2); to be denoted by B1 and B2, respectively. The whole procedure is repeated 10 times, providing 210 pseudo minimum values in each case of B1 and B2. The probability plots of these values over a representative run are shown in Figs. 8.15 and 8.16. The R^2 value for linear regression is .7631 for B1 and .7664 for B2, these are quite high indicating the possibility of extreme value distribution as a model.

However, anticipated superiority of the estimates (8.2) over (8.1) could not be established in terms of better fit measured by R^2 over different runs.

Next, to have a nonparametric solution of the problem we may compute the estimates given by (8.1) from 21 observed yearly minimum ozone level and obtain the *standardized value of lowest ozone level* observed so far as $y = \frac{89 - \hat{\mu}}{\hat{\sigma}} = -1.1227$.

The relative proportion of observations *above* this value in Figs. 8.15 and 8.16 are 0.909 and 0.869, respectively, indicating that future minimum will be higher than the minimum already observed up to the year 2006, with a high probability nearing 0.9.

The minimum of yearly total ozone in DU for the 6 years 2007–2012 are, respectively, as follows: 125, 107, 98, 122, 102, 136.

See e.g., http://www.esrl.noaa.gov/gmd/dv/spo_oz/spmin.html.

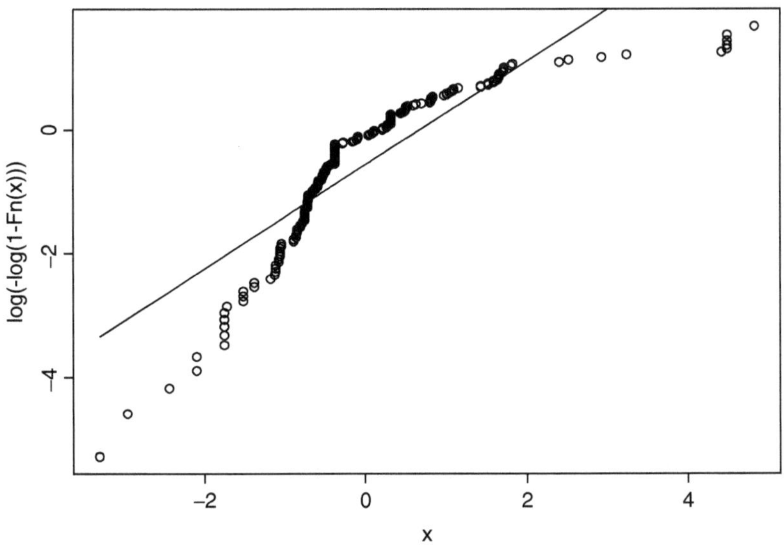

Fig. 8.15 Bootstrap 1 extreme value fit of ozone level for 21 years

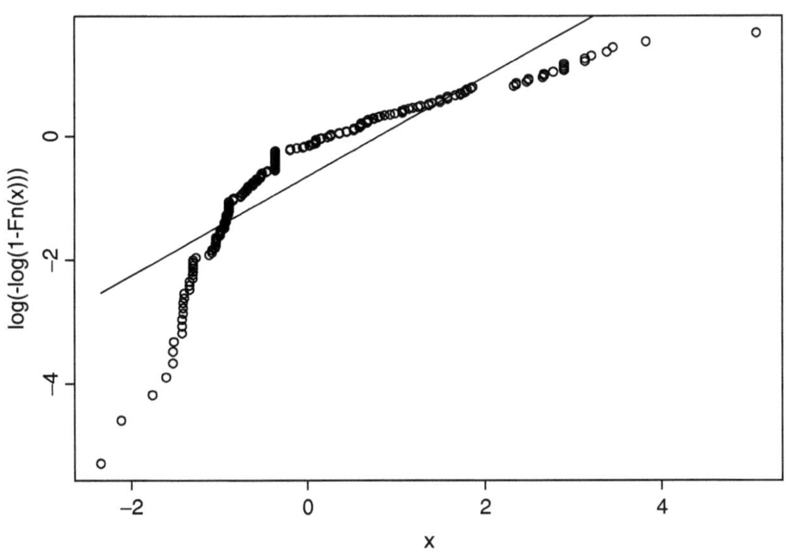

Fig. 8.16 Bootstrap 2 extreme value fit of ozone level for 21 years

All these values lie above the lower tolerance limit predicted from model.

Recently Antarctic ozone hole is getting smaller possibly due to regulation adherence, the average area covered by the Antarctic ozone hole in the year 2012 was the second smallest in the last 20 years.

Acknowledgment Thanks are due to Prof. J.K. Ghosh for his valuable comments. Data analyzed are obtained from NOAA web site.

Appendix

A.1 Empirical Distribution of Observations on Nonstationary Gaussian Process Realized on Non-equispaced Time Points

Here we prove Theorem 1. We follow the steps (5.2)–(5.5), pp. 32–33 of Dasgupta (2006). As the condition (5.5.3) of Cramér and Leadbetter (1967) is satisfied, from (5.6) Dasgupta (2006),

$$\frac{1}{T}\int_0^T [I(X(t) < d) - EI(X(t) < d)]dt \to 0 \tag{8.3}$$

with probability one, as $T \to \infty$.

(In Table 8.1, the observations are not equispaced, there are weekly observations taken at a stretch, observations are taken biweekly, or at a gap of 2/3/4/5 days.)

Subdivide the interval $[0, T)$ into k subintervals, where the time gap is homogeneous between two consecutive observations in each subdivision. In the jth time segment $[T^{(j)}, T^{(j+1)}), j = 0, 1, \cdots, k-1$, there are n_j observations, where $T^{(0)} = 0, T^{(k)} = T$. We may obtain from (5.8), p. 34 of Dasgupta (2006) the following:

$$\frac{1}{n_j}\sum_{i=0}^{n_j-1} I\{X(T^{(j)} + iT^{(j+1)}/n_j) < d\} \to \Phi(d/\sigma) = P(X(\infty) < d) \text{ a.s.} \tag{8.4}$$

In other words,

$$\sum_{i=0}^{n_j-1} I\{X(T^{(j)} + iT^{(j+1)}/n_j) < d\} - n_j\Phi(d/\sigma) = o(n_j) \text{ a.s.}$$

Consider all the components with inter-homogeneous time gaps; weekly, biweekly, etc. A few "isolated" observations which may not fall in a group have negligible contribution to the above sum of indicator variables, as the divisor is n. Combining, it follows that

$$\sum_{i=0}^{n-1} I\{X_i < d\} - n\Phi(d/\sigma) = o(n), \text{ a.s.}$$

as $n \to \infty$, where n is the total number of observations, X_i, $i = 0, 1, 2, \cdots, n-1$. Hence,

$$\frac{1}{n}\sum_{i=0}^{n-1} I\{X_i < d\} \to \Phi(d/\sigma) \text{ a.s.} \tag{8.5}$$

In our application, X_i are the residual ϵs of ozone recordings from the mean response $\mu(t)$ at n time points.

A.2 Construction of Lower Tolerance Limit

Here we prove Theorem 2. Denote

$$L_{\mu,\sigma}(x) = P(w \leq x) = 1 - \exp(-e^{\frac{x-\mu}{\sigma}}) \tag{8.6}$$

Let z_i be distributed as $L_{\mu,\sigma}$ approximately and $z_{(1)} = \min(z_1, \cdots, z_k)$.

A reasonable tolerance limit for a future observation Z with approximate distribution L would be $z_{(1)}\delta(\mu, \delta)$ where δ is a positive function of μ and σ. Thus we need

$$P_{z(1)}[P_L\{Z \geq z_{(1)}\delta(\mu, \sigma)\} \geq \beta] = \gamma \tag{8.7}$$

Hence at least $100\beta\%$ of the future observations would be above $z_{(1)}\delta(\mu, \sigma)$ with a high probability γ. From (8.7) we get,

$$P[1 - L(z_{(1)}\delta(\mu, \delta)) \geq \beta] = \gamma, \text{ i.e., } P[z_{(1)}\delta(\mu, \sigma) \leq L^{-1}(\overline{\beta})] = \gamma,$$

where $\overline{\beta} = 1 - \beta$ and L^{-1} is the inverse function of L.
i.e., $P[z_{(1)} > L^{-1}(\overline{\beta})/\delta(\mu, \sigma)] = 1 - \gamma$
i.e., $G^k[L^{-1}(\overline{\beta})/\delta(\mu, \sigma)] \simeq 1 - \gamma$, from Theorem A,
where $G = 1 - L$, i.e., $L[L^{-1}(\overline{\beta})/\delta(\mu, \sigma)] \simeq 1 - (1 - \gamma)^{1/k}$
i.e.,

$$\delta(\mu, \sigma) \simeq L_{\mu,\sigma}^{-1}(\overline{\beta})/L_{\mu,\sigma}^{-1}(1 - (1 - \gamma)^{1/k}) \tag{8.8}$$

In the present case, $L_{\mu,\sigma}^{-1}(y) = \mu + \sigma \log\{-\log(1 - y)\}$, from (8.6) and the yearly minimum ozone, $Z \xrightarrow{D} \mu^* + \sigma^* W \sim L_{\mu^*,\sigma^*}$, where $P(W \leq x) = 1 - \exp(-e^x), -\infty < x < \infty$. Now the right hand side of (8.8) is a continuous function of μ, σ. The result follows as $\hat{\mu}^*$ and $\hat{\sigma}^*$ are consistent estimates of μ^* and σ^*, respectively.

A.3 Estimate of $\mu = E(Y_t)$ for Random t

Let $t \in [a,b]$ be uniformly distributed on the interval. Then, $E(Y_t) = \frac{1}{b-a}\int_a^b E(Y_s)ds \simeq \frac{1}{m}\sum_{i=1}^m \mu_i$, where i refers to the day numbers lying on the time interval $[a,b]$ and μ_i is mean ozone level of ith day. In our application, $m = 17$ and $i \in$ [September 26, October 12]; μ_i may be taken as value of ith day mean ozone level given by the response curve, interpolated by the method of smoothing spline from the whole data set. This provides $\hat{\mu} = 105.86$.

One may estimate the above directly from the ozone observations in the segment $[a,b]$. We assume that the correlation function of the Gaussian process decays polynomially. Following the same arguments of (8.3) in A.1, $\frac{1}{b-a}\int_a^b [Y_s - E(Y_s)]ds \simeq 0$; see also A2 of Dasgupta (2006). Hence, $E(Y_t) = \frac{1}{b-a}\int_a^b E(Y_s)ds \simeq \frac{1}{b-a}\int_a^b Y_s ds \simeq \frac{1}{m}\sum_{j=1}^n w_j Y_j$, the weighted mean of observations in that time period where the weights w_j are proportional to the time gap between consecutive recorded ozone observations $Y_j,\ j = 1,\cdots,7$. For example, $w_1 = 2, w_2 = 2, w_3 = 4, w_4 = 2, w_5 = 2, w_6 = 2, w_7 = 3$; $m = 17, n = 7, m = \sum_{j=1}^n w_j$. Thus we get, $\hat{\mu} = 107.94$.

In the first method, the whole data set is used to estimate μ, whereas observations falling in the specific time interval $[a,b]$ are used in the latter method of estimation.

References

Cramér, H., & Leadbetter, M. R. (1967). *Stationary and related stochastic process*. New York: Wiley.

Dasgupta, R. (2006). Modeling of material wastage by Ornstein–Uhlenbeck process. *Calcutta Statistical Association Bulletin, 58*(229–230), 15–35.

Dasgupta, R., & Bhaumik, D. K. (1995). Upper and lower tolerance limits of atmospheric ozone level. *Sankhyā B, 57*, 182–199.

Galambos, J. (1987). *The asymptotic theory and extreme order statistics* (2nd ed.). Malabar: Krieger.

Chapter 9
A New Technique for Estimating Population Distribution of Growth Curve Parameters with Longitudinal and Cross-sectional Data

Sedigheh Mirzaei Salehabadi and Debasis Sengupta

Abstract In this paper, we present a new approach for estimating the population distribution of biological parameters related to individual growth. The most attractive feature of the proposed method is that, while some amount of longitudinal data is required, information contained in sparse longitudinal as well as completely cross-sectional data can also be harnessed. Although the method is not Bayesian, it can be implemented through recursions based on Gibb's sampling. Computer simulations in the special case of the Preece–Baines growth model show that inclusion of some cross-sectional data indeed reduces the mean squared errors of the estimators. The method is then used to compare the population distribution of human growth parameters among the male and female subjects of a study conducted some years ago by the Indian Statistical Institute.

9.1 Introduction

Consider longitudinal growth data having the form (t_{i1}, y_{i1}), (t_{i2}, y_{i2}), $\ldots$, (t_{in_i}, y_{in_i}), $i = 1, \ldots, n$, where n is the number of individuals, n_i is the number of observations for the ith individual, $1 \le i \le n$, and y_{ij} is the observed size variable at age t_{ij}. For such data, one often postulates a parametric random effects model of the form

$$y_{ij} = h(t_{ij}; \boldsymbol{\tau}_i) + \varepsilon_{ij}, \quad j = 1, 2, \ldots, n_i, \; i = 1, 2, \ldots, n, \tag{9.1}$$

where the function h has a known functional form, with a random vector parameter $\boldsymbol{\tau}_i$ controlling its shape, ε_{ij}, $j = 1, 2, \ldots, n_i$, $i = 1, 2, \ldots, n$ are samples from

S. Mirzaei Salehabadi • D. Sengupta (✉)
Applied Statistical Unit, Indian Statistical Institute, Kolkata, India
e-mail: sedigheh_r@isical.ac.in; sdebasis@isical.ac.in

R. Dasgupta (ed.), *Advances in Growth Curve Models*, Springer Proceedings in Mathematics & Statistics 46, DOI 10.1007/978-1-4614-6862-2_9,

a zero-mean error distribution with density φ with vector parameter $\boldsymbol{\eta}$, and $\boldsymbol{\tau}_i$ (independent of the errors) are samples from a distribution with density f with vector parameter $\boldsymbol{\theta}$. The function h is referred to as the individual growth curve, while the ε_{ij}s are regarded as measurement errors. The $\boldsymbol{\tau}_i$s account for the variation of the growth curve from one individual to another. The distribution (f) of these individual-specific parameters, captured here through the parameter $\boldsymbol{\theta}$, is a matter of general interest in a wide variety of applied areas, including biology, psychology, economics, and sociology. In this paper, we consider the problem of estimating this parameter.

Potthoff and Roy (1964) considered a model of h that is possibly nonlinear (e.g., polynomial) in the age variable, but is linear in the parameter $\boldsymbol{\tau}_i$. Since this model falls into the framework of multivariate linear models, Potthoff and Roy's (1964) work was followed up by many other researchers. However, when there are only a handful of observations per individual, a parsimonious model can be fitted to individual growth data only if the model is allowed to be nonlinear in the parameters. Simplest examples of such models include the exponential growth model and the logistic growth model, while more complex models with larger number of parameters have also been considered (see Falkner and Tanner 1986, Hauspie et al. 2004). We consider a popular model, proposed by Preece and Baines (1978) in the next section.

Work in this area had begun with the Potthoff–Roy model (see Rao 1965) and has continued ever since (see Laird and Ware 1982, Cnaan et al. 1997, Huggins and Loesch 1998, Duncan et al. 2006, and Donnet et al. 2010). The methods proposed by these authors rely heavily on the structure of the data. In particular, a healthy number of observations per individual are needed. Such data may be obtained through longitudinal studies (Diggle et al. 2002), which are time consuming and expensive. On the other hand, data from cross sectional studies are not suitable for the methods mentioned above. Because of this difficulty, some studies are designed to track different individuals over different age ranges. The different age ranges used in the study may have only partial overlap. This way, the duration of the study can be shorter. Huggins and Loesch (1998) considered analysis of this type of data.

However, many large scale studies on human growth happen to be entirely cross-sectional. Thus, one might seek to combine the strengths of the two types of data, for solving longitudinal data problems. We present in this paper a general method for estimating $\boldsymbol{\theta}$ of (9.1) on the basis of a combination of longitudinal and cross-sectional data. We illustrate it by analysing a classic data set on human growth through a popular model (see Dasgupta and Hauspie 2001, Falkner and Tanner 1986).

There have been some attempts to use a combination of cross-sectional and longitudinal data for various types of analysis. For example, Verbeke et al. (2001) consider a linear mixed model and go into the issue of estimation of the model parameters. However, their focus is on the estimation of overall trend (a shared attribute in all individuals). They regard the population-specific parameters as nuisance parameters and do not go into the question of the distribution of these parameters across different individuals in the population. We focus on the latter problem, for which there is no comparable method available as yet.

Independence of the errors $\varepsilon_{ij},\ j = 1, 2, \ldots, n_i,\ i = 1, 2, \ldots, n$ is crucial to the methodology developed in this paper. While various researchers have attempted to model in different ways the possible dependence of observations for a single individual, the model used here amounts to assuming that such dependence is adequately taken into account through the shared random parameter, $\boldsymbol{\tau}_i$.

9.2 A Growth Model

The following model proposed by Preece and Baines (1978) has been one of the most popular models for human growth:

$$h(t; \boldsymbol{\tau}) = h_{max} - \frac{2(h_{max} - h_\theta)}{e^{s_0(t-\theta)} + e^{s_1(t-\theta)}}. \tag{9.2}$$

Here, $\boldsymbol{\tau}$ consists of five parameters. Out of these, s_0 and s_1 are parameters controlling the rates of growth at different stages, h_{max} is the final size, and h_θ is the size at a threshold age θ. For the sake of identifiability, it is assumed that $s_0 < s_1$.

The derivative of the growth function, known as the velocity function, has the following form for the Preece–Baines model:

$$h'(t; \boldsymbol{\tau}) = -h(t; \boldsymbol{\tau}) \frac{s_0 e^{s_0(t-\theta)} + s_1 e^{s_1(t-\theta)}}{e^{s_0(t-\theta)} + e^{s_1(t-\theta)}}. \tag{9.3}$$

There are several biological parameters of interest that can be linked to the five mathematical parameters of the model:

(a) Peak Height Velocity (PV) is a measure of the maximum rate of growth in size during a growth spurt.
(b) Age at PV is the age corresponding to maximum velocity of growth.
(c) Take off Velocity (TO) is a measure of the minimum rate of growth in size during a growth spurt.
(d) Age at TO is the age corresponding to minimum velocity of growth.
(e) Final size is the limiting h for large age.

Figure 9.1 shows the plot of a typical growth curve and the corresponding velocity curve, in the special case where the size in question is the stature of a person (in cm), and age is measured in years. The biological parameters are also indicated in the plot. The final size is the largest value attained by the growth curve. The age at TO and the age at PV are the locations of first minimum and the unique maximum of the velocity function, respectively. The TO velocity and PV are the values of the velocity function at these points of inflection.

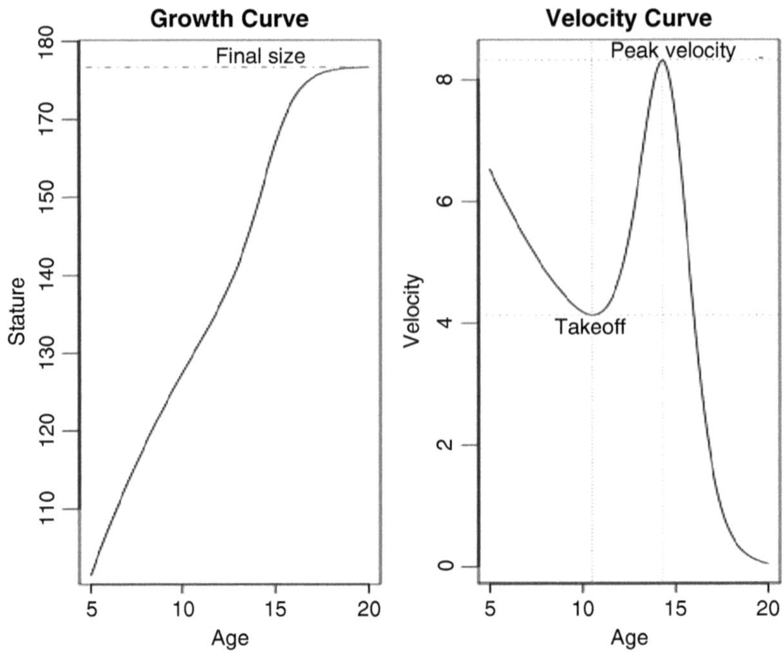

Fig. 9.1 Biological parameters of interest

All the parameters of the Preece–Baines curve are required to be positive. There are further constraints. In order that the growth function is strictly increasing, it is necessary that

$$h_{max} > h_\theta. \tag{9.4}$$

Another constraint is necessary in order that the velocity curve has two distinct points of inflection. To see this, let $\phi(t;\boldsymbol{\tau}) = -h'(t;\boldsymbol{\tau})/h(t;\boldsymbol{\tau})$, so that we have from (9.3)

$$h''(t;\boldsymbol{\tau}) = h(t;\boldsymbol{\tau})\left[\phi^2(t;\boldsymbol{\tau}) - \phi'(t;\boldsymbol{\tau})\right].$$

The points of inflection of the velocity function satisfy the condition $\phi^2(t;\boldsymbol{\tau}) - \phi'(t;\boldsymbol{\tau}) = 0$, which simplifies to

$$\begin{aligned} 2\left(s_0 e^{s_0(t-\theta)} + s_1 e^{s_1(t-\theta)}\right)^2 &= \left(s_0^2 e^{s_0(t-\theta)} + s_1^2 e^{s_1(t-\theta)}\right)\left(e^{s_0(t-\theta)} + e^{s_1(t-\theta)}\right), \\ \text{i.e., } \left(s_0 e^{s_0(t-\theta)} + s_1 e^{s_1(t-\theta)}\right)^2 &= (s_1 - s_0)^2 e^{(s_0+s_1)(t-\theta)}, \\ \text{i.e., } \left(s_0 e^{s_0(t-\theta)} + s_1 e^{s_1(t-\theta)}\right) &= (s_1 - s_0) e^{\frac{(s_0+s_1)(t-\theta)}{2}}, \\ \text{i.e., } s_0 + s_1 e^{(s_1-s_0)(t-\theta)} &= (s_1 - s_0) e^{\frac{(s_1-s_0)(t-\theta)}{2}}. \end{aligned}$$

The last equation is a quadratic one in $e^{\frac{(s_1-s_0)(t-\theta)}{2}}$, which has distinct solutions if and only if $(s_1 - s_0)^2 - 4s_0s_1 > 0$, i.e.,

$$\frac{s_1}{s_0} > 3 + 2\sqrt{2}. \tag{9.5}$$

Thus, the parameter space of the Preece–Baines model (9.2) is the orthant corresponding to positive values of the five parameters, subject to the restrictions (9.4) and (9.5).

We end this section by relating the five biological parameters mentioned above to the mathematical parameter $\boldsymbol{\tau}$. The final size has already been identified as h_{max}. It also follows from the quadratic equation given above that

$$\text{Age at TO} = \frac{2}{s_1 - s_0} \log\left(\frac{(s_1 - s_0) - \sqrt{(s_1 - s_0)^2 - 4s_1s_0}}{2s_1}\right) + \theta,$$

$$\text{Age at PV} = \frac{2}{s_1 - s_0} \log\left(\frac{(s_1 - s_0) + \sqrt{(s_1 - s_0)^2 - 4s_1s_0}}{2s_1}\right) + \theta.$$

The remaining two parameters, viz., TO velocity and PV, are obtained by substituting the above ages in (9.3).

9.3 A Data Example

This work is motivated by an anthropometric study conducted by the Indian Statistical Institute under the leadership of Professor S.R. Das during the 1950s and 1960s, from the Sarshuna–Barisha (S–B) region of Kolkata (Das et al. 1985). The data set of male subjects obtained from this study reflects 298 individuals and that of female subjects represents 253 individuals, many of whom were tracked over the said period for different durations. The variables include age, stature, and a few other anthropometric characteristics. The number of observations per individual range from 1 to 21, for the age interval 0.5–21. Lack of samples in the 19-or-more and 10-or-less age ranges come in the way of fitting a reasonable parametric model in most of the cases. The fitting of individual-specific growth curves is further restricted by convergence problems in some cases. Fitting of the Preece–Baines model is possible in the cases of only 36 male and 15 female subjects. On the other hand, the total number of observations is substantial, and these may be tapped for improved estimation.

9.4 Estimating Population Distribution of Parameters

The parameter $\boldsymbol{\tau}_i$ in (9.1) completely determines the growth function h. Therefore, any functional of h, such as the biological parameters mentioned in Sect. 9.2, can be written as a function of $\boldsymbol{\tau}_i$. The population distribution of such parameters can be determined from the population distribution of $\boldsymbol{\tau}_i$, namely, $f(\boldsymbol{\tau}_i; \boldsymbol{\theta})$. It transpires that the task of estimating the population distribution of any functional of h reduces to the problem of estimating $\boldsymbol{\theta}$, in the presence of the nuisance parameter $\boldsymbol{\eta}$.

The likelihood for $\boldsymbol{\theta}$ and $\boldsymbol{\eta}$ is

$$\prod_{i=1}^{n} f(\boldsymbol{\tau}_i; \boldsymbol{\theta}) \prod_{j=1}^{n_i} \varphi\left([y_{ij} - h(t_{ij}; \boldsymbol{\tau}_i)]; \boldsymbol{\eta}\right). \tag{9.6}$$

Since $\boldsymbol{\tau}_1, \boldsymbol{\tau}_2, \ldots, \boldsymbol{\tau}_n$ are unobserved, these can be treated as nuisance parameters. Maximizing the likelihood in the presence of the nuisance parameters is generally rather difficult. A standard approach to this problem is to maximize the likelihood (9.6) with respect to $\boldsymbol{\theta}$, $\boldsymbol{\eta}$, and $\boldsymbol{\tau}_1, \boldsymbol{\tau}_2, \ldots, \boldsymbol{\tau}_n$. The other approach is to integrate the likelihood with respect to the nuisance parameters, i.e., to maximize the integrated likelihood

$$\prod_{i=1}^{n} \int f(\boldsymbol{\tau}_i; \boldsymbol{\theta}) \prod_{j=1}^{n_i} \varphi\left([y_{ij} - h(t_{ij}; \boldsymbol{\tau}_i)]; \boldsymbol{\eta}\right) d\boldsymbol{\tau}_i. \tag{9.7}$$

The EM algorithm and Gibb's sampling (see McLachlan and Krishnan 1997) provide computational methods for solving such problems. Even so, the nonlinear nature of the function h complicates the optimization problem that needs to be solved at each step of an iterative procedure.

As a computationally feasible alternative, we use a heuristic approach. Let $\boldsymbol{\theta}$ be a functional parameter, i.e., well defined for any given distribution with appropriate support. Further, let the observations described in Sect. 9.1 be generated through a random sampling mechanism, which may involve successive stages of generating (a) the parameter $\boldsymbol{\tau}_i$ for an individual, (b) the number of observations n_i for that individual, (c) the times $t_{i1}, \ldots, t_{in_i}$ of size measurement of that individual, and (d) the corresponding size measurements $y_{i1}, \ldots, y_{in_i}$. Consider the "posterior distribution" of the random parameter $\boldsymbol{\tau}_i$ obtained from the usual formula by using the correct (but unknown) distribution as prior. The expected value of this posterior, taken with respect to all the distributions underlying the sampling mechanism described above, is equal to the prior. Therefore, the functional parameter $\boldsymbol{\theta}$ obtained from the prior distribution f is identical to that obtained from the expected posterior.

The expected value of the posterior can be approximated, for the purpose of estimation, by the sample average. Thus, one can define an estimator of $\boldsymbol{\theta}$ by the estimating equation, which equates the functional parameter obtained from the sample average of posteriors to the parameter value used in the prior:

$$\boldsymbol{\theta}(f) = \boldsymbol{\theta}\left(\frac{\sum_{i=1}^{n} n_i g_i\left(\boldsymbol{\tau}_i | y_{i1}, \ldots, y_{in_i}\right)}{\sum_{i=1}^{n} n_i}\right). \tag{9.8}$$

In the above equation, $\boldsymbol{\theta}(\cdot)$ is the functional representation of the parameter $\boldsymbol{\theta}$, and g_i is the posterior distribution based on the ith set of observations (t_{i1}, y_{i1}), ..., (t_{in_i}, y_{i,n_i}),

$$g_i\left(\boldsymbol{\tau}_i | y_{i1}, \ldots, y_{in_i}\right) \propto f(\boldsymbol{\tau}_i; \boldsymbol{\theta}) \prod_{j=1}^{n_i} \varphi\left([y_{ij} - h(t_{ij}; \boldsymbol{\tau}_i)]; \boldsymbol{\eta}\right). \tag{9.9}$$

The estimating equation (9.8) defines a frequentist estimator, even though the argument used to motivate it is based on a posterior representation. The argument also differs from the empirical Bayes approach (see Carlin and Louis 2000), as there is no difference between the prior and the posterior parameters in (9.8).

The estimating equation (9.8) can be solved iteratively. In data sets such as the one described in Sect. 9.3, there would be a longitudinal part, where fitting of individual-specific growth functions would be possible. We can obtain preliminary estimates of $\boldsymbol{\tau}_i$ for these individuals, use these estimated $\boldsymbol{\tau}_i$s to estimate $\boldsymbol{\eta}$ and $\boldsymbol{\theta}$, and substitute the latter in f to get an empirical version of the prior density. We can then iterate over this entire process, by treating the average posterior distribution at a particular step as the prior distribution at the next step, until the "prior" and the average of the posteriors come sufficiently close.

The proposed frequentist method can benefit from the Markov Chain Monte Carlo technique, a tool developed primarily for Bayesian computations. The crux of the problem is to avoid computing the proportionality constant of (9.9), even though the samples need to be drawn from an average of the posterior densities (and not the posterior densities themselves). In order to make this possible, the average of the posterior densities is viewed as a mixture distribution, so that the samples from the targeted density can be obtained by judiciously pooling samples from the posterior densities of the individuals.

The steps to be used, adapted from the Metropolis–Hastings algorithm (Albert 2007), are as follows:

Step I. For those individuals i with sufficiently large n_i (i.e., exceeding the dimension of $\boldsymbol{\tau}_i$), estimate $\boldsymbol{\tau}_i$ through nonlinear least squares (Bates and Watts 2007). Estimate $\boldsymbol{\theta}$ by using these estimates as observed data and denote the estimator by $\boldsymbol{\theta}^{(0)}$. Also estimate $\boldsymbol{\eta}$ from the mean-corrected data pooled from the above individuals and denote the estimator by $\hat{\boldsymbol{\eta}}$. Set the index of iteration $k = 0$.

Step II. Generate samples from the posterior density of $\boldsymbol{\tau}_i$, defined for each individual i by (9.9) with $\boldsymbol{\theta}$ and $\boldsymbol{\eta}$ replaced by $\boldsymbol{\theta}^{(k)}$ and $\hat{\boldsymbol{\eta}}$, respectively, as follows. Generate M samples from a proposal distribution, say $\boldsymbol{\tau}_1^*, \boldsymbol{\tau}_2^*, \ldots, \boldsymbol{\tau}_M^*$. For $j = 1, 2, \ldots, M$, compute

$$r_{ij} = \frac{g_i\left(\boldsymbol{\tau}_j^*|y_{i1},\ldots,y_{in_i};\boldsymbol{\theta}^{(k)},\hat{\boldsymbol{\eta}}\right)}{g_i\left(\boldsymbol{\tau}_0|y_{i1},\ldots,y_{in_i};\boldsymbol{\theta}^{(k)},\hat{\boldsymbol{\eta}}\right)},$$

where $\boldsymbol{\tau}_0$ is the mean of the distribution f for $\boldsymbol{\theta} = \boldsymbol{\theta}^{(k)}$. If $r_{ij} > 1$, accept the sample $\boldsymbol{\tau}_j^*$; else, accept it with probability r_{ij}. Let M_i be the number of selected samples.

Step III. Draw N samples from the average posterior as follows. Let

$$p_i = \frac{n_i I(M_i > 0)}{\sum_{j=1}^n n_j I(M_j > 0)}, \quad \text{for } i = 1, 2, \ldots, n,$$

and $(m_1, m_2, \ldots, m_n)$ be multinomial with parameters $N, p_1, p_2, \ldots, p_n$. Then, for $i = 1, 2, \ldots, n$, the desired sample would consist of m_i samples selected with replacement from the M_i samples generated from the posterior density of $\boldsymbol{\tau}_i$, as mentioned in Step II.

Step IV. Define the updated estimate $\boldsymbol{\theta}^{(k+1)}$ as that obtained from the sample of size N generated in Step III.

Steps II–IV are iterated until the estimates of $\boldsymbol{\theta}$ from successive steps come sufficiently close. The population distribution of any function of $\boldsymbol{\tau}_i$ can be obtained from f evaluated at the converged value of $\boldsymbol{\theta}$.

While stochastic convergence of the iterative procedure has not been established as yet, no instance of non-convergence was found in any of the simulations or data analysis reported in the next two sections.

9.5 Simulation Results

We examine the performance of the proposed estimator in the following special case. We assume that the density φ of the measurement errors is normal with mean 0 and variance σ^2. As for the growth function h, we work with the Preece–Baines model (9.2) having five parameters. The conditions for the simulation study are largely determined by the characteristics of the S–B data mentioned in Sect. 9.3.

Any computational method for the model parameters may be affected by the different orders of their magnitude and the various constraints. It follows from a preliminary analysis of the S-B data that an unconstrained set of parameters of somewhat uniform magnitude is

$$\psi_1 = \log\left(\frac{3s_0}{1-3s_0}\right), \qquad \psi_3 = \frac{\theta}{10},$$

$$\psi_2 = \log\left(\frac{\frac{(3+2\sqrt{2})s_0}{s_1}}{1-\frac{(3+2\sqrt{2})s_0}{s_1}}\right), \qquad \psi_3 = \log(h_{max} - h_\theta),$$

$$\psi_5 = \log(h_\theta).$$

The distribution of these transformed parameters, obtained from the nonlinear least squares fit of the longitudinal part of the S–B data, appeared to be normal, with a rank 2 variance–covariance matrix. Accordingly, the first two principal components, with empirically determined coefficients, were used for the simulations. Thus, the random parameter used here is a vector with two components, such that the parameters $\psi_1, \ldots, \psi_5$ are linear functions of these, and the model parameters s_0, s_1, θ, h_{max} and h_θ are nonlinear functions thereof. Independent normal distributions $N(0, 0.1634)$ and $N(0, 0.0391)$ for the two principal components ν_1 and ν_2 are assumed. The parameter σ^2 of the measurement error distributions is chosen as 1. The equations relating the transformed mathematical parameters to the principal components are as follows:

$$\begin{pmatrix} \psi_1 \\ \psi_2 \\ \psi_3 \\ \psi_4 \\ \psi_5 \end{pmatrix} = \begin{pmatrix} -0.3289 & 0.8108 \\ -0.9365 & -0.3406 \\ 0.0378 & 0.0573 \\ 0.1142 & -0.4710 \\ -0.0186 & 0.0380 \end{pmatrix} \begin{pmatrix} \nu_1 \\ \nu_2 \end{pmatrix} + \begin{pmatrix} -0.8457 \\ -0.0167 \\ 1.4687 \\ 2.5090 \\ 5.0386 \end{pmatrix}.$$

As for estimation of parameters of the distribution of $\boldsymbol{\tau}$, we assume that the distribution is bivariate normal and estimate its parameters through the sample mean and the sample variance–covariance matrix. The parameter σ^2 is estimated from the longitudinal part of the data by averaging over the error sum of squares, after the parameter $\boldsymbol{\tau}$ has been estimated through nonlinear regression, separately for each individual that permits such fitting. The proposal distribution is considered to be the bivariate normal distributions, with mean and dispersion matrix given by the current mean and dispersion matrix of the components of $\boldsymbol{\tau}$.

We seek to address the following questions while evaluating the proposed method for a mixture of longitudinal and cross-sectional data. (a) Is there any value addition to the original estimate (obtained from the longitudinal part of the data) from the cross-sectional part of the data? (b) Would the performance be substantially better if the cross-sectional part of the data are replaced by equivalent amount of longitudinal data? In order to answer these questions, we repeatedly (100 times) generate three types of data. The first type of data consists of 50 individuals each with 10 data points (height at ages 4–21). The second type comprises 10 individuals each with 10 data points and 400 individuals with only one data point (i.e., cross-sectional data). The third type of data is a subset of the second one, where only the longitudinal part of the data are included.

By using the above method, we can find the estimated parameters of the distribution (f) of individual specific biological parameters and study their bias and standard deviation. Table 9.1 gives a comparison of the bias, the standard deviation and the root mean square error (RMSE) of the biological parameters estimated from the three types of data. Table 9.2 gives a similar comparison for the parameters of the mathematical model. As expected, the bias, the standard deviation, and the mean square error for the second type of data (mixture of 20% longitudinal and

Table 9.1 Bias, standard error, and RMSE of estimated biological parameters

Data type	Property of estimator	Age at takeoff (year)	Takeoff velocity (cm/year)	Age at peak velocity (year)	Peak velocity (cm/year)	Final size (cm)
I	Bias	0.080	0.028	−0.027	−0.231	−0.146
	Standard error	0.049	0.024	0.018	0.035	0.079
	RMSE	0.094	0.037	0.032	0.234	0.166
II	Bias	0.605	−0.086	0.521	0.286	1.560
	Standard error	0.494	0.106	0.364	0.088	1.300
	RMSE	0.781	0.136	0.636	0.299	2.031
III	Bias	0.681	−0.119	0.554	0.292	1.927
	Standard error	0.518	0.155	0.368	0.142	1.786
	RMSE	0.856	0.195	0.665	0.325	2.627

Table 9.2 Bias, standard error, and RMSE of estimated mathematical parameters

Data type	Property of estimator	s_0 (cm/year)	s_1 (cm/year)	θ (year)	h_{max} (cm)	h_θ (cm)
I	Bias	−0.001	−0.035	0.018	−0.146	−0.190
	Standard error	0.001	0.017	0.015	0.079	0.149
	RMSE	0.001	0.039	0.023	0.166	0.241
II	Bias	0.004	0.063	0.496	1.560	1.832
	Standard error	0.003	0.028	0.364	1.300	1.485
	RMSE	0.005	0.069	0.615	2.031	2.358
III	Bias	0.006	0.072	0.534	1.927	2.315
	Standard error	0.004	0.054	0.382	1.786	1.925
	RMSE	0.007	0.090	0.657	2.627	3.011

80% cross-sectional data) lie in between those of the other two types. Substantial reduction in bias and variance is found to have occurred from the use of the cross-sectional part of the data, though an equivalent amount of additional longitudinal data would have produced even greater improvement.

9.6 Data Analysis

As for the male subjects of the Sarshuna–Barisha data, there are 36 cases with 10–18 data points in the range 7–18 years, where estimation of the model parameters subject to the constraint (9.5) is possible. For the remaining 262 cases, there are many with only a few observations, including 16 cases with only one observation. Among the female subjects, there are 30 cases where initial model fitting is possible. For the remaining 223 cases, there are many with only a few observations, including 16 cases with only one observation. This makes it difficult to estimate the population distribution of individual specific biological parameters, using conventional

Table 9.3 Mean and standard error of estimated biological and mathematical parameters, using full data and some part of data set for boys

	All 298 cases		36 cases of longitudinal data	
Parameter	Mean	Standard error	Mean	Standard error
Age at takeoff (year)	10.27	0.0813	10.49	0.5531
Takeoff velocity (cm/year)	4.012	0.0419	4.115	0.1805
Age at peak velocity (year)	14.33	0.0298	14.32	0.1761
Peak velocity (cm/year)	7.908	0.0528	8.370	0.2454
Final size (cm)	166.2	0.1020	166.5	0.4662
s_0 (cm/year)	0.0934	0.0020	0.1007	0.0125
s_1 (cm/year)	1.100	0.0210	1.176	0.1445
θ (year)	14.75	0.0299	14.73	0.1545
h_{max} (cm)	166.2	0.1020	166.5	0.4662
h_θ (cm)	153.4	0.2453	154.3	1.3462

Table 9.4 Mean and standard error of estimated biological and mathematical parameters, using full data and some part of data set for girls

	All 253 cases		30 cases of longitudinal data	
Parameter	Mean	Standard error	Mean	Standard error
Age at takeoff (year)	9.180	0.4207	8.407	0.9094
Takeoff velocity (cm/year)	3.864	0.1293	3.983	0.3971
Age at peak velocity (year)	12.42	0.1220	11.81	0.7370
Peak velocity (cm/year)	7.786	0.1886	7.331	0.6789
Final size (cm)	149.0	0.3054	147.9	0.7496
s_0 (cm/year)	0.122	0.0080	0.092	0.0128
s_1 (cm/year)	1.334	0.0923	1.165	0.2622
θ (year)	12.85	0.1066	12.21	0.1965
h_{max} (cm)	149.0	0.3054	147.9	0.7496
h_θ (cm)	138.8	0.9393	136.7	1.4124

methods. Application of the method proposed in this paper gives rise to the summary of mean and standard error of biological and mathematical parameters, reported in Table 9.3 for boys data and in Table 9.4 for girls data. For comparison, the summary from the longitudinal part of the data is also reported alongside. The standard deviations show substantial improvement when the additional cases (262 for boys; 223 for girls) are included in the analysis.

As a further illustration of the utility of the proposed approach, we test for significance of the mean difference between boys and girls in the Sarshuna–Barisha data for the different parameters. Table 9.5 summarizes the results of the tests carried out from the 36 longitudinal samples for boys and the 30 longitudinal cases for girls. The corresponding results for the entire data set (298 boys and 253 girls) are reported alongside. It is found that the mean difference for two parameters (s_0 and s_1) are found to be significant only when the full data set is used. For the mean differences of all the parameters, the p-values computed from the full data set are found to be smaller.

Table 9.5 Results of Z test of mean difference between boys and girls in respect of the five biological and mathematical parameters, using full data and some parts of data

	All data			Selected longitudinal data		
Parameter	Mean difference	Z value	2 sided p value	Mean difference	Z value	2 sided p value
Age at TO (year)	1.087	2.537	0.0111	2.087	1.960	0.0500
TO velocity (cm/year)	0.148	1.091	0.2751	0.132	0.303	0.7616
Age at PV (year)	1.913	15.25	0.0000	3.132	4.134	0.0000
PV (cm/year)	0.122	0.625	0.5320	1.039	1.439	0.1503
Final size (cm)	17.28	53.68	0.0000	18.58	21.04	0.0000
s_0 (cm/year)	−0.029	−3.449	0.0006	0.009	0.489	0.6248
s_1 (cm/year)	−0.234	−2.471	0.0135	0.011	0.036	0.9713
θ (year)	1.902	17.18	0.0000	2.514	10.06	0.0000
h_{max} (cm)	17.28	53.68	0.0000	18.57	21.44	0.0000
h_θ (cm)	14.66	15.10	0.0000	17.60	9.021	0.0000

References

Albert, J. (2007). *Bayesian computation with R*. New York: Springer.

Bates, D. M., & Watts, D. G. (2007). *Nonlinear regression analysis and its applications*. New York: Wiley.

Carlin, B. P., & Louis, T. A. (2000). *Bayes and empirical Bayes methods for data analysis*. Boca Raton: Chapman and Hall.

Cnaan, A., Laird, N. M., & Slasor, P. (1997). Using the general linear mixed model to analyse unbalanced repeated measures and longitudinal data. *Statistics in Medicine, 16*, 2349–2380.

Das, S. R., Mukherjee, D. P., Bose, L., & Das, A. (1985). *Mixed longitudinal growth data for 22 measures: The Sarsuna Barisha series*. Calcutta: Anthropological Survey of India.

Dasgupta, P., & Hauspie, R. (2001). *Perspectives in human growth, development and maturation*. Dordrecht, The Netherlands: Kluwer.

Diggle, P., Heagerty, P., Liang, K. -Y., & Zeger, S. L. (2002). *Analysis of longitudinal data*. Oxford: Oxford University Press.

Donnet, S., Foulley, J. L., & Samson, A. (2010). A Bayesian analysis of growth curves using mixed models defined by stochastic differential equations. *Biometrics, 66*, 733–741.

Duncan, T. E., Duncan, S. C., Strycker, L. A., Li, F., & Alpert, A. (2006). *An introduction to latent variable growth curve modeling: Concepts, issues and applications*. Mahwah: Lawrence Erlbaum Associates.

Falkner, F., & Tanner, J. M. (1986). *Human growth: A comprehensive treatise* (Vol. 3, 2nd ed.). New York: Plenum Press.

Hauspie, R. C., Cameron, N., & Molinari, L. (2004). *Methods in human growth research*. Cambridge: Cambridge University Press.

Huggins, R. M., & Loesch, D. Z. (1998). On the analysis of mixed longitudinal growth data. *Biometrics, 54*, 583–595.

Laird, N. M., & Ware, J. H. (1982). Random effects models for longitudinal data: An overview of recent results. *Biometrics, 38*, 963–974.

McLachlan, G., & Krishnan, T. (1997). *The EM algorithm and extensions*. New York: Wiley.

Potthoff, F., & Roy, S. N. (1964). A generalized multivariate analysis of variance model useful specially for growth curve problems. *Biometrika, 51*, 313–326.

Preece, M. A., & Baines, M. J. (1978). A new family of mathematical models describing the human growth curve. *Annals of Human Biology, 5*, 1–24.
Rao, C. R. (1965). The theory of least squares when the parameters are stochastic and its application to the analysis of growth curves. *Biometrika, 52*, 447–458.
Verbeke, G., Spiessen, B., & Lesaffre, E. (2001). Conditional linear mixed models. *The American Statistician, 55*, 25–34.

Chapter 10
Tuber Crop Growth and Pareto Model

Ratan Dasgupta

Abstract Pareto distribution has wide applications in natural sciences. We show that yield of some crops may be modeled by a Pareto like density from growth viewpoint of additional tubers. Extending a result of Dasgupta (Discrete distributions with application to lifestyle data. In International conference on productivity, quality, reliability, optimization and modeling proceedings (Vol. 1, pp. 502–520), New Delhi: Allied Publishers, 2011), we obtain an analytic expression for density of the variable that is a mixture of exponential densities relevant in real life situations; a beta prior induced on exponential of intensity function for variables with memoryless property (viz., exponential random variables) results in a heavy tailed distribution much like a Pareto variable. This may be appropriate for modeling some real life situations when memoryless property of a variable may hold only in subgroups. The results are applied to model yield of tuber crops with real data sets.

MS subject classification 2010: Primary: 62E15, secondary: 62P99

10.1 Introduction

Pareto model may be used to explain real data set arising out of different branches of science, especially in economics and actuarial studies, e.g. see Krishnaji (1970) for modeling underreported income. In Dasgupta (2011), it is shown that exponential distribution in different subgroups with exponential of intensity following a beta distribution may result in a heavy tailed Pareto like distribution for aggregated data. Such a situation may arise when an exponential model may seem appropriate

R. Dasgupta (✉)
Theoretical Statistics and Mathematics Unit, Indian Statistical Institute, Kolkata, India
e-mail: ratandasgupta@gmail.com

R. Dasgupta (ed.), *Advances in Growth Curve Models*, Springer Proceedings in Mathematics & Statistics 46, DOI 10.1007/978-1-4614-6862-2_10,

from theoretical consideration, although a heavy tailed distribution may fit real data set more closely. Yield data of Elephant foot yam crossing a threshold is seen to exhibit Pareto like polynomial decay towards the tail of the distribution, see Dasgupta (2013). Small value of the estimated Pareto index indicates a heavy tailed distribution for yield i.e., abundance of crops.

An important tuber crop is potato. The above ground biomass of potato plant is an indicator of final yield after harvest. Following centuries of selective breeding from the wild variety, there are now over a thousand different types of potatoes. Of these subspecies, a variety that at one point grew in the Chiloe Archipelago left its germplasm on over 99% of the cultivated potatoes worldwide.

Potato plant growth may be divided into five phases. During the first phase, sprouts emerge from the seed potatoes and root growth at the base of emerging sprouts begins. During the second, photosynthesis begins as the plant develops leaves and branches, roots and stolons develop at below ground nodes. In the third phase stolons develop further and tubers develop as swellings at stolon tips. At the fourth step tuber cells expand with accumulation of water, nutrients and carbohydrates, as their sugar content convert to starches. In the final phase of maturation, photosynthesis decreases, plant canopy dies, tuber growth slows, tuber dry matter content reaches to a maximum, and the tuber skins harden. New tubers may arise at the soil surface. Since exposure to light leads to greening of the skins and the development of toxic solanine, such tubers need to be covered either by piling additional soil around the base or by mulching with straw, plastic, etc., as these continue emerging.

Growth of additional tubers from a well-nourished plant at later stages irrespective of those already developed resembles memoryless property of a distribution. Plants may grow more tubers underground to store carbohydrates, independent of number of tubers already developed, when sufficient nutrients are available. Such a model is plausible, as groomed potato plants are seen to have more tubers. For continuous distributions this memoryless property is shared only by exponential distribution.

Potato plant is a low-growing, branching perennial herb with weak stems, a leaf is divided into five to nine oval leaflets; these constitute above ground biomass, an indicator of underground yield. In a related study, yield of Elephant foot yam is seen to be highly correlated with stem-growth above ground.

The potatoes are large tubers food-storing bodies that grow from the end of underground stems, below the fibrous roots. Each tuber bears several buds, i.e., eyes of the potato, from which new plants grow.

In the eighteenth century potato began to replace cereals in the diet of the poor. The failure of this staple food in Ireland in the years 1845–1846 due to plant decease resulted in a severe famine. Pareto index estimated from the tail of yield data may reveal a crop production scenario when such a theoretical model fits appropriately to real data set. Lower values of index indicate better production scenarios.

As already mentioned, an exponential random variable with random intensity following a beta prior results in a heavy tailed distribution much like a Pareto variable. Here memoryless property holds for an individual with fixed intensity.

The property is lost when aggregated over individuals having different intensities, resulting in a heavy tailed distribution. The model is seen to be appropriate in some specific cases like yield of Elephant foot yam with weight exceeding a threshold value, among others including environmental data.

In this paper we obtain an expression of the above-mentioned mixture density in terms of digamma function, providing a compact analytic form of resultant density in a nice form. We further check the appropriateness of the mixture model with real data set arising out of growth experiments on potato conducted at Indian Statistical Institute's Giridih farmhouse in Jharkhand. Comparison of two crop production scenarios is also made by the proposed model.

The paper is organized as follows. In Sect. 10.2 we discuss modeling problem related to yield of potato and obtain an expression of the density having decay like a Pareto density. Section 10.3 deals with Pareto fit for two sets of yield data on a tuber crop potato. Section 10.4 provides a discussion on the results obtained along with applications.

10.2 Modeling Crop Yield and Derivation of Compound Density

The number and size of tubers arising out of underground stems vary over potato plants. However, for a particular plant the sizes of tuber are more or less similar. Additional underground stems may develop in a potato plant at a later stage irrespective of how many such stems the plant already developed. Total size of tubers in a particular plant may then follow an exponential variable with an intensity specific to the plant. This gives rise to the possibility that the yield of potato over plants is a mixture of exponential variables having intensities specific to the plants.

We prove the following result extending Theorem 2 of Dasgupta (2011).

Theorem 1. *Let the random variable X be exponentially distributed with density function $g(x|\theta) = (-\log\theta)\theta^x,\ x > 0, 0 < \theta < 1$, where θ has a prior beta density*

$$f_{\alpha,\beta}(\theta) = \frac{\Gamma(\alpha+\beta)}{\Gamma(\alpha)\Gamma(\beta)}\theta^{\alpha-1}(1-\theta)^{\beta-1},\ \alpha > 0,\ \beta > 0.$$

Then the marginal distribution of X is given by

$$f(x) = \frac{\Gamma(\alpha+\beta)}{\Gamma(\alpha)}\frac{\Gamma(\alpha+x)}{\Gamma(\alpha+x+\beta)}[\psi_0(\alpha+x+\beta) - \psi_0(\alpha+x)]$$

where $\psi_0 = \Gamma'/\Gamma$ is a digamma function.

Proof. Write,

$$\frac{d}{dx}\int_0^1 \theta^{\alpha-1}(1-\theta)^{\beta-1}\theta^x d\theta = \frac{d}{dx}\frac{\Gamma(\alpha+x)\Gamma(\beta)}{\Gamma(\alpha+x+\beta)}$$

$$= \Gamma(\beta)\frac{\Gamma'(\alpha+x)\Gamma(\alpha+x+\beta)-\Gamma(\alpha+x)\Gamma'(\alpha+x+\beta)}{(\Gamma(\alpha+x+\beta))^2} \quad (10.1)$$

Thus the required density

$$f(x) = \frac{\Gamma(\alpha+\beta)}{\Gamma(\alpha)\Gamma(\beta)}\int_0^1 \theta^{\alpha-1}(1-\theta)^{\beta-1}g(x|\theta)d\theta$$

$$= \frac{\Gamma(\alpha+\beta)}{\Gamma(\alpha)\Gamma(\beta)}\int_0^1 \theta^{\alpha-1}(1-\theta)^{\beta-1}(-\log\theta)\theta^x d\theta$$

$$= \frac{\Gamma(\alpha+\beta)}{\Gamma(\alpha)}\frac{\Gamma(\alpha+x)\Gamma'(\alpha+x+\beta)-\Gamma'(\alpha+x)\Gamma(\alpha+x+\beta)}{(\Gamma(\alpha+x+\beta))^2}$$

$$= \frac{\Gamma(\alpha+\beta)}{\Gamma(\alpha)}\frac{\Gamma(\alpha+x)}{\Gamma(\alpha+x+\beta)}[\psi_0(\alpha+x+\beta)-\psi_0(\alpha+x)] \quad (10.2)$$

where $\psi_0(x) = \frac{\Gamma'(x)}{\Gamma(x)} = \frac{d}{dx}\log\Gamma(x)$ is digamma function and $\psi_0(x)$ is increasing for $x > 0$. See e.g., Arfken (1985) for properties of digamma and polygamma functions that appear in the expression of f and its derivatives. The density function f in (10.2) is continuous.

Remark 1. One may consider variate values crossing a threshold $x > c$ in the above formulation. Exponential distribution possesses memoryless property, and the form of the resultant density in (10.2) remains same, with x replaced by $x - c(> 0)$. The initial value of c may be estimated from the observed minimum in the data set. The values may then be modified sequentially to have a better fit towards the tail of the distribution.

In Dasgupta (2011) by a series expansion it is shown that (10.2) has decay similar to Pareto density, viz.,

$$f(x) = \frac{\Gamma(\alpha+\beta)}{\Gamma(\alpha)\Gamma(\beta)}\frac{\Gamma(\alpha+x)\Gamma(\beta+1)}{\Gamma(\alpha+x+\beta+1)} + \frac{1}{2}\frac{\Gamma(\alpha+\beta)}{\Gamma(\alpha)\Gamma(\beta)}\frac{\Gamma(\alpha+x)\Gamma(\beta+2)}{\Gamma(\alpha+x+\beta+2)}$$
$$+\frac{1}{3}\frac{\Gamma(\alpha+\beta)}{\Gamma(\alpha)\Gamma(\beta)}\frac{\Gamma(\alpha+x)\Gamma(\beta+3)}{\Gamma(\alpha+x+\beta+3)} + \ldots = O_e((x+\alpha)^{-(\beta+1)}), x > 0 \quad (10.3)$$

Remark 2. Mode of the distribution (10.2) is at the starting point. Differentiating (10.2) with respect to $x(> 0)$, one gets

$$f'(x) = -\frac{\Gamma(\alpha+\beta)}{\Gamma(\alpha)\Gamma(\beta)}\int_0^1 \theta^{\alpha-1}(1-\theta)^{\beta-1}(\log\theta)^2\theta^x d\theta \quad (10.4)$$

$f'(x) < 0$, for all $x > 0$. From (10.3) one gets,

$$f(0) = \frac{\beta}{\alpha+\beta}[1 + \frac{1}{2}\frac{\beta+1}{\alpha+\beta+1} + \frac{1}{3}\frac{\beta+1}{\alpha+\beta+1}\frac{\beta+2}{\alpha+\beta+2} + \frac{1}{4}\frac{\beta+1}{\alpha+\beta+1}\frac{\beta+2}{\alpha+\beta+2}\frac{\beta+3}{\alpha+\beta+3} + \ldots] \quad (10.5)$$

One may estimate the unknown parameters appearing in the prior used in mixture density from observed data.

Parameter β, the Pareto index may be estimated from the tail behavior of the observed frequency distribution. An estimate of α may then be obtained from expression in (10.2) or, (10.5) coupled with height of the observed histogram near the starting point as origin (or, c).

10.3 Fitting the Model and Analysis of Yield Data

We analyze two sets of potato yield data from Indian Statistical Institute, Giridih. Seed size varies from 10 to 30 g. Plantation date for the first set of 100 seeds in a fertile piece of farmland in Indian Statistical Institute was 19 January 2012 and harvesting date was 15 April 2012. The bunch of tubers (potatoes) harvested in each plant are properly developed, except for plantations nos. 4 and 37 having nil yield possibly due to infection incurred.

The second set of 100 seed were planted to a different piece of barren unfertile land in main Rosevilla campus of Indian Statistical Institute, cultivated for the first time. Plantation date for this set is 29 January 2012 and harvesting date is 16 April 2012. The tubers developed in this unfertile land are undernourished and relatively smaller in size than the first set. Also the period of growth is comparatively short for the second set (about 10 days less than that for first set). Weather in April is harsh in Giridih, Jharkhand. Most of the plant canopies in two experiments dried up at the time of harvest.

The two set of yield observations in grams are as follows:

Data set 1

142, 242, 128, 0, 66, 310, 178, 170, 128, 156, 92, 74, 136, 206, 244, 24, 184, 4, 122, 96, 490, 312, 86, 242, 254, 136, 68, 192, 56, 1, 182, 4, 22, 244, 34, 62, 0, 112, 94, 280, 108, 520, 288, 66, 418, 156, 238, 126, 110, 224, 192, 350, 318, 30, 382, 184, 234, 48, 222, 66, 2, 106, 170, 18, 164, 70, 44, 236, 1, 212, 32, 134, 18, 12, 44, 10, 84, 154, 202, 90, 104, 28, 104, 192, 254, 304, 152, 78, 162, 166, 104, 246, 278, 1, 336, 142, 140, 128, 80, 492

Data set 2

32, 70, 30, 20, 16, 104, 64, 16, 52, 48, 4, 80, 70, 50, 46, 70, 66, 32, 6, 18, 28, 14, 42, 50, 32, 80, 40, 82, 26, 10, 82, 32, 22, 2, 46, 68, 26, 22, 32, 10, 44, 16, 88, 46, 12, 6, 25, 10, 10, 96, 58, 94, 66, 98, 62, 42, 52, 8, 1, 2, 12, 60, 92, 1, 60, 12, 66, 42, 22, 28, 20, 8, 46, 70, 72, 24, 6, 16, 26, 24, 56, 14, 28, 34, 26, 42, 6, 18, 66, 32, 62, 52, 12, 6, 64, 48, 8, 30, 30, 46

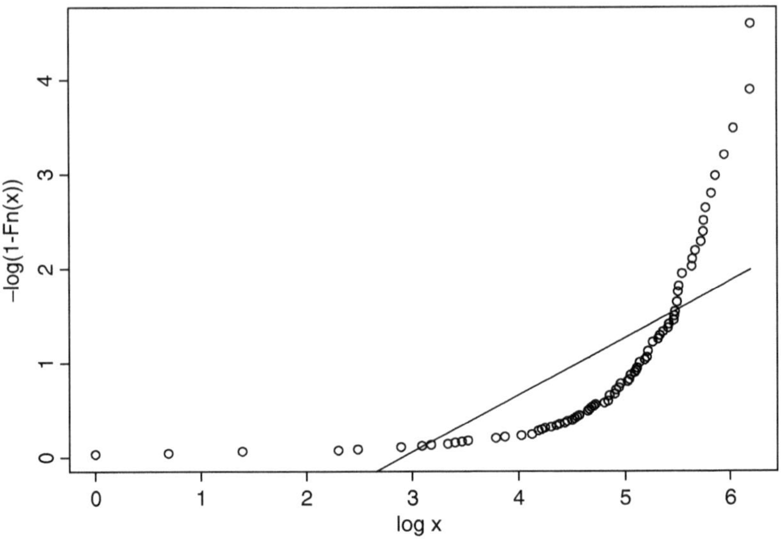

Fig. 10.1 Pareto fit for potato yield Expt. 1

Figure 10.1 shows the data set 1 (of Experiment no. 1) along with linear regression fit for Pareto model; R^2 of linear regression with 98 non zero yield (two nil yield excluded) is low, $R^2 = 0.5127$.

Next we fit a Pareto model to the yield values crossing a threshold, we check whether these relatively high yield do conform to the assumption on exponential distribution with random intensity specific to the plant. Such a possibility arises in view of the fact that homogeneous and fully developed tubers are seen to be present in well-nourished plants in a farmland in the middle of season, with a possibility of growing more tubers in plants. In presence of conducive environmental factors such as darkness and humidity for development of additional stolons at a later stage, additional tubers in a plant may grow in conformity of the memoryless property of an exponential distribution. In a recent experiment it is seen that such a memoryless property holds in case of potato tubers.

In Fig. 10.2, R^2 of Pareto fit in linear regression for the values $\{x : \log x > 5\}$ in Experiment no. 1 is quite high, $R^2 = 0.9508$. Estimated value of the slope is $\beta = 2.8125$.

To obtain maximum likelihood estimate (mle) of the index one may use the result that if X is Pareto-distributed with minimum a and index β, then $Y = \log(X/a)$ is exponentially distributed with intensity β. One may estimate the intensity from the mean of transformed variables. Thus the mle of the Pareto index based on the data set $\{x\colon \log x>5\}=(152, 154, 156, 156, 162, 164, 166, 170, 170, 178, 182, 184, 184, 192, 192, 192, 202, 206, 212, 222, 224, 234, 236, 238, 242, 242, 244, 244, 246, 254, 254, 278, 280, 288, 304, 310, 312, 318, 336, 350, 382, 418, 490, 492, 520)$ after

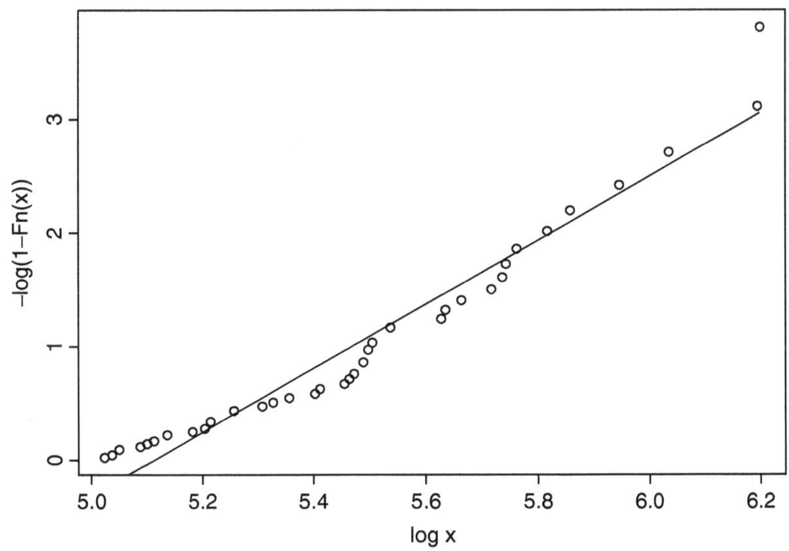

Fig. 10.2 Pareto fit for potato yield Expt. 1: log x > 5

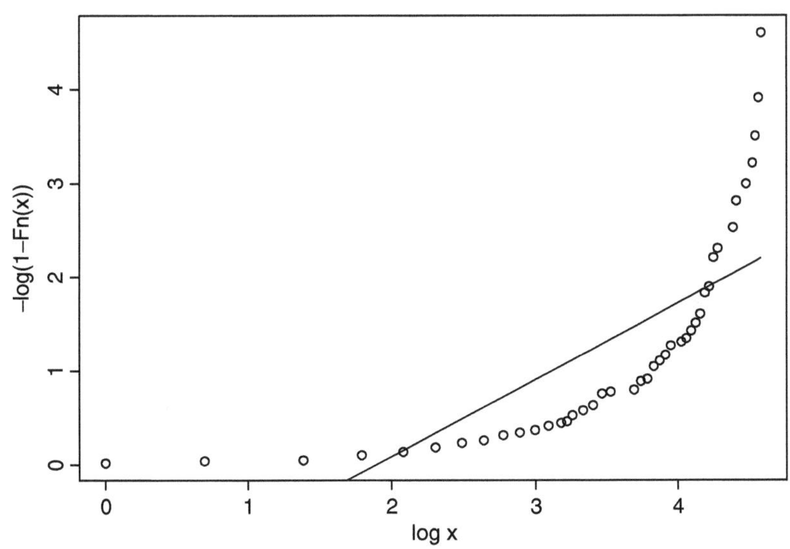

Fig. 10.3 Pareto fit for potato yield Expt. 2

trimming the upper one and lower four values i.e., deleting the extreme observations 152, 154, 156, 156, and 520 is $\beta = 1/0.384784 = 2.598861$, with $a = 162$.

Data set of Experiment no. 2, along with linear regression fit for Pareto model, is shown in Fig. 10.3. R^2 of linear regression with 100 nonzero yield is low, $R^2 = 0.5638$.

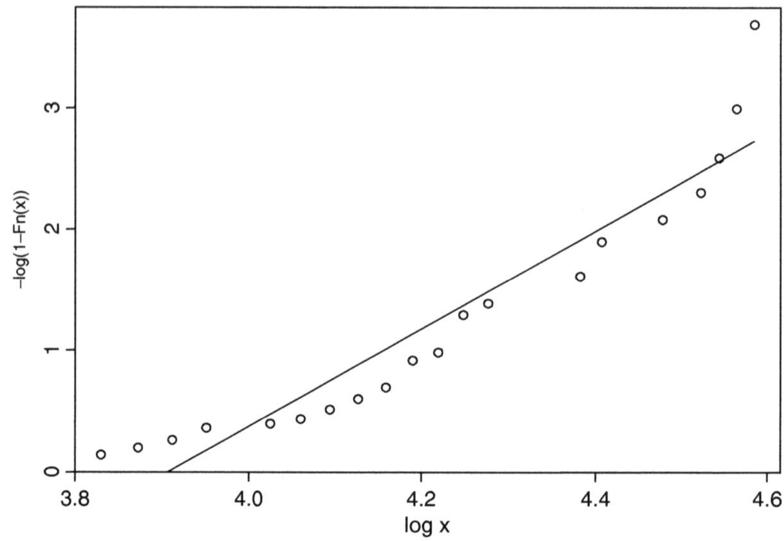

Fig. 10.4 Pareto fit for potato yield Expt. 2: log x > 3.8

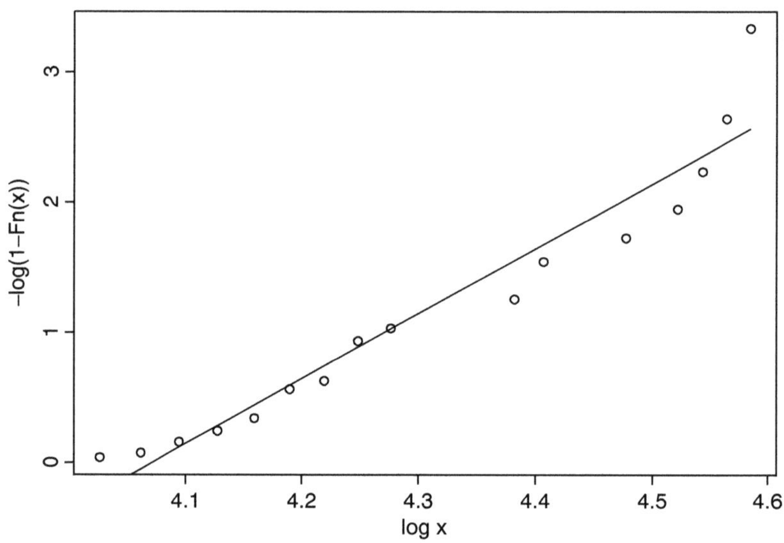

Fig. 10.5 Pareto fit for potato yield Expt. 2: log x > 4

In Fig. 10.4, R^2 of linear regression for the values $\{x : \log x > 3.8\}$ in Experiment no.2 is 0.898.

The value of R^2 improves with slight increase in the threshold value. In Fig. 10.5, R^2 of linear regression for the values $\{x : \log x > 4\}$ in Experiment no. 2 is 0.9296. Estimated value of the slope is $\beta = 4.9834$.

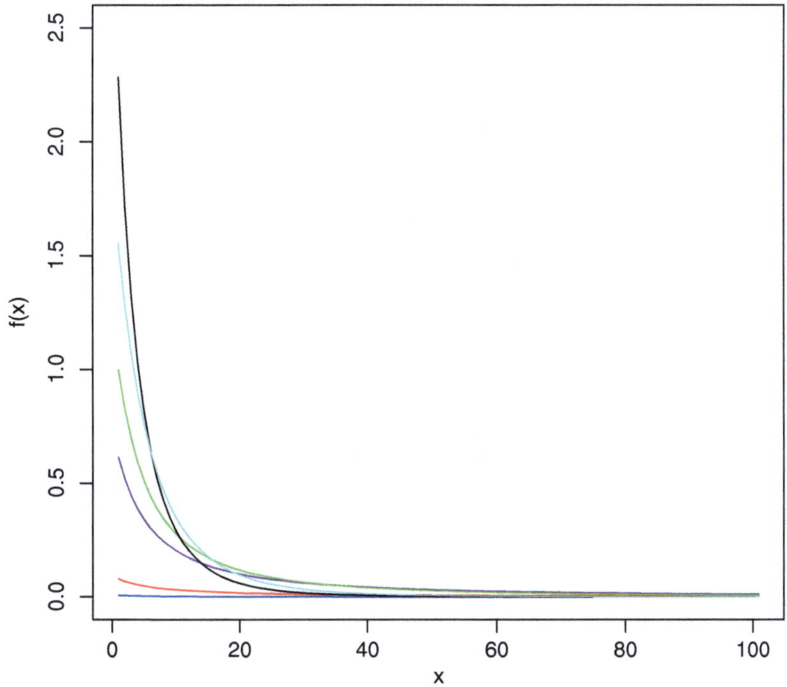

Fig. 10.6 Mixture density function

In a similar manner the mle of the Pareto index based on the data set $\{x : \log x > 4\}$=(56, 58, 60, 60, 62, 62, 64, 64, 66, 66, 66, 66, 68, 70, 70, 70, 70, 72, 80, 80, 82, 82, 88, 92, 94, 96, 98, 104) after trimming the upper two and lower two values i.e., deleting the extreme observations 56, 58 and 98, 104 is $\beta = 1/0.184562 = 5.418233$, with $a = 60$.

The two indices corresponding to two different yield scenarios may be compared to judge for the better yield. The Pareto index is unchanged with a different initial point of start, the conditional probability distribution of a Pareto-distributed random variable, given the event that it is greater than or equal to a particular number a_1 exceeding a, is a Pareto distribution with the same Pareto index, but with minimum a_1 instead of a.

For the first data set the mle of the Pareto index is 2.598861, which is lower than that for the second set with index 5.418233. Thus the first production scenario is superior in a Pareto model-based analysis, as yield distribution has a heavier tail in the first case. An index of performance may be considered as the ratio of Pareto exponents, viz., $e_{1,2} = \beta_2/\beta_1$. For the present case $\hat{e}_{1,2} = 2.084849$. Value of threshold c is also higher in the first data set.

Figure 10.6 plots some representative densities of the form (10.3) for different values of (α, β). The graph nearest to the x axis is with $(\alpha = 1, \beta = .005)$. Peak of the densities sequentially increases from second to sixth graphs with the following

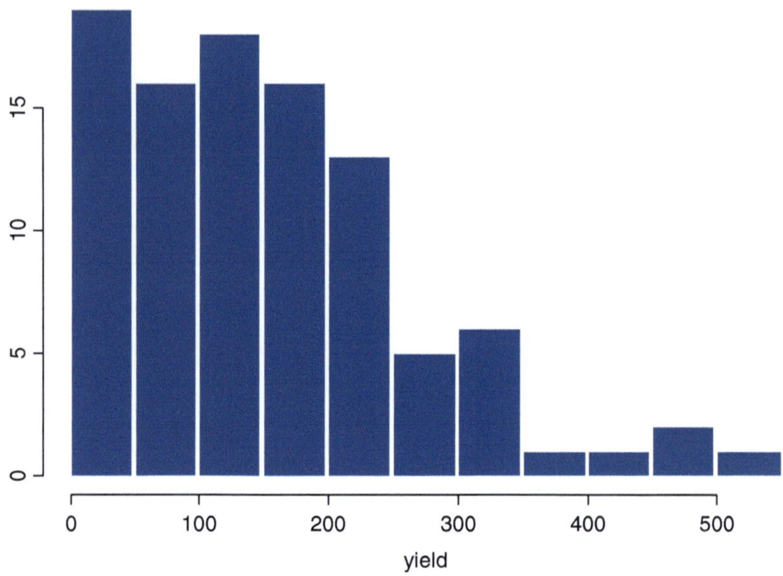

Fig. 10.7 Histogram of yield: data set 1

changes in the values of parameters: $(1, .05), (1, .5), (1, 1), (1.8, 5), (1, 5)$. The case of equal ignorance for the intensity i.e., ($\alpha = 1, \beta = 1$) corresponds to the fourth graph from below.

Histograms of data sets 1 and 2 are shown in Figs. 10.7 and 10.8, respectively. Once the values of β are obtained from the tail behavior of the distribution, the other parameter α may be obtained by equating the height of the relative histogram observed with $f(0)$ given in (10.3)/(10.5). Figures 10.7 and 10.8 suggest that $\hat{f}(0)$ is smaller for data set 1 compared to that for data set 2; the values are 0.35 and 1.55, respectively. This indicates the modeled yield-distribution is of heavier tail for first experiment than the second, suggesting that the yield scenario in first experiment is better.

We next proceed to see the appropriateness of the mixture model (10.2)/(10.3) over entire data sets. As mentioned before, fixing the value of β from the tail behavior, we start with an initial value α by equating the heights of observed relative histogram with the theoretical value and iteratively proceed to obtain an estimate of α that minimizes the Kolmogorov–Smirnov distance.

For data set 1 with 98 nonzero observations, the estimated value of α so obtained is $\hat{\alpha} = 349$ and the value of K–S statistic is $\sqrt{n} D_n = \sqrt{98}(0.1162808) = 1.151121$ that is insignificant at 10% level. Empirical cdf and theoretical cdf with estimated parameters $\alpha = 349, \ \beta = 2.598861$ are shown in Fig. 10.9.

Similarly, for data set 2 all the 100 observations are nonzero and the estimated value of α to minimize K–S distance is $\hat{\alpha} = 83$ and the resultant value of K–S statistic is $\sqrt{n} D_n = \sqrt{100}(0.12361) = 1.2361$, is barely significant at 10% level;

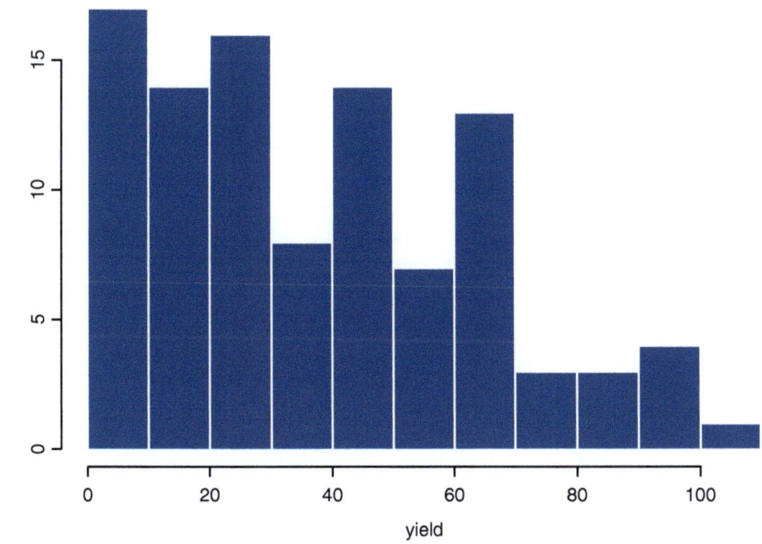

Fig. 10.8 Histogram of yield: data set 2

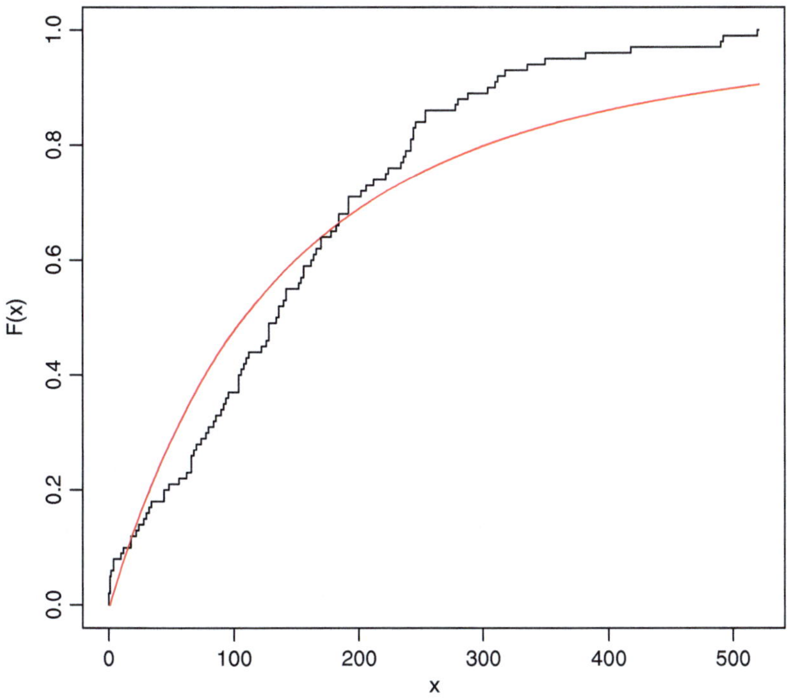

Fig. 10.9 Empirical and theoretical cdf: data set 1

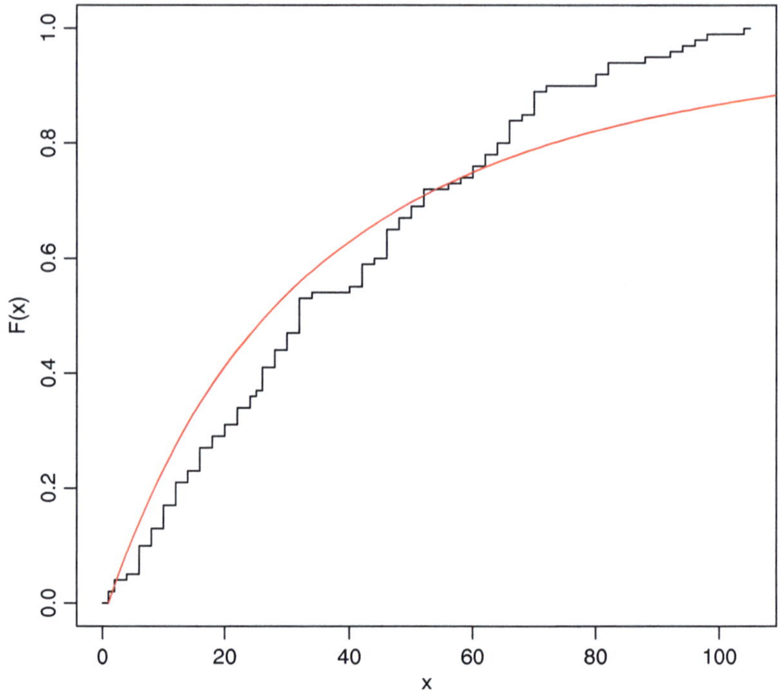

Fig. 10.10 Empirical and theoretical cdf: data set 2

K–S statistic at 10% level is 1.23. Empirical cdf and theoretical cdf with estimated parameters $\alpha = 83,\ \beta = 5.418233$ are shown in Fig. 10.10.

The mixture model provides satisfactory fit to observed data sets 1 and 2.

10.4 Discussion

In some real-life situations exponential distribution having memoryless property may seem to be an appropriate model. However, realized data may sometimes suggest that a heavy tailed distribution could provide a better fit. In this paper we derive a precise expression, in terms of digamma function, for the density function of a variable which is a mixture of exponential density having random intensity related to a beta distribution. The derived family of distributions start from origin (or some positive point c) like an exponential random variable and have a relatively slower rate of decay towards the tail; much like polynomial decay of a Pareto random variable. The peak of the density $f(x) = O_e((x+\alpha)^{-(\beta+1)}), x > 0$, increases at origin; $f(x) \uparrow \infty$ as $x \downarrow 0$ and $\alpha \downarrow 0$. This includes the possibility of modeling variables with high concentration near the starting point, and polynomial decay of density towards tail.

From (10.3) it is clear that from a large value of truncation point onward, the decay of the mixture density is polynomial, like Pareto. And Pareto has this nice property: truncation towards tail is once again Pareto with same index β, but different initial point. So, if the mixture model is appropriate for a data set, then gradually moving towards the tail, there will be some stage where Pareto fit via straight line of regression on remaining data points exceeding a threshold will show a very high value of R^2, providing a reliable estimate of index β for Pareto. The same index also appears in the mixture model in (10.2)/(10.3). This is what is achieved in Figs. 10.1–10.5, for gradually moving truncation points. One may not choose a very high threshold, which may reduce the number of data points exceeding that high value, thereby reducing the efficacy of β estimate based on small number of observations. Now the value of other parameter (α) in the mixture model is estimated via minimizing the K–S statistics. The benefit is twofold, it shows if the mixture model is good over the whole range of data (when K–S value is low) and provides corresponding reliable estimate of α; thus specifying the model.

For some data we had to go for a high threshold, for some not so high. The mixture densities are shown in Fig. 10.6. There are many cases where Pareto has a good fit towards tail (may be due to some change in the behavior of random variable, like a phase change at high values), although for moderate values the variable may behave differently; such phenomena also occurs for some high temperature states.

The results may be applied to symmetric random variables via a mirror image of the density considered. This has polynomially decaying density towards both tails, in contrast to exponentially decaying Laplace density. Such modeling of tuber-crop yield data leads to a comparison of production scenarios in two different locations in terms of Pareto index associated with the model.

References

Arfken, G. (1985). Digamma and polygamma functions. In *Mathematical methods for physicists*, Sect. 10.2 (3rd ed., pp. 549–555). Orlando: Academic.

Dasgupta, R. (2011). Discrete distributions with application to lifestyle data. In *International conference on productivity, quality, reliability, optimization and modeling proceedings* (Vol. 1, pp. 502–520). New Delhi: Allied Publishers.

Dasgupta, R. (2013). Characterization theorems based on conditional quantiles with applications. Proceedings of TIES 2012 conference. *Journal of Environmental Statistics*, *4*(6), in press.

Krishnaji, N. (1970). Characterization of the Pareto distribution through a model of underreported incomes. *Econometrica, 38*, 251–255.

Chapter 11
Effect of Past Demographic Events on the mtDNA Diversity Among the Adi Tribe of Arunachal Pradesh, India

S. Krithika and T.S. Vasulu

Abstract The effect of past demographic upheavals of population growth and evolutionary changes in population size on the genetic diversity of the regional sub-tribes of *Adi,* a Tibeto-Burman speaking population of central Arunachal Pradesh, India, are studied. Demographic expansion or population growth experienced in the past from a state of equilibrium is examined by comparison of the Mitochondrial DNA sequences via different statistical techniques including Tajima's D statistic.

11.1 Introduction

Population genetics deals with change in gene frequency and its influencing factors over generations. Type and mode of such evolutionary changes may be examined from Hardy-Weinberg (HW) law. Deviation from the law takes into account the changes that can be expected under each of the evolutionary forces and their interaction. In addition, the genetic diversity may also be affected by population growth or variation in population size (Tajima 1989a; Rogers and Harpending 1992). In general, changes in population size and other demographic variables exhibit similar pattern of genetic diversity that is indistinguishable from the influences of selection and evolutionary factors (Tajima 1989a; Stajich and Hahn 2005; Li 2011). To understand the relative roles of demographic events and the influences of evolutionary forces on the genetic diversity it is important to get an insight into the antiquity and origin of Man at the global level and to investigate microevolution

S. Krithika (✉)
University of Toronto, Toronto, Mississagua, ON, Canada
e-mail: krithika.sundararaman@utoronto.ca

T.S. Vasulu
Indian Statistical Institute, Kolkata, India
e-mail: vasulu@gmail.com

R. Dasgupta (ed.), *Advances in Growth Curve Models*, Springer Proceedings in Mathematics & Statistics 46, DOI 10.1007/978-1-4614-6862-2_11,

of local and regional human populations (Tajima 1989b; Harpending et al. 1993; Kimmel et al. 1998; Rogers and Harpending 1992; Harpending 1994; Harpending and Rogers 2000; Stajich and Hahn 2005; Li 2011).

The DNA sequence (both nuclear and mitochondrial genome) contains the biological (genetic) information encrypted in each cell that is responsible for the ability of the organism to survive in a given eco-niche. In addition, it is also a storehouse of evolutionary information including its putative origin, affinity, and diversity within and between species (Cann et al. 1987; Rogers and Harpending 1992; Rogers and Jorde 1995). Both the nuclear and mtDNA genome contain the functional "coding part" that is essential for the genotypic and phenotypic expression and therefore are under the influence of evolutionary forces of selection. The noncoding part does not involve in the genotypic or phenotypic expression and is little influenced by selection. By comparing the DNA sequences in a population, one can investigate the past genetic history and detect the signatures of demographic effects of migration, bottleneck effect and population expansion and the effect of different evolutionary forces e.g., selection etc. on the gene pool. It can also help to detect the influences of past population structure (e.g., fission-fusion) and its genetic consequences in a regional population.

11.2 Mitochondrial Genome: mtDNA—Non-coding Region

The theoretical models that help us to examine the evolutionary implications in different populations vary with respect to the type of genetic information. In case of nuclear genome (DNA sequences pertaining to genes), the influence of evolutionary forces and the confounding factors could be different when compared to the noncoding regions of the mitochondrial genome. In case of coding regions, selection could be one of the basic evolutionary forces which can lead to micro-evolutionary changes in the gene pool of the population.

In case of mtDNA noncoding regions, the selection is of little value, since these are basically influenced by segregating and mutational process and recombination is almost absent or a very rare event. To infer the influence of demographic factors on the distribution pattern of noncoding mitochondrial genome one has to take care of segregating and mutational process. The mtDNA is maternally inherited and the mutations and the segregating sites are passed on from the ancestral individuals (populations) to their descendents over generations (Anderson et al. 1981). By comparing the sequence information about the mutations and the segregating sites in a population it is possible to trace their mtDNA maternal lineage and putative ancestors who possibly had lived in the remote past (Kumar et al. 2008). From the extant pairwise distribution of sequence information from a suitable population, it is possible to estimate the origin of these mutations or to trace back the remote ancestral population and also examine the estimates of past population size—"the effective size" of the ancestral populations and the effects of past changes in population growth (demographic changes that the population had experienced in recent past) on the nucleotide sequence diversity. The basis of such inferences is based on theoretical models on pairwise comparison of mtDNA sequences; "mismatch distribution" (MMD) and its influence on the genetic diversity in a population (Saltkin and Hudson 1991; Rogers and Harpending 1992).

Using population genetic models, this study investigates variation in mitochondrial genome (specific to hypervariable regions I and II: HVRI and HVRII) of a few extant human regional populations to infer the influence and significant effect of earlier demographic events in the recent past as against the evolutionary factors that had resulted or shaped the distribution of the mitochondrial genome of the extant regional populations. The Adi, a Tibeto-Burman speaking tribe in central Arunachal Pradesh, offers a suitable example to investigate the influence of recent changes in population size on the genetic diversity among its subpopulations.

Adi tribe retains folklore tradition which describes their putative origin, migration, and dispersal from upper northeastern Himalayan region towards southern lower Himalayan region around Siang river valley for four to five hundred years. The folklore information also describes past events concerning recent history of upheaval in population size and formation of splinter groups as a result of inter-tribal warfare and feuds, etc. This is characteristic of fission–fusion process of population structure commonly observed among tribes practicing hunting-gathering and shifting cultivation. A detailed account of the population structure and the genetic consequences of fission and fusion process has been described by Neel and others among South American Indian tribes (Neel 1970, 1973, 1978; Neel and Salzano 1967; Rosenberg and Morton 1970; Fleishman 1980) and among Semai Senoi of Malaysia (Fix 1975, 1978; Fix and Lie-Injo 1975). The internal tribal warfare, in the recent past, is characterized by events of sudden and drastic change in the demographic size and subsequent population growth over generations among its splinter groups or subpopulations. Such events affect their genetic diversity, leave their mark in the genome, and pose a test case to investigate the effects of fission and fusion population structure on the genetic diversity of the sub-tribes of *Adi*. The study investigates the effects of past demographic upheavals of population growth and changes in population size on the genetic diversity of the regional sub-tribes of *Adi*, a Tibeto-Burman speaking population of central Arunachal Pradesh, India. This study, perhaps for the first time, investigates the genetic status of six sub-groups of *Adi* tribe based on the mitochondrial DNA markers.

11.3 Materials and Methods

11.3.1 Theoretical Considerations

The theoretical models that describe the effect of demographic factors from the segregating and/or mutational changes assume a variety of population structure models. Those are: the Tajima' D, Fu's F_S and Fay and Wu's H. Each of these models has its advantages in terms of efficiency and power of detecting the effect of demographic factors on the genetic diversity in a population.

11.3.1.1 Tajima's D

Let us assume that "*t*" generations ago, a population at equilibrium of size N_0 experiences a demographic expansion or population growth and let N_1 be the size after expansion (after "*t*" generations). Then under "infinite site mutation model" (Kimura 1969), that assumes that every new mutation occurs at a different site, Kimura has derived an expression for the probability of observing "i" differences between two sequences (genes) taken at random is expressed:

$$F^I_i(\theta_1, \theta_0, \tau) = F_i(\theta_1) + \exp[-\tau(\theta_1 + 1)/\theta_1)]x$$
$$\Sigma^i_{j=0}\left(\tau^j/j!\right)\left[F_{i-j}(\theta_0) - F_{i-j}(\theta_1)\right]$$

where $\theta_0 = 2N_0\mu$, $\theta_1 = 2N_1\,\mu$, $\tau = 2\,\mu t$ and μ is the total mutation rate per generation per site, "t" is generations.

In a population, at equilibrium, $F_i(\theta)$ is also the probability of observing "i" differences between two sequences (sites/genes).

Interestingly Waterson (1975) has shown that if the population remains constant over time, then $F_i(\theta)$ converges toward an equilibrium distribution.

$$F_i(\theta) = \theta^i/(\theta + 1)^{i+1}$$

The mean and variance of the distribution, respectively, are

$$E[F(\theta)] = \theta; \quad V[F(\theta)] = \theta(\theta + 1)$$

The population genetics theory of molecular evolution concerning the nucleotide sequences specifies:

The *number* of nucleotide positions of a sequence which is polymorphic at the position or the number of segregating sites κ,

The average nucleotide diversity per site or sequence is defined as $\pi = \Sigma\, x_i\, x_j\, \delta_{ij}/N$, where x_i is the frequency of *i*th haplotype, δ_{ij} is the number of nucleotide differences between haplotypes *i* and *j*, and N is the total length of the sequence.

Under infinite-site model of DNA sequence evolution, it can be shown that

$$E(\pi) = \theta \quad \text{and} \quad E(\kappa) = \theta\Sigma_i^{n-1}1/i$$

Therefore, $\theta_\pi = \acute{\kappa}$ and $\theta_\kappa = \kappa/\Sigma_i^{n-1}1/i$ where π is the average heterozygosity at nucleotide sites in the sample and κ is the observed number of segregating sites.

Under neutral equilibrium and if the population is at equilibrium w.r.t drift and mutation, then the two estimates of θ_π and θ_κ are statistically indistinguishable from one another, or

$D = E(\theta_\pi - \theta_\kappa)$ should be distinguishable from zero.

This is Tajima's D statistic (Tajima 1989b). D can be either positive or negative.

This can serve as a statistic to assess if the given data is consistent with the theoretical expectation of mutation-drift equilibrium. In short, D provides an insight to investigate the evolutionary history of a particular nucleotide sequences in a population. In case estimated,

Tajima's D $= 0$: It indicates no change in population size or role of selection acting on the locus;
Tajima's D < 0: It indicates the population size may be increasing or expanding or it might also indicate purifying selection;
Tajima's D > 0: It indicates the population might have experienced bottleneck or fluctuation of size or might suggest over dominant selection.

Though Tajma's D is very useful and can help us to investigate if the given nucleotide sequence is under mutation-drift equilibrium and/or search for other possible scenarios of molecular evolution. It has been suggested by Zheng et al. (2006) that it does not account for some aspects of population structure variables, as such may not be powerful to detect evolutionary history of sequence of a particular locus in a population. Therefore, to augment the results of Tajima's D, it is necessary to consider a statistics proposed by Fu's $\dot{F}s$ (Fu 1997), which was based on the "infinite sites mutation" model.

11.3.1.2 Fu's $\dot{F}$s

In general, excess of rare alleles or recent mutations can behave in a pattern different to the pattern of polymorphisms that can result from background selection than logistic population growth or genetic hitchhiking. In this regard, Fu's F_s test is more powerful for detecting the presence of evolutionary forces resulting from a population of excess of young mutations. Fu (1997) proposed a statistic, based on "estimating the probability of observing a random sample with a number of allele equal to or smaller than the observed value under assumption of selective neutrality criteria" (Holsinger 2006). If S is the above-mentioned probability, then Fu's Fs statistic is defined as:

$$\dot{F}\text{s} = \ln\{\text{s}/\left(1 - \text{s}\right)\}$$

A negative value of the statistic $\dot{F}$s is an indication of an excess number of alleles as would be expected from a recent population expansion or from genetic hitchhiking. A positive value is an indication for deficiency alleles as that can be expected from a recent population bottleneck or from over dominant selection. Fu's $\dot{F}$s is more sensitive to population expansion and genetic hitchhiking. Further Fay and Wu (2000) have proposed a statistic H to detect departures from neutrality for the DNA sequence evolutionary scenario:

$$H = \theta_\pi - \theta_H$$

H measures departures from neutrality that are a result of differences between high-frequency and intermediate-frequency alleles. Thus D is sensitive to population expansion as the number of segregating sites responds more rapidly to changes in population size than nucleotide heterozygosity. H will not detect this. Both tests therefore help us to distinguish between the effects of population expansion from that of purifying selection.

11.3.1.3 Harpending's Raggedness Index "r"

The mismatch distribution obtained from a high resolution nonrecombinant mtDNA pairwise data is generally expected to show a smooth distribution for the range of differences corresponding to the nucleotide diversity of a population; however, in case of low-resolution nonrecombinant mtDNA sometimes one may not obtain characteristic smooth distribution and might show several modes. As one can infer from the Saltkin and Hudson (1991), theoretical distribution or simulated mismatch distribution shows different pattern of distribution under different demographic scenarios of population expansion, stationary for a long time, or experienced short and sudden burst of population expansion. These show characteristic smooth curve in case of expanding population scenarios and show "raggedness" for others. Harpending et al. (1993) have developed an index: "raggedness index 'r'", a statistic to distinguish the observed MMD from a population as against the theoretical expected distribution under the assumption of expanding population.

If there are a maximum *d* differences in the mismatch distribution x, then the raggedness index defined by Harpending (1994) is:

$$\mathbf{r} = \sum_{i=1}^{d+1} (X_i - X_{i-1})^2$$

where x_i is the pairwise differences and d is the number of differences between pair of sequences.

11.4 Population Samples

11.4.1 ADI Tribe of Arunachal Pradesh

Adi, one of the largest tribes of Arunachal Pradesh, is a collection of regional tribes speaking Adi language and is distributed in the temperate and sub-tropical regions within the districts of West Siang, East Siang, Upper Siang, Upper Subansiri, and Dibang Valley in central Arunachal Pradesh (Rapson 1955; Elwin 1959; Gordon 2005; Singh 1998; APHD Report 2006; Tabi 2006). They share similar physical features of East Asian populations and speak Adi language (was a dialect a decade ago), which belongs to North-Assam branch of Tibeto-Burman sub-linguistic family (Majumdar 1980; Ruhlen 1991; Gordon 2005). The ethno-history suggests their origin from southern regions of Tibet (China). Migration and settlement history of their ancestors (the "Tani" group) at different time periods during about fifth to seventh century AD can also be traced (Tabi 2006; Lego 2005; Nath 2000; Roy 1997). There are about 12 sub-tribes of Adi, categorized under two major clusters based on their dialect, culture, clan-structure or kinship criterion, ethno-historical migration and distribution along the Siang river valley. While one cluster consists of seven regional tribes; the Minyong, Padam, Shimong, Milan, Pasi, Panggi, and Komkar sub-tribes, the rest five form another cluster that includes: Gallong, Ramo, Bokar, Pailobo, and Bori sub-tribes. Among these 12 sub-tribes, Minyong and Padam are numerically large (about several thousands), the rest being small (in hundreds) (Nath 2000; Roy 1997) and live in isolation in upper mountain regions. Their ethno-historical records suggest that these sub-tribes are as a resultant of fission–fusion processes due to inter-tribal warfares in the recent past of their settlement history (Lego 2005; Tabi 2006). Some of the remotely located (in mountainous terrains) Adi sub-populations practice hunting-gathering, while some others located over plain lands, in close proximity to urban area, practice settled agriculture. Those living in towns are educated and work in various offices and business enterprises. These wide variations in population structure, coupled with the vast diversity in their culture including religious beliefs, customs, dialect, house types, food habits, dress and ornaments' pattern and demography, render importance to these groups from a population genetic perspective (Nath 2000; Blackburn 2003/2004). There are a few studies on the culture and social organization, and however biological studies among these tribes are hardly found, except for reports of ABO blood group data in some of the sub-tribes and hemoglobin types and recent molecular genetic studies among a few individuals of Adi (Dhani 1960; Das et al. 1980; Roychoudhury 1981; Walter et al. 1986; Deka et al. 1988; Bhasin and Walter 2001; Krithika et al. 2006, 2007; Cordaux et al. 2004).

For the molecular genetics study among the sub-tribes of the Adi, we have collected 836 blood samples from six regional Adi populations. A total of 530 DNA samples were tested for polymorphism for 15 autosomal microsatellite loci and the results were published elsewhere (Krithika et al. 2006, 2008). For the mitochondrial genome analysis, a total of 97 samples were selected so that at least one individual represents the various clans in each of the regional populations. The sample size for the sub-tribes for mitochondrial region varies from 17 in Adi Padam to 23 in Adi Panggi (Table 11.1). The location of the studied populations is shown in Fig. 11.1.

Based on the mtDNA sequences of the studied six populations, the gene diversity, nucleotide diversity, and the mean number of pairwise differences along with standard deviations were estimated (Nei and Jin 1989) to understand the extent of genetic variation and the within-population genetic heterogeneity of the Adi subpopulations. These were computed using software package ARLEQUIN 3.01 (Excoffier et al. 2005). With the help of the software package Fst distances (Mega

Table 11.1 Diversity parameters deduced from mtDNA HV1 and HV2 sequences of Adi sub-populations

Population	Number of sequences	Number of polymorphic sites	Gene diversity	Nucleotide diversity	Mean pairwise differences
Adi Pasi	19	49	1.0000 ± 0.0171	0.207185 ± 0.110673	10.152047 ± 4.852679
Adi Minyong	14	47	1.0000 ± 0.0270	0.235806 ± 0.127870	11.318681 ± 5.467784
Adi Panggi	19	34	1.0000 ± 0.0171	0.298747 ± 0.159287	10.456140 ± 4.988735
Adi Komkar	18	40	1.0000 ± 0.0185	0.207875 ± 0.112763	8.522876 ± 4.134276
Adi Padam	13	37	1.0000 ± 0.0302	0.255128 ± 0.140121	10.205128 ± 4.986200

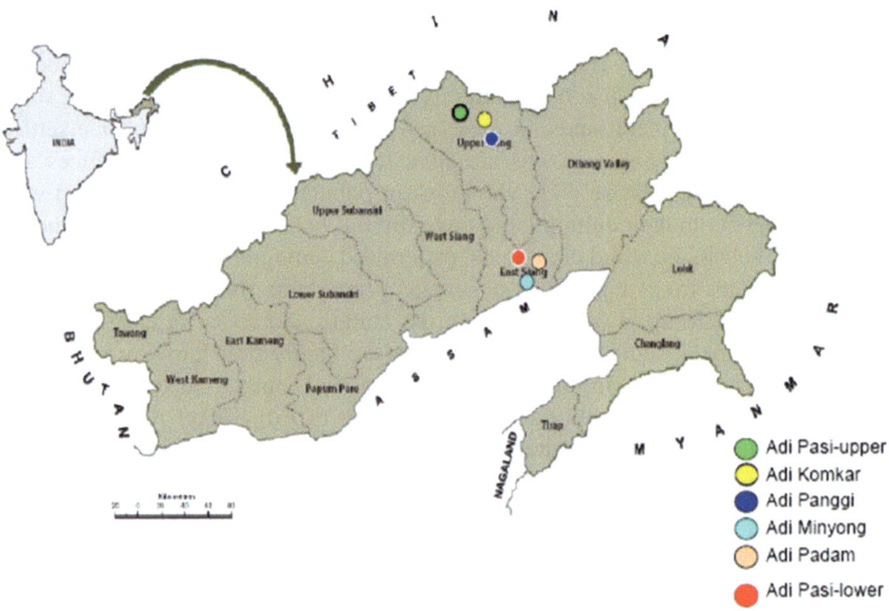

Fig. 11.1 Map of Arunachal Pradesh showing the geographical distribution of the studied Adi sub-tribes

3.1 software package) and associated *P*-values based on 1,000 simulations were computed for the studied populations. Subsequently the Fst distance matrix was used to construct the neighbor-joining (NJ) tree-rectangular and radiation forms by employing Mega 3.1 software package. Further to characterize the clustering trends, the data dimensionality was reduced by performing a covariance analysis between factors (principal component analysis—PCA). Based on Fst distance matrix PCA was performed using SPSS (version 11.0) Chicago, IL, USA. Analysis of molecular variance (AMOVA) based on HV1 and HV2 sequences of the studied groups was performed to understand the possible role of geography and ethno-history towards the genetic differentiation of Adi. However the results of the PCA plot and AMOVA are not included in this report.

11.4.2 Population Genetic Models: Parameter Estimates

Various demographic parameters of the populations including mismatch distributions, Fu's "F_S" statistic and associated *P*-values based on 1,000 simulations, raggedness index "r" and Tajima's *D* value were also computed by employing ARLEQUIN (version 3.01). Signatures of past population expansions were examined from the above obtained demographic parameters. Unimodal mismatch distributions were interpreted as signs of demographic expansion while multimodal distributions were interpreted as signs of constant population size over time (Harpending et al. 1993). Also, values of raggedness index "r" lower than 0.05 and negative values, differing significantly from zero of Fu's "F_s" statistic were inferred as signs of population demographic expansion (Fu 1997; Harpending et al. 1993).

The strength of genetic drift depends on the population size. If a population is at a constant size with constant mutation rate, the gene frequencies for several of the loci or the DNA sequences at different sites will reach equilibrium. This equilibrium has important properties, including the number of segregating sites *S* and the number of nucleotide differences between pairs sampled (these are called pairwise differences). To standardize the pairwise differences, the mean or "average" number of pairwise differences is used. This is simply the sum of the pairwise differences divided by the number of pairs and is signified by π. The clarification of the effects of demographic effects can be examined by statistic Tajima's D.

A negative Tajima's D signifies an excess of low frequency polymorphisms, indicating population size expansion. A positive Tajima's D signifies low and high frequency polymorphisms, indicating a decrease in population size and/or balancing selection. This test is based on the fact that under the neutral model, estimates of the number of segregating sites and of the average number of nucleotide differences are correlated. The bioinformatics software DnaSP calculates the confidence limits of D (two-tailed test) assuming that this statistic follows a beta distribution. Molecular evolutionary software MEGA (Tamura et al. 2007) is used to estimate the rates of molecular evolution, inferring ancestral sequences, and testing evolutionary hypotheses. Based on the mismatch distribution of the six regional populations of the Adi, we tested deviation from neutrality expectations by the statistics: the Tajima's D and Fu's F_s.

11.5 Results

11.5.1 Mitochondrial DNA Diversity

The extent of diversity measured by estimators such as gene diversity, nucleotide diversity, and the mean pairwise differences (MPD) among the Adi regional populations is shown in Table 11.1. In general, the sub-groups show a high level of gene (haplotype) diversity, similarly a high range of nucleotide diversity (0.207–0.298). Pasi and Komkar show the least and Panggi shows the highest nucleotide diversity values among the six Adi populations. Similar high range of nucleotide diversity has been reported from a composite sample of Adi (Cordaux et al. 2004). Mean pairwise differences (MPD) among the Adi vary from 8.523 among Adi Kokmar to 11.32 among the Adi Minyong, the average MPD of the six Adi groups is 10.13.

11.5.2 Genetic Affinity Among Adi Sub-groups

The dendrogram (figure not shown) obtained by rooted and unrooted methods depicts a close cluster of three populations: Panggi, Minyong, and Pasi, among the six studied Adi sub-groups. Padam and Komkar show a longer branch length and stand away from the major cluster. The PCA plots (two and three components) also show close clustering of the three sub-groups and distant location of Padam and Komkar groups (figures not shown).

11.5.3 Mismatch Distributions of the Adi Sub-groups

Mismatch distribution for the six studied *Adi* sub-tribes is shown in Fig. 11.2. Examining the distribution of pairwise nucleotide differences is expected to give considerable insight into the recent demographic history of the Adi sub-groups.

The size, shape, and the pattern of the MMD observed among each of the five Adi sub-groups show considerable variation and also significantly differ from the theoretical expectations. We find that Adi Pasi and Adi Minyong tend to show unimodal distribution, though there is slight variation in the shape. On the contrary, Adi Komkar exhibit bimodal tendency and Adi Panggi show multimodal distribution (one crest followed by two small crests). Adi Padam display three major crests of increasing intensity. Although, Pasi and Minyong and also Komkar (to an extent) appear to correspond to the expected MMD in case of demographic expansion model, the expected distributions do not fit well with the observed distributions. In view of the nature of the distribution curves, the statistical estimates of the mean and variance hardly describe the characteristics of the distribution except in case of

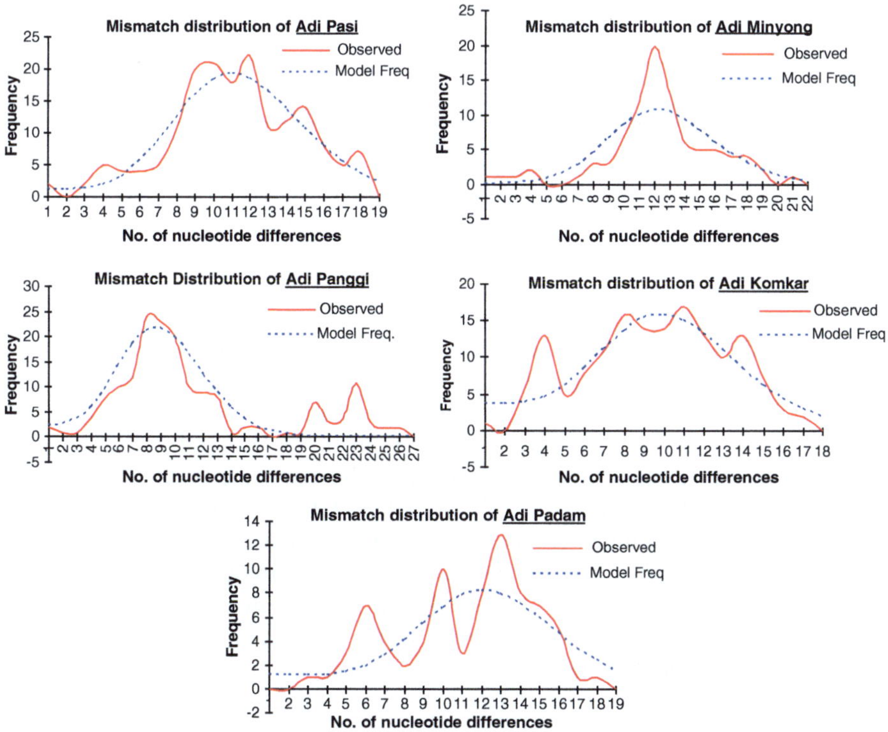

Fig. 11.2 Mitochondrial mismatch distributions for the six studied Adi sub-tribes

Pasi and Minyong groups. The different distributions suggest the past demographic upheavals that had recorded its signatures in the extant tribes—Padam and Panggi and to certain extent in Komkar. The observed mismatch mean for the populations is around 10 except in case of Komkar (8.5) and the observed mismatch variance falls within 12–13 range excepting for Panggi (35.838).

11.5.4 Demographic Parameters of the Studied Adi Sub-groups

Based on the mtDNA HV1 and HV2 sequences, an attempt is made to understand the past demographic events among the *Adi* sub-tribes. The different demographic parameters, including the Fu's F_S statistic; raggedness index "r" and Tajima's D, estimated to obtain considerable insight into the population expansion of the studied Adi groups is shown in Table 11.2.

The studied sub-tribes of Adi show significantly large negative Fu's "F_S" values, with an approximate range from -5 (in Minyong and Padam groups) to -11 (in Pasi, Komkar, and Panggi groups). Negative values of F_S among the five sub-groups

Table 11.2 Demographic indices deduced from mtDNA HV1 and HV2 sequences of Adi sub-populations

Population	Number of sequences	Fu's F_S (*P* value)	Raggedness Index "r" (*P* value)	Tajima's D (*P* value)
Adi Pasi	19	−11.21824 (0.000)	0.0132 (0.576)	−1.1216 (0.092)
Adi Minyong	14	−5.97461 (0.011)	0.0272 (0.490)	−1.02473 (0.116)
Adi Panggi	19	−10.98509 (0.000)	0.0176 (0.454)	0.2991 (0.602)
Adi Komkar	18	−11.53593 (0.000)	0.0127 (0.855)	−1.08973 (0.096)
Adi Padam	13	−5.64377 (0.005)	0.0362 (0.394)	−0.63823 (0.242)

of Adi are indicative of their demographic expansion in the past. We observe that the raggedness index "r" is lower than 0.05 in all the studied Adi sub-populations. The values of "r" range from 0.012 to 0.03, with high *p* values denoting nonsignificance. Padam and Minyong, the two largest populations of the Adi and were the earlier settlers of the region, show higher "r" values than the remaining smaller three groups (0.012 to 0.017). These values of "r" lower than 0.05 also suggest demographic expansion among the studied groups. Except for Panggi, the remaining populations show negative values for Tajima's "D" indicative of low frequency polymorphisms and population size expansion. Positive Tajima's "D" value signifies high levels of intermediate frequency polymorphisms in the population, probably indicating population bottleneck in Panggi.

11.6 Discussion

The values of mean number of pairwise differences (MPD) too indicate the diversity among the Adi sub-groups. In fact Adi sub-groups show higher MPD values when compared to that reported by Cordaux et al. (2004) among the groups of north-east India [5.73 ± 2.79 (Apatani) to 7.17 ± 3.51 (Tipperah)]. The highest value of MPD (11.318681 ± 5.467784) in Adi Minyong reflects their high diversity when compared to other groups. This scenario in case of Adi Minyong may probably be attributed to their large size and distribution over plain region in proximity to the urban locality. As a consequence of this, the population might plausibly have experienced external gene flow leading to high diversity within the group. On the contrary, the lowest MPD value is exhibited by Adi Panggi (8.522876 ± 4.134276) which is likely considering their small size and isolated location in remote mountainous terrain. However, the higher F_{ST} value 0.1333 obtained through Analysis of Molecular Variance (AMOVA) indicates considerable genetic differentiation among Adi (the results of AMOVA is not presented). This suggests the probable involvement of different processes, governed by the population structure including the marriage patterns, during the differentiation of the females of Adi sub-groups.

The mitochondrial DNA mismatch distribution shows an overall unimodal distribution, accompanied by some discrepancy, among Pasi and Minyong groups indicating their recent population expansion. Adi Padam which, like Minyong, is also settled over the plains near the urban area, however, shows multimodal distribution. This is unexpected in view of their large size and urban settlement. However this requires further investigation in the background of their history of settlement and details of their population structure. Panggi and Komkar are small populations comprising of a couple of thousand individuals residing in remote mountainous terrains and surviving on hunting-gathering economy. The population structure is characteristic of preferential marriages among close kin and high endogamy. This is expected to reflect in their genetic profiles. The observed multimodal and bimodal distributions among Panggi and Komkar sub-groups and their larger negatives values of Fu's F_S statistic and lower values of raggedness index indeed reflect their population structure.

In Panggi, about 70% of the husbands and wives in general belong to seven surnames (Maji et al. 2007; Maji and Vasulu 2008). However in recent years there are other non-Panggi surnames entering into the population through marriages with non-Panggi females among the younger generations living in the Geku town. Possibly if the population samples analyzed for mtDNA variation consist of both non-Panggi and Panggi females, the resultant mismatch distribution is expected to be multimodal with higher variance and also positive Tajima's D value. That is what we observe in case of Adi Panggi. The bimodal distribution observed among Adi Komkar needs to be explained in view of their population structure.

11.7 Conclusions

The considerable cultural differences and geographic isolation, exhibited by the different sub-tribes of *Adi*, are reflected (to an extent) in their mitochondrial DNA profile, where these groups tend to show differentiation to a certain extent. The genetic affinity among the *Adi* sub-tribes is in agreement with their ethno-historical records and folklore traditions that describe the antiquity and the settlement history of the formation of sub-groups. The mtDNA study shows signatures of demographic upheavals undergone by the sub-tribes in the recent past. The results indicate the possibility of bottleneck effect among the small and isolated Panggi group as a result of internal tribal warfare, where the Panggi possibly had significant effect on their group size and history of settlement. The validity of folklore narration about the fission–fusion population structure and its possible genetic consequences are illustrated in mtDNA mismatch distribution among the regional *Adi* tribes.

Acknowledgements The authors acknowledge the helpful comments and suggestions from the unknown referees and the editor.

References

Anderson, S., Bankier, A. T., Barrell, B. G., de Bruijn, M. H., Coulson, A. R., Drouin, J., et al. (1981). Sequence and organization of the human mitochondrial genome. *Nature, 290*, 457–465.

APHD Report. (2006). *Arunachal Pradesh human development report 2005 (2006)*. Itanagar, Arunachal Pradesh: Department of Planning, Government of Arunachal Pradesh.

Bhasin, M. K., & Walter, H. (2001). *Genetics of castes and tribes of India*. Delhi: Kamla-Raj Enterprises.

Blackburn, S. (2003/2004). Memories of migration: notes on legends and beads in Arunachal Pradesh, India. *European Bulletin of Himalayan Research, 25/26*, 15–60.

Cann, R. L., Stoneking, M., & Wilson, A. C. (1987). Mitochondrial DNA and human evolution. *Nature, 325*(1), 31–36.

Cordaux, R., Weiss, G., Saha, N., & Stoneking, M. (2004). The northeast Indian passageway: a barrier or corridor for human migration? *Molecular Biology and Evolution, 21*, 1525–1533.

Das, B. M., Deka, R., & Das, R. (1980). Haemoglobin E in six populations of Assam. *Journal of Indian Anthropological Society, 15*, 153–156.

Deka, R., Reddy, A. P., Mukherjee, B. N., Das, B. M., Banerjee, S., et al. (1988). Haemoglobin E distribution in ten endogamous population groups of Assam, India. *Human Heredity, 38*, 261–266.

Dhani, A. H. (1960). *Prehistory and protohistory of eastern India*. Calcutta: Firma KL Mukhopadhyay.

Elwin, V. (1959). *A philosophy for NEFA*. Shillong: North-East Frontier Agency.

Excoffier, L., Laval, G., & Schneider, S. (2005). ARLEQUIN ver 3.01: an integrated software package for population genetics data analysis. *Evolutionary Bioinformatics Online, 1*, 47–50.

Fay, J. C., & Wu, C.-I. (2000). Hitchhiking under positive Darwinian selection. *Genetics, 155*, 1405–1413.

Fix, A. G. (1975). Fission-fusion and lineal effect: aspects of the population structure of the Semai Senoi of Malaysia. *American Journal of Physical Anthropology, 43*, 295–302.

Fix, A. G. (1978). The role of kin-structural migration in genetic differentiation. *American Journal of Human Genetics, 41*, 329–339.

Fix, A. G., & Lie-Injo, L. E. (1975). Genetic microdifferentiation in the Semai Senoi of Malaysia. *American Journal of Physical Anthropology, 43*, 47–56.

Fleishman, M. L. (1980). An unusual distribution for height among males in a Warao Indian village: a possible case of lineal effect. *American Journal of Physical Anthropology, 53*, 397–405.

Fu, Y. X. (1997). Statistical tests of neutrality of mutations against population growth, hitchhiking and background selection. *Genetics, 147*, 915–25.

Gordon, R. G. (Ed.). (2005). *Ethnologue: languages of the world* (15th ed.). Dallas: SIL International.

Harpending, H. (1994). Signature of ancient population growth in a low resolution mitochondrial DNA mismatch distribution. *Human Biology, 66*(4), 591–600.

Harpending, H., & Rogers, A. (2000). Genetic perspective on human origins and differentiation. *Annual Review of Genomics and Human Genetics, I*, 361–385.

Harpending, H. C., Sherry, S. T., Rogers, A. R., & Stoneking, M. (1993). The genetic structure of ancient human populations. *Current Anthropology, 34*, 483–496.

Holsinger, K. E. (2006). *Lecture notes in population genetics*. Storrs-Mansfield: Dept. Ecology and Evolutionary Biology, University of Connecticut. http://darwin.eeb.uconn.edu/eeb348/lecture-notes/notes.html.

Kimmel, M., Chakraborty, R., King, J. P., Bamshad, M., Watkins, W. S., & Jorde, L. B. (1998). Signatures of population expansion in microsatellite repeat data. *Genetics, 148*, 1921–1930.

Kimura, M. (1969). The rate of molecular evolution considered from the standpoint of population genetics. *Proceedings of the National Academy of Sciences of the United States of America, 63*, 1181–1188.
Krithika, S., Maji, S., & Vasulu, T. S. (2006). Genetic heterogeneity among three Adi tribes of Arunachal Pradesh, India. *Human Biology, 78*(2), 221–227.
Krithika, S., Maji, S., & Vasulu, T. S. (2007). Intertribal and temporal allele-frequency variation at the ABO locus among Tibeto-Burman-speaking Adi subtribes of Arunachal Pradesh, India. *Human Biology, 79*(3), 355–362.
Krithika, S., Maji, S., Vasulu, T. S. (2008). A microsatellite guided insight into the genetic status of Adi, an isolated hunting-gathering tribe of northeast India. *Plos One*:e2549.
Kumar, S., Padmanabham, P. B., Ravuri, R. R., Uttaravalli, K., Koneru, P., Mukherjee, P. A., et al. (2008). The earliest settlers' antiquity and evolutionary history of Indian populations: evidence from M2 mtDNA lineage. *BMC Evolutionary Biology, 8*(1), 230.
Lego, N. (2005). *History of the Adis of Arunachal Pradesh*. Itanagar, Arunachal Pradesh: Peregrine Graphics.
Li, H. (2011). A new test for detetcting recent positive selection that is free from the confoudning impacts of demography. *Molecular Biology and Evolution, 28*(1), 365–375.
Maji, S., Krithika, S., & Vasulu, T. S. (2007). Genetic kinship among an isolated Adi tribe of Arunachal Pradesh: isonymy in the Adi Panggi. *Human Biology, 79*(3), 321–337.
Maji, S., & Vasulu, T. S. (2008). Genetic structure of an isolated sub-tribe of the Adi people of Arunachal Pradesh state in Northeast India: isonymy analysis and selective neutrality of surname distribution in Adi Panggi. *Journal of Genetic Genealogy, 4*(1), 1–11.
Majumdar, D. N. (1980). Northeast India: a profile. In T. C. Sharma & D. N. Majumdar (Eds.), *Eastern Himalayas: a study on anthropology and tribalism*. New Delhi: Cosmo.
Nath, J. (2000). *Cultural heritage of tribal societies* (Vol. 1 (The Adis)). New Delhi: Omsons Publications.
Neel, J. V. (1970). Lessons from a primitive people. *Science, 170*(3960), 815–822.
Neel, J. V. (1973). 'Private' genetic variants and the frequency of mutation among South American Indians. *Proceedings of the National Academy of Sciences of the United States of America, 70*, 3311.
Neel, J. V. (1978). Rare variants, private polymorphism and locus heterozygosity in Amerindian Indians. *American Journal of Human Genetics, 30*, 455.
Neel, J. V., & Salzano, F. M. (1967). Further studies on the Xavante Indians X. Some hypothesis, generalizations resulting from these studies. *American Journal of Human Genetics, 19*, 554–574.
Nei, M. N., & Jin, L. (1989). Variances of the average number of nucleotide substitutions within and between populations. *Molecular Biology and Evolution, 6*(3), 290–300.
Rapson, E. J. (1955) People and languages. In: E. J. Rapson (Ed.), *Cambridge history of India*, vol. 1 (pp. 33–57). Delhi: S. Chand.
Rogers, A. R., & Harpending, H. (1992). Population growth makes waves in the distribution of pairwise genetic differences. *Molecular Biology and Evolution, 9*(3), 552–569.
Rogers, A. R., & Jorde, L. B. (1995). Genetic evidence on modern human origins. *Human Biology, 67*(1), 1–36.
Rosenberg, I., & Morton, N. E. (1970). Population structure of blood groups in Central and South American Indians. *American Journal of Physical Anthropology, 32*, 373–376.
Roy, S. (1997). *Aspects of Padam Minyong culture*. Itanagar, Arunachal Pradesh: The Director of Research.
Roychoudhury, A. K. (1981). The genetic composition of the people in Eastern India. *Journal of Indian Anthropological Society, 16*, 153–170.
Ruhlen, M. (1991). *A guide to the world's languages: volume 1, classification*. Stanford: Stanford University Press.
Saltkin, M., & Hudson, R. (1991). Pairwise comparison of mitochondrial DNA sequences in stable and exponentially growing population. *Genetics, 129*, 555–562.

Singh, K. S. (1998). *People of India: India's communities* (Vol. 1). New Delhi: Oxford University Press.
Stajich, J. E., & Hahn, M. W. (2005). Disentangling the effects of demography and selection in Human history. *Molecular Biology and Evolution, 22*(1), 63–73.
Tabi, T. (2006). *The Adis*. Pasighat, Arunachal Pradesh: Siang Literary Forum.
Tajima, F. (1989a). The effect of change in population size on DNA polymorphism. *Genetics, 105*, 437–460.
Tajima, F. (1989b). Statistical methods for testing the neutral mutation hypothesis by DNA polymorphism. *Genetics, 123*, 597–601.
Tamura, K., Dudley, J., Nei, M., & Kumar, S. (2007). MEGA4: Molecular Evolutionary Genetics Analysis (MEGA) software version 4.0. *Molecular Biology and Evolution, 24*, 1596–1599.
Walter, H., Mukherjee, B. N., Gilbert, K., Lindenberg, P., Dannewitz, A., et al. (1986). Investigations on the variability of haptoglobin, transferrin and Gc polymorphisms in Assam, India. *Human Heredity, 36*, 388–396.
Waterson, G. A. (1975). On the number of segregating sites in genetical models without recombination. *Theoretical Population Biology, 7*, 256–276.
Zheng, K., Fu, Y.-X., Shi, S., & Wu, C.-I. (2006). Statistical tests for detecting positive selection by utilizing high-frequency variants. *Genetics, 174*, 1430–1439.

Chapter 12
Growth Curve Model in Relation to Extremal Processes Based on Stationary Random Variables

Ratan Dasgupta

Abstract We study the growth models of Elephant foot yam in relation to a Markov process and explain the adequacy of the model. The said Markov process may be bounded by two extremal processes based on stationary sequence of random variables. Properties of such extremal processes are investigated. Parameters of the model are estimated from observed growth by minimising the maximum distance between the realised growth curve and simulated theoretical process. The proposed Markov process depends on the behaviour of basic variables crossing a threshold. A characterisation for uniform distribution based on specified linear regression of conditional quantiles on unrestricted quantiles, while crossing a threshold is proved.

MS classification 2010: Primary 62P10; Secondary 60J99, 62J02

12.1 Introduction

Consider the problem of studying effect of seed weight and seed skin texture through growth experiments on Elephant foot yam when the size of the seed corm is around the critical weight. Indian Statistical Institute's Giridih farmland consists mainly of lateritic soil full of gravels, and 200g weight of cut yam seed is seen to be critical for sprouting and subsequent survival of the plant therein. Farmers sometimes use cut and underweight yam corm rather than the whole corm, in absence of adequate supply of yam seed to cover a large area of farmland. The yam plants grown may then turn out to be relatively tender, slim and undernourished.

R. Dasgupta (✉)
Theoretical Statistics and Mathematics Unit, Indian Statistical Institute,
Kolkata - 700 108, India
e-mail: rdgupta@isical.ac.in

R. Dasgupta (ed.), *Advances in Growth Curve Models*, Springer Proceedings in Mathematics & Statistics 46, DOI 10.1007/978-1-4614-6862-2_12,

Yam is considered to be a staple food among tribals. Yam stems and petioles of young unexpanded yam leaves are good source of leaf protein and minerals. These tender yam leaves are edible when thoroughly cooked. Dasgupta (1995) proposed locally more accurate confidence intervals of leaf protein content in plants.

If the seed corms are already in sprouting stage before plantation, then the critical corm weight for subsequent yam-plant survival may be lowered further. Experience shows that sprouted corms may later fully develop as plants having more than one stem and has higher chances of survival. Experiments are now being conducted with cut-seed weights 100g, 200g, 300g, 400g and 500g along with five grading of the seed-skin textures in a 5×5 Graeco-Latin square design. The seeds are cut from yam-corms, which were already sprouting from "main eye". The lowest weight viz., 100g of cut corm is then seen to be critical for plant survival in a preliminary investigation.

This scenario of undernourished plants may be rectified by appropriate aftercare of plants from tender stage onward by periodically administering different fertilisers and manual care so as to increase above ground vegetative mass that has a positive impact on the yam yield. Plant hormones may also be used as growth regulators that determine formation of leaves, stems, flowers and underground yam. Kim et al. (2005) suggested combined treatment of Chinese yam plants with gibberalic acid and jasmonic acid with low concentration (GA 50ppm and JA 5ppm) to promote tuber yield, by a single treatment. Higher doses may produce contrary results.

Manual care such as loosening the soil, weeding and mulching, may produce higher vegetative growth. Casual farm labourers gradually become experienced in their work as the plants grow over time. Gradual growth of yam over different stages of time may be compared with a non-decreasing process, e.g., an extremal process based on a stationary sequence of random variables.

A question may arise whether variation of nutrients administered by different inexperienced farm-labours over large area of farmland and subsequent heterogeneous absorption of nutrients by plants at initial stages be ignored in data analysis part when we are mainly interested in distinguishing the effects of seed size and seed skin texture on yield.

We propose a growth model, which is close to the lowess curve of observed growth, and we show that under the proposed model it is possible to answer the above question in affirmative.

In the next section we propose a transient Markov process as growth curve model for yam. In general, this process has upper and lower bounds in terms of two extremal processes, see Dasgupta et al. (1981). We find that if the expertise of the labourers develop fast over the lifetime of yam plants, then one may ignore the variation arising out of heterogeneous nourishment administered by inexperienced labourers in the beginning, thereby allowing the effect of the main parts viz., seed weight and seed skin texture to be tested.

In general, problems where one may compensate for deficiency in initial inputs by subsequent remedial measures, a similar analysis may be adopted.

It may not be always possible to sequentially order the impact period of causal variables in a production session as time progresses. One may then consider a model based on day to day additional increments on yield, e.g., the Markov model (12.6) of next section.

The growth curve of the yam stems is simulated by a transient Markov process based on uniform random variables for comparison with observed growth. All the variate values are equally likely for uniform random variables, these are used as basic inputs in the present growth curve simulation.

Similarity of patterns noted in simulated and observed growth curves indicates that the model is satisfactory. Parameters associated with the model is estimated from observed data so as to minimise the maximum absolute distance between observed and theoretical growth curves, much like the idea of minimising Kolmogorov–Smirnov distance. A similar model may be used for underground yam deposition over time as above ground vegetative growth is found to be highly correlated with yam yield.

The proposed Markov model depends on the behaviour of the random variables crossing a threshold. This behaviour may be described in terms of conditional quantiles. A characterisation for uniform distribution based on linear regression of conditional quantiles on unrestricted quantiles, while crossing a threshold is proved.

12.2 Growth Model and a Characterisation of Uniform Distribution

We state a theorem on extremal process for stationary sequence of random variables. The proof is given in Appendix of Dasgupta (2011).

Theorem A. Let $(X_i, i \geq 1)$ be a sequence of stationary random variables and there exist constants $a_m,\ b_m$ such that the standardised maximum,

$$Z_m = b_m^{-1}(\max_{1\leq i\leq m} X_i - a_m) = O_p(1), \tag{12.1}$$

i.e., the distribution of $\{Z_m, m \geq 1\}$ is tight. Also let c_i; $(0 < c_i < 1,\ c_i \to 1)$ be a sequence of non-decreasing constants and let $i_o = i_o(n)$ be a sequence of positive integers satisfying

$$n^{-1} i_o(n) \to 0,\ \ c(i_o) = c(i_o(n)) = 1 - o(|\, a_n^{-1} b_n|);\ n \geq 1. \tag{12.2}$$

Then for any Y_m satisfying

$$\max_{1\leq i\leq m} c_i X_i \leq Y_m \leq \max_{1\leq i\leq m} X_i;\ \text{one has,} \tag{12.3}$$

$$b_m^{-1}(Y_m - \max_{1\leq i\leq m} X_i) \to 0,\ \text{in probability.} \tag{12.4}$$

Further if,

$$f(i) = |\, a_i b_i^{-1}| \ (\to \infty) \tag{12.5}$$

is non-decreasing in i and $\overline{\lim}_{i\to\infty} \frac{f(i\delta)}{f(i)} < \infty$, for every fixed $\delta > 0$, then the condition (12.2) on c is implied by $c(i) = 1 - o(|a_i^{-1} b_i|)$.

Remark 1. From the proof given in Dasgupta (2011) it is seen that stationarity assumption of the random variables may be relaxed by a weaker assumption, which should ensure that maximum value of an expanding set of random variables is attained in an index with large value. As for example, it is enough to have a weaker assumption that for a fixed sequence $i_o = i_o(n) = o(n) \to \infty$, $P\{\cup_{i=i_o+1}^{n}(X_i = \max_{1\leq j\leq n} X_j)\} \geq 1 - \frac{i_o}{n}$; $\forall n \geq 2$, to replace stationarity assumption for validity of Theorem A.

This mild assumption excludes the possibility that maximum of a sequence of random variables be confined only within a finite segment in the beginning of the sequence. It should also be possible to locate maximum beyond a finite range of the sequence (i.e., towards tail) as well.

Later we shall identify the variables X_i to be fresh random input of growth or, ideal possible growth in a time segment in presence of a large number of causal variables at optimal level. While modeling growth curves, relaxation of stationarity assumption of X_i thus provides a wide coverage of real life situations when growth inputs or, ideal possible growth in different time segments may not be homogeneous (or stationary) due to non-homogeneity of basic causal variables.

Consider a yam seed-corm with fixed weight and skin texture. In the above representation identify $X \equiv X_i$ to be the ideal possible growth of a response variable e.g., growth of stem or, above ground biomass due to photosynthesis in presence of $i = 1, \cdots, m$ auxiliary variables, like climatic conditions e.g., sunshine/cloudy weather/rain, humidity, temperature, etc.; absorption level of nutrients administered, manual care e.g., appropriate weeding, irrigation, fungicide, insecticide, plant hormone treatment, etc.; that are relevant for growth experiment, when all these characteristics remain at perfect level for ideal growth of response in a time segment; m being large when the number of these causal and uncertain characteristics is large. Sometimes these may have simultaneous effect on proper growth of the crop.

We shall assume the following. Deviation from the ideal situation may at most dampen the ideal growth during plant lifetime by a scale c_i due to dominant i-th characteristic in a time segment, resulting in a lower bound of growth $c_i X_i$ as in (12.3), when the status of other characteristics are within permissible limit. It therefore follows from (12.4) that the maximum growth variation due to these uncertain variables are of negligible order $o_p(b_m)$, compared to the main part $O_p(b_m)$ as mentioned in (12.1); for all the plants when $c_i \to 1$ at an appropriate rate. In some occasions it is possible to sequentially order the impact period of causal variables, growth deficiency due one causal variable may be compensated by possible subsequent adjustment of the others, as a result lower bound of growth due to i-th characteristic may be improved upon by favorable j-th characteristic (i<j) at a later stage. Different stages of growth may be seen in most of the plants' life cycle.

As for an example, the life cycle of cannabis is usually complete in four to nine months. First stage is germination during the warmth of spring. Water is absorbed and the embryo's tissues swell and grow, splitting the seed along its edge. Temperature and water are two important factors at this stage. Anchored by the roots and receiving water and nutrients, the embryonic leaves (cotyledons) unfold.

The formation of the second pair of leaves begins the seedling stage. This stage is complete when the plant has reached the maximum leaves per blade, usually within four to six weeks. Depending on variety, they reach their maximum number. Next is period of vegetative growth which depend on the availability of nutrients. In the preflowering stage, a period of one to two weeks, the growth slows considerably till the appearance of the first flowers. At the flowering and seed set stage causal variables conducive to pollination is most important.

The actual time of life cycle is regulated by local growing conditions, specifically the photoperiod (length of day vs. night). Cannabis is a long night (or short day) plant. When exposed to a period of two weeks of long nights (13 h or more of continuous darkness each night), the plants respond by flowering. This allows the grower to control the life cycle of the plant and adapt it to local growing conditions or unique situations. Since one can control flowering, one can control maturation and, hence, the age of the plants at harvest. A few causal variables assume dominant role depending on a particular stage of life cycle.

In the beginning of life cycle when growth of Yam-plant is fast, frequent recordings are taken compared to sparse recording towards end. Rainy season may provide adequate supply of water to yam plants in the beginning, which has to be supplemented by regular irrigation at later stages of plant growth. There are secondary sprouting of the plants resulting in higher vegetative growth as the experiment proceeds. The total number of stems may therefore increase over time. Above ground biomass takes into account overall growth of all the stems. The random variable Y_m may be taken as a good predictor like "above ground biomass" for underground yam deposition.

The assumption $c_i \uparrow 1$ is reasonable in view of the fact that the temporarily employed farm workers gradually gain expertise of plant care over time.

There are situations when it is not possible to sequentially order the impact period of causal variables in a production session as time progresses. Modeling may then be done by viewing the growth over successive time segments of daytime when a plant starts a new cycle of photosynthesis after a night rest. Suppose in ideal scenario, on i-th daytime of the production session, the plant gets a fresh (random) input for growth represented by $X = X_i$. An alternative model for total growth in terms of sequentially ordered time intervals may then be made as follows. The cutting model described in Dasgupta et al. (1981) is Markov.

$$T_1 = c_1 X_1^+, T_2 = T_1 + c_2(X_2 - T_1)^+, \cdots, T_n = T_{n-1} + c_n(X_n - T_{n-1})^+ \ldots \quad (12.6)$$

where $X^+ = X \vee 0$. The sequence T_n is non-decreasing, additional contribution at n-th stage is $c_n(X_n - T_{n-1})^+$, to be interpreted as additional growth at n-th time period. This is justified in the following fashion. If $X_n > T_{n-1}$, then a part of surplus input after maintaining the present requirement of plant at state T_{n-1} may be added to growth at n-th stage. A part of excess carbohydrates gathered from photosynthesis is deposited in underground yam. Under this model additional growth of stem/above ground biomass $c_n(X_n - T_{n-1})^+$ decreases as the magnitude T_{n-1} of these variables at $(n-1)$-th time interval increases. The fraction $c = c_n \uparrow 1$ may partly represent appropriate care for growth of the plant by experienced labours towards mature stage of plant. The above representation (12.6) provides a similar bound (12.3) for T_n, see Dasgupta et al. (1981). The component due to perturbation in causal variables is once again of negligible order.

The variables $X_1, \cdots, X_n, \cdots$ are assumed to be stationary and the constants c_i may be assumed to represent the combined dampening effect, being off the target, of the causal variables on the response variable at i-th time period, $0 < c_i < 1$; small values of c indicate more dampening, whereas values of c near 1 indicate less dampening of the additional part $(X_n - T_{n-1})^+$.

A similar model may be proposed to the underground yam deposition Z_i, over time segments $i = 1, \cdots, n, \cdots$ with a different sequence of constants $d_i \in (0,1)$, d_i converging to 1; thus uncontrolled and random perturbations in causal variables on yam deposition is of negligible order as before under the model.

Remark 2. Instead of a single sequence $\{c_i\}$ of dampening factor, there may be a number of sequences of dampening factors $\{c_{i1}\}, \cdots, \{c_{ip}\}$ over different regions of land and associated with different farm workers nourishing the plants in these regions over time, where each sequence of c satisfies the conditions of the theorem. These sequences of factors may be of different magnitudes. In such a situation, let the variable Y_i for i-th characteristic be either dampened by one of the sequence of c's depending on that *particular* region of land, or let it be the full magnitude X_i. The resultant variable Y then satisfies the following.

$$\min_{j=1,\cdots,p} \max_{1 \le i \le m} c_{ij} X_i \le Y_m \le \max_{1 \le i \le m} X_i. \tag{12.7}$$

Towards the end of Appendix in Dasgupta (2011), a modified proof is outlined extending Theorem A for Y_m satisfying the bound (12.7). Once again error term is of order $o_p(b_m)$ compared to the main part $O_p(b_m)$.

Based on a stationary sequence of random variables, we consider transient Markov process (12.6) and check whether this has behaviour similar to observed growth curve. The process resembles actual growth of yam-stem for some specific choice of constants and that provides a good visual approximation in growth curve simulation.

Figures 12.1 and 12.2 show realised growth curves (i.e., observed points joined by straight lines) of yam-stems and corresponding non-parametric lowess regression fit marked by small circles for two different stems of same seed corm.

These observed patterns are simulated in Figs. 12.3 and 12.4 by transient Markov process (12.6) based on $U(0,1)$ random variables with $c = c_i = 1 - i^{-.04}$ and $c =$

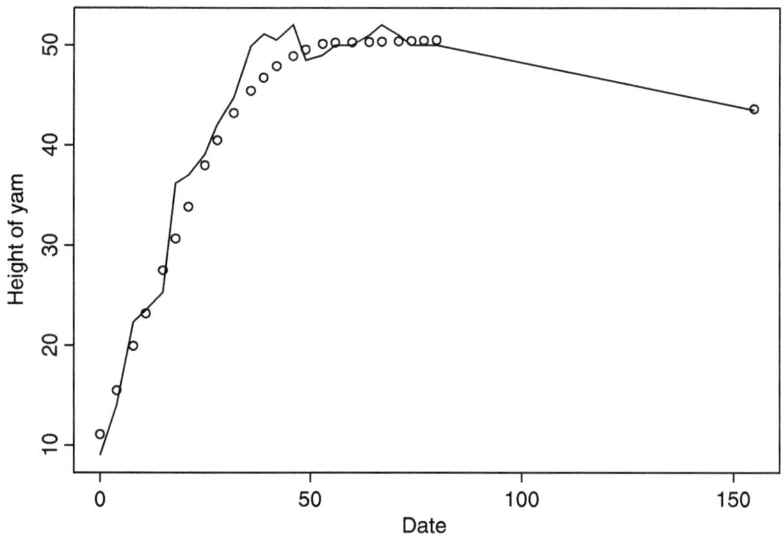

Fig. 12.1 Lowess fit for height curve of Yam plant 9, stem 1

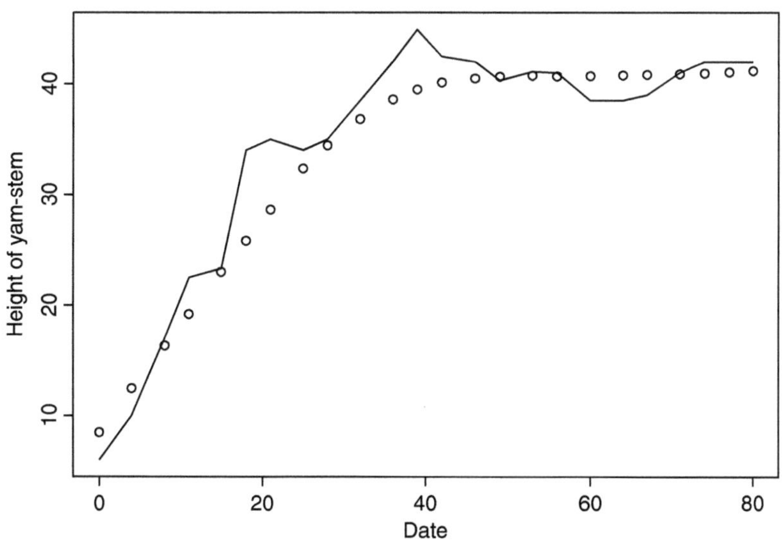

Fig. 12.2 Lowess fit for height curve of Yam plant 9, stem 2

$c_i = 1 - i^{-.007}$ respectively, for $i = 1, \cdots, 200$. In absence of precise information of soil and other field conditions, these uncertain variables may be assumed to generate uniform random impulse within equispaced time points towards plant growth.

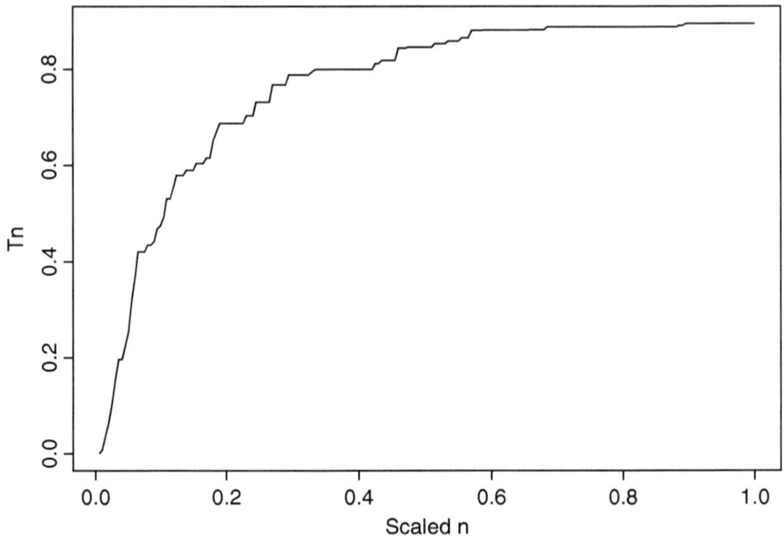

Fig. 12.3 Transient Markov process

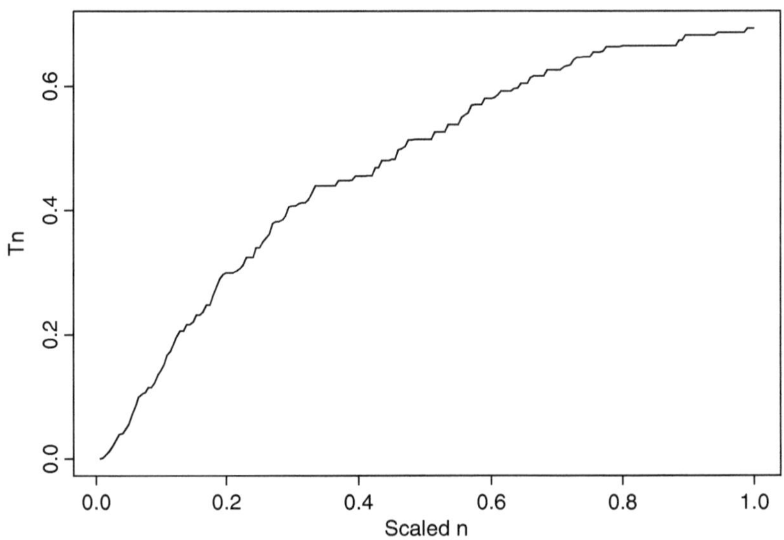

Fig. 12.4 Transient Markov process

Choice of $m = 200$ is made from the fact that Yam plant is harvested in about six months; in this particular case harvest was made after 200 days. Thus the number of time segments was taken to be $m = 200$. The concavity of the realised growth curve provides an appropriate choice of constants. Fast growth means that the coefficients c approach 1 fast. The growth observed at the half lifetime (100-th day) in the smooth lowess curve may be equated with theoretical growth $(1-i^{-\alpha})(X_n-T_{n-1})^+$ around the point of time $i = 100$ that has a positive growth. The second part $(X_n - T_{n-1})^+$ is obtainable from simulation, thus the product term when equated to growth in lowess curve estimated from observed data around that time point provides an initial value of α. Adjusting this value to minimise the maximum distance between the lowess growth curve of observed data and the simulated process where the jumps occur at 200 points, an appropriate choice of α may be made.

It was observed that a slight change in the choice of α from .04 (.007) increases the maximum distance between the lowess and simulated theoretical model with specific choice of seed value made, for the first (second) growth curve.

In other words, the value of α is so selected as to minimise the absolute distance between the realised lowess and simulated growth curves, much like the idea of minimising Kolmogorov–Smirnov statistics.

For the first process simulating the observed pattern of Stem 1 (Fig. 12.1), the growth is faster as reflected in the rate at which $c = c_i = 1 - i^{-.04} \uparrow 1$, compared to that for Stem 2 with estimated $c = c_i = 1 - i^{-.007} \uparrow 1$. An index for relative growth of Stem 1 and Stem 2 may then be defined as the ratio of the power indices, viz. $I_{1,2} = .04/.007 \approx 6$. Note that for uniform random variables X the scaling factor for $\max_{1,\cdots,i} X_j$ is $b_i = i^{-1}$. For $c = c_i = 1 - o(i^{-1}) \uparrow 1$, the limiting behaviour of T_i is similar to that of $\max_{1,\cdots,i} X_j$.

It was shown in Dasgupta et al. (1981) that for a fixed constant $c \in (0,1)$, limiting behaviour of T_i is different from that of $\max_{1,\cdots,i} X_j$. Here we see that for c with magnitude in between viz., $c = c_i = 1 - O(i^{-\alpha})$, $\alpha \in (0,1)$, it is possible to approximate the growth curve of yam-stem by the Markov model of T_i in (12.6).

Since the empirical study shows that the growth of Stem and Yam deposition has high correlation, one may explore similar model for predicting yam yield.

In the Markov model (12.6) one may take into account the situation where expertise-gain of the farm workers may have a random component. In other words, one may consider a more general model where $c_i \uparrow 1$, are random and independent of X.

Instead of assuming $c_i = 1-i^{-.04}$ assume that $c_i = 1-w_i^{-.04}$, where w_i are non-decreasing with $E(w_i) = i$. This construction is achieved by considering successive increments in w values to be iid uniform random variables with expectation 1, rather than a fixed constant 1, as considered for drawing Figs. 12.3 and 12.4. In Fig. 12.5 the process (12.6) is simulated with random c as suggested above and with same set of uniform X values as considered in Fig. 12.3. There is no significant difference in Figs. 12.3 and 12.5, indicating that random gain in expertise may not affect the growth curves significantly.

Similar patterns are observed between Figs. 12.4 and 12.6, where in the latter figure we consider random coefficients $c_i = 1 - w_i^{-.007}$, keeping the choice of variables X in (12.6) to be same as that for Fig. 12.4, where we consider constant coefficients $c_i = 1 - i^{-.007}$ associated with increments X.

Simulations are done by SPLUS software with assigned seed value 1234.

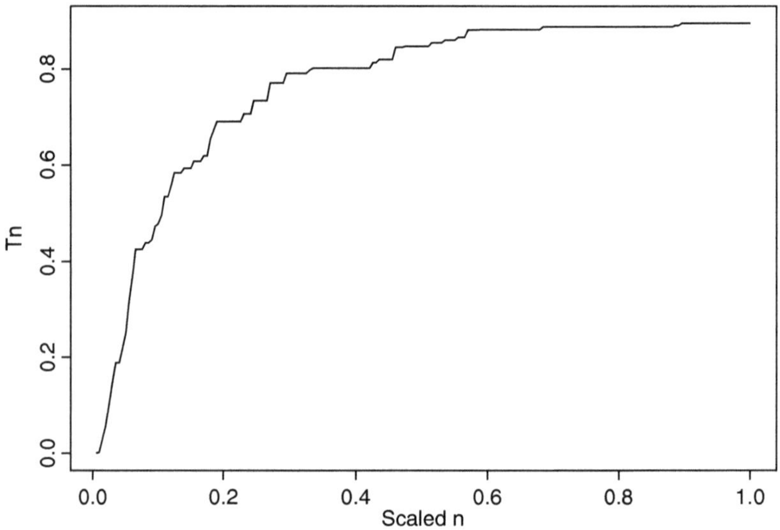

Fig. 12.5 Transient Markov process (Fig. 12.3) with random c

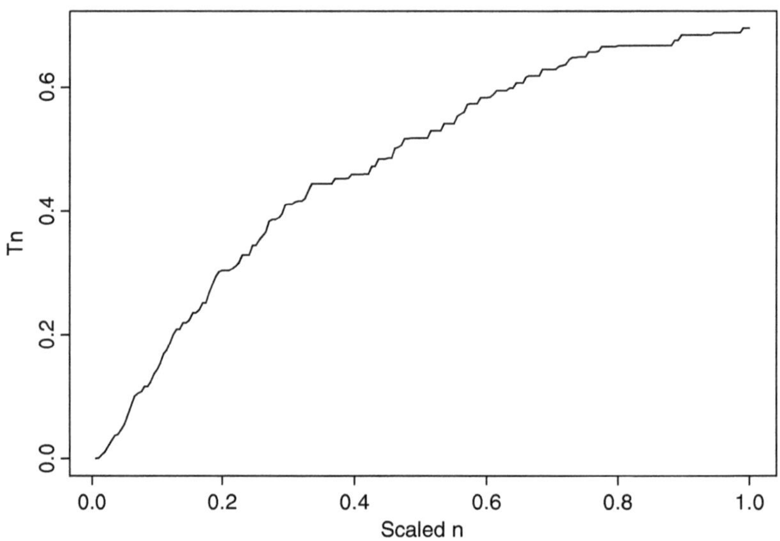

Fig. 12.6 Transient Markov process (Fig. 12.4) with random c

In the formulation (12.6) note that growth at n-th stage depends on whether the realised value of the variable X_n crosses a threshold T_{n-1}. It is therefore important to study the behaviour of the random variable X exceeding a threshold.

Uniform distribution has the property that the conditional p-th quantile, given that the variable exceeds a threshold u_0, is linear in u_0.

Without loss of generality consider uniform distribution on $[0, 1]$. The equation

$$P(U > u|U > u_0) = (1-u)/(1-u_0) = 1-p, \ p \in (0,1), \ 0 \leq u_0 \leq u \leq 1 \quad (12.8)$$

provides the value of conditional p-th quantile as $u = p + u_0(1-p)$, expressible in terms of unconditional p-th quantile plus shift u_0 multiplied with diminishing slope $(1-p)$. Such a property characterises uniform distribution as seen in the following theorem.

Theorem 1. *Let X be a non-negative and non-degenarate random variable with distribution function F that is continuously differentiable except possibly at the end points where $F = 0$ and $F = 1$. Let $x_0 \in (0, \epsilon)$, where $\epsilon > 0$ may be taken arbitrarily small. Let the p-th quantile of the distribution under the restriction $X > x_0$, be of the form of a linear regression $\alpha + \beta x_0$, where α is the unrestricted p-th quantile of X and the slope β does not depend on x_0. The above property holds for a dense choice of $p \in (0, 1)$ with $\beta = (1-p)$ iff F is a uniform distribution function on an interval.*

Proof. First part of the theorem follows easily as seen from (12.8) above. To see the "necessary part" observe that

$$P(X > x|X > x_0) = (1 - F(x))/(1 - F(x_0)) = 1 - p, \ x = \alpha + \beta x_0 \quad (12.9)$$

provides the relation,

$$F(\alpha + \beta x_0) = F(\alpha) + (1 - F(\alpha))F(x_0) \quad (12.10)$$

Differentiating both sides with respect to x_0, one gets the following relationship for the density f, which is assumed to be continuous.

$$\beta f(\alpha + \beta x_0) = (1 - F(\alpha)) f(x_0) \quad (12.11)$$

Let $x_0 \downarrow 0$, to get $\beta f(\alpha) = (1-p) f(0)$ over a dense set of $p \in (0, 1)$, where $\alpha = F^{-1}(p)$.

In other words $f(\alpha) = (1-p) f(0)/\beta$, as $\beta \neq 0$; since X is non-degenarate. This provides a characterisation of uniform distribution $f(\alpha) = f(0)$, $\forall \alpha \in (0, \delta)$, for $\delta = F^{-1}(1)$ iff $\beta = (1-p)$.

Remark 3. The form of linear regression $\alpha + \beta x_0$ for conditional quantiles given $X > x_0$, where α is the unrestricted p-th quantile of X may hold for distributions other than uniform with a slope $\beta \neq (1-p)$. As for example, consider $F(x) = 1-(1-x)^{\gamma}$, where $\gamma > 0$ is rational and $x \in [0, 1]$. With an application of binomial theorem for rational index, linear regression of conditional p-th quantile hold for $p \in (0, 1)$, where $\beta \neq (1-p)$; the slope β then involves higher powers of p.

References

Dasgupta, R., Ghosh, J.K., & Ranga Rao, N.T.V. (1981). A cutting model and distribution of ovality and related topics. In *Proc. of the ISI Golden Jubilee Conference* (pp. 182–204).
Dasgupta, R. (1995). Locally more accurate confidence sets. *Frontiers in Probability and Statistics (Proc. of the 2nd triennial Calcutta symp. Special Volume of CSA)* (pp. 130–140).
Dasgupta, R. (2011). On the distribution of burr with applications. *Sankhya B, 73*(1), 1–19.
Kim, S.K., Kim, J.T., Jang, S.W., Lee, S.C., Lee, B.H., & Lee, I.J. (2005). Exogenous effect of gibberellins and jesmonate on tuber enlargement of *Daoscorea opposita*. *Agronomy Research, 3*(1), 39–44.

Chapter 13
Projection of Indian Population by Using Leslie Matrix with Changing Age Specific Mortality Rate, Age Specific Fertility Rate and Age Specific Marital Fertility Rate

Prasanta Pathak and Vivek Verma

13.1 Introduction

The importance of projection in national and state level planning and policy formulation is quite well recognized in any country which attempts at achieving sustainability in human development and improving the quality of life. Population projection exercises are basically a part of forecasting growth of human population in the future years over a time horizon. There are various means of extrapolating past trend of change in human population over the future years. These means are determined by the assumptions that are made on the determinants of population change such as time, pattern of changes in fertility, mortality, and migration and other associated factors. The success of population projection depends not only on the technique of projection but also on the proximity of the assumptions to reality so that changes in the future years get estimated with least possible errors. Projections, however, might not suffice when there are significant deviations from the assumption that prevailing conditions would continue unchanged in the future. Also, the projection might not be satisfactory due to failure to incorporate adequately the changes in the policy parameters, technological changes, changes in the migration pattern, etc. Forecasting attempts at overcoming these drawbacks by incorporating the elements of judgment in the projection exercise. Forecasting enjoys the advantage of being based upon one or more assumptions that are likely to be realized in the future years. Thus, forecasts give more realistic picture of the future.

There are various mathematical models for population projection as discussed in Lee and Carter (1992), Islam (2009), and Gayawan et al. (2010). These are also used for finding out the determinants of population growth. The models are either stochastic with certain distributional assumptions on the variables or deterministic

P. Pathak (✉) • V. Verma
Indian Statistical Institute, Kolkata, India

R. Dasgupta (ed.), *Advances in Growth Curve Models*, Springer Proceedings in Mathematics & Statistics 46, DOI 10.1007/978-1-4614-6862-2_13,

as discussed in Zakria (2009). For forecasting population under the assumption that migration plays an insignificant role, models have been derived in de Jong and Tickle (2006), Yashin et al. (2000), Islam (2009, 2011), and Gayawan et al. (2010), incorporating fertility and mortality determinants. Two common forecasting methods are exponential smoothing, discussed in Brown et al. (1961) and Trigg and Leach (1967) and regression, discussed in any standard statistical textbook.

This study attempts at projecting the Indian population by applying the Leslie matrix, discussed in Lewis (1942) and Leslie (1945). The use of such matrix as an operator for projection of distribution of population by ages in the future has not been reported so far in the Indian context. Previously, the Indian population got projected by the World Bank and the Census Directorate by using predominantly the cohort-component method. The method and the references on the history of its development are given in O'Neill et al. (2001). Differences in assumptions based on expert opinions have made them obtain different projections. These assumptions are not known in detail, but both have done the projection with some assumptions on the targeted life expectancy and total fertility rate at the end of the projection period. Time series data on life expectancy and total fertility were assumed to follow the logistic and the Gompertz forms, respectively. The logical guidelines of cohort-component method have been followed also in the application of the Leslie matrix. The use of such matrix has been found predominantly in the context of various living organisms and animals as discussed in Jensen (1974), Werner and Caswell (1977), Cheke (1978), Van Groenendael et al. (1988), and Gauthier et al. (2007). It has been used in demography and health by Gross et al. (2006), Kajin et al. (2012), Thomas and Clark (2008), and few others. Changes in the policy parameters over time could be incorporated in a model based on Leslie matrix by modifying the matrix elements appropriately. Dependence of population growth on population density, however, is not taken into account in such a model. An attempt is made to use the matrix for age group wise projection of Indian population and compare the projection with the ones obtained by the World Bank and the Census. It does not assume any targets for life expectancy and total fertility rate at the end of the projection period nor does it impose any standard functional form like Gompertz or logistic function to describe the temporal change in the fertility and the mortality. It uses the best fitted statistical models to capture the temporal changes in the age specific fertility, the mortality, and the sex ratio while projecting population.

13.2 Methodology

The sources of data for this study are mainly the following web sites of the Census of India, the Ministry of Statistics and Programme Implementation, the Central Bureau of Health Intelligence, the World and the Ministry of Health and Family Welfare. http://www.indiastat.com/demographics/7/population/217/16662/census.aspx (Census of India, MOSPI), http://cbhidghs.nic.in/content%282002%29.asp (Central Bureau Of Health Intelligence—India), http://data.worldbank.

org/data-catalog/ (The World Bank), http://nrhm-mis.nic.in/UI/Public%20Periodic/Population_Projection_Report_2006.pdf, http://nrhm-mis.nic.in/ (Ministry of Health & Family Welfare).

It is assumed that the female population is the only contributor of the population growth and their childbearing ages are between 15 and 49 years. It is also assumed that births are given only by the married females. All females are considered in 5-year age interval namely 0–4, 5–9, 10–14, 15–19, 20–24, 25–29, 30–34, 35–39, 40–44, 45–49, 50–54, 55–59, 60–64, 65–69, 70–74 and projections are also made for the same intervals. The population at ages above 74 is ignored due to nonavailability of sufficiently reliable data. They constitute about 3% of the total population. It is assumed that each age specific rate remains constant over each 5-year interval and there is a perfect balance in emigration and immigration over the period of projection.

Under the above assumptions, the cohort-component method of population projection gives rise to the following 16 equations.

$$\begin{aligned} &\frac{1}{2}\left({}_{15}^{5}\mathrm{P}^{t} + {}_{20}^{5}\mathrm{P}^{t+5}\right){}_{15}^{5}\mathrm{F}\left(\frac{{}_{0}^{5}\mathrm{L}}{\mathrm{l}_0}\right) + \frac{1}{2}\left({}_{20}^{5}\mathrm{P}^{t} + {}_{20}^{5}\mathrm{P}^{t+5}\right){}_{20}^{5}\mathrm{F}\left(\frac{{}_{0}^{5}\mathrm{L}}{\mathrm{l}_0}\right) + \\ &\cdots + \frac{1}{2}\left({}_{45}^{5}\mathrm{P}^{t} + {}_{45}^{5}\mathrm{P}^{t+5}\right){}_{45}^{5}\mathrm{F}\left(\frac{{}_{0}^{5}\mathrm{L}}{\mathrm{l}_0}\right) \\ &= {}_{0}^{5}\mathrm{P}^{t+5} \end{aligned} \tag{13.1}$$

$${}_{0}^{5}\mathrm{P}^{t}\left(\frac{{}_{5}^{5}\mathrm{L}}{{}_{0}^{5}\mathrm{L}}\right) = {}_{5}^{5}\mathrm{P}^{t+5} \tag{13.2}$$

$$\cdots$$

$$\cdots$$

$$\cdots$$

$$\cdots$$

$${}_{70}^{5}\mathrm{P}^{t}\left(\frac{{}_{75}^{5}\mathrm{L}}{{}_{70}^{5}\mathrm{L}}\right) = {}_{75}^{5}\mathrm{P}^{t+5} \tag{13.16}$$

where ${}_{x}^{n}\mathrm{P}^{t}$ = Female Population in the age interval $(x, x+n)$ at time t. ${}_{x}^{n}\mathrm{F}$ = Age Specific Fertility Rates (or Age Specific Marital Fertility Rates) in the age interval $(x, x+n)$. ${}_{x}^{n}\mathrm{L}$ represents the person-years lived by the cohort between years x to $(x+n)$ and l_0 represents the number of persons surviving at exact age 0.

The above set of equation can be written as the following matrix equation.

$$\begin{pmatrix} {}_{0}^{5}P^{t+5} \\ {}_{5}^{5}P^{t+5} \\ \cdots \\ {}_{75}^{5}P^{t+5} \end{pmatrix} = \begin{pmatrix} 0 & 0 & 0 & \left[\frac{1}{2}\left(\frac{{}_{0}^{5}L}{l_0}\right)\left(\frac{{}_{15}^{5}L}{{}_{10}^{5}L}\,{}_{15}^{5}F\right)\right] & \left[\frac{1}{2}\left(\frac{{}_{0}^{5}L}{l_0}\right)\left({}_{15}^{5}F + \frac{{}_{20}^{5}L}{{}_{15}^{5}L}\,{}_{20}^{5}F\right)\right] & \cdots & \left[\frac{1}{2}\left(\frac{{}_{0}^{5}L}{l_0}\right){}_{45}^{5}F\right] & 0 & 0 & 0 & 0 & 0 \\ \frac{{}_{5}^{5}L}{{}_{0}^{5}L} & 0 & \cdots & \cdots & \cdots & \cdots & 0 & 0 & 0 & 0 & 0 & \\ 0 & \frac{{}_{10}^{5}L}{{}_{5}^{5}L} & 0 & \cdots & \cdots & \cdots & \cdots & 0 & 0 & 0 & 0 & 0 \\ 0 & 0 & 0 & \cdots & \cdots & \cdots & \cdots & \cdots & \cdots & \cdots & \cdots & \frac{{}_{75}^{5}L}{{}_{70}^{5}L} \end{pmatrix} \times$$

$$\begin{pmatrix} {}_{0}^{5}P^{t} \\ {}_{5}^{5}P^{t} \\ \vdots \\ {}_{75}^{5}P^{t} \end{pmatrix}$$

Equation (13.1) provides the number of female babies in the age class (0–4) at time (t + 5) and they are those daughters who are born between time (t) and (t + 5). It is the sum of products of number of females in the childbearing age classes, i.e. from (15–19) to (45–49) at time (t), fecundities in their respective childbearing age classes, and their survival chance from previous age at time (t) to current age at time (t + 5). Equation (13.2) represents the number of females in the age class (5–9) at time (t + 5) and they are those females in the age class (0–4) at time (t) who continue to survive at time (t + 5). In general, Eqs. (13.3)–(13.16) represent the number of females in the age class labeled (m + 1) at time (t + 5), and they are those females in age class labeled (m) at time (t) who continue to survive at time (t + 5). The label m stands for {(5–9), (10–14), (15–19), (20–24), ..., (70–74)} for m = 1, 2, 3, 4, ..., 14.

The above matrix equation in abbreviated form becomes

$$P^{t+5} = LP^{t}$$

where P^{t+5} and P^{t} represent the population by age vector of order (16×1) in the year (t + 5) and t, respectively, and L is the Leslie Matrix consisting of elements which are functions of fertility and mortality parameters.

The age specific mortality rates (ASMR) are available intermittently for the 20 different years in the period 1981–2009 and the age specific fertility rates (ASFR) are available intermittently for 12 different years in the period 1980–2009. Age specific marital fertility rates (ASMFR) are available intermittently for 13 different years in the period 1984–2009. Inconsistencies in the available data on ASFR and ASMFR prior to 2000 made the study depend only on 7-year data, available in the period 2000–2009.

The ASMRs are projected till 2020 by using either polynomial regression models or exponential smoothing algorithm. The projection is done for each age interval. The choice of the appropriate technique has been made based on the comparison of the root mean square errors (RMSE) over all the age intervals. The same strategy has been adopted for projecting the ASFR and ASMFR. A brief description on the algorithm of the exponential smoothing is given in the appendix.

13.3 Findings

Table 13.1 below gives for different age groups the degrees of fit of polynomial regression models along with degrees of best fitted polynomials and the degrees of fit obtained by exponential smoothing to estimates on ASMR in different years along with RMSE and their differences. It shows that exponential smoothing is a better choice for projection of ASMR. Table 13.2 gives age group wise ASMR projections for 2010, 2015, and 2020 by exponential smoothing. The projections, in general, indicate downward trend of the ASMRs for all the age groups except the age groups like 30–34 years, 35–39 years, 50–54 years, and 70–74 years. The near constancy of the projections for these age groups needs to be investigated at greater depth.

Table 13.3 above gives for different reproductive age groups the degrees of fit of polynomial regression models along with degrees of best fitted polynomials and the degrees of fit obtained by exponential smoothing to estimates on ASFR in different years along with RMSE and their differences. It shows that polynomial regressions are better choices for projection of ASFR. Table 13.4 below gives the ASFR projections for 2010, 2015, and 2020 for each reproductive age group by using polynomial regressions. The projections for each age group show a downward trend with varying rate of falling.

Table 13.5 below gives for different reproductive age groups, the degrees of fit of polynomial regression models along with degrees of best fitted polynomials and

Table 13.1 The degrees of fit of polynomial regression models along with degrees of polynomials and the degrees of fit for exponential smoothing to estimates on age specific mortality rates in different years along with RMSEs and their differences

Age interval (year)	RMSE using polynomial regression [RMSE(reg.)]	RMSE using exponential smoothing [RMSE(exp)]	RMSE(reg.) − RMSE(exp)
0–4	1.314 [1] (0.967)	1.27 (0.970)	0.044
5–9	0.225 [1] (0.917)	0.209 (0.928)	0.016
10–14	0.080 [1] (0.895)	0.073 (0.913)	0.007
15–19	0.129 [1] (0.827)	0.102 (0.890)	0.026
20–24	0.155 [1] (0.833)	0.138 (0.868)	0.017
25–29	0.170 [1] (0.734)	0.189 (0.672)	−0.019
30–34	0.268 [1] (0.337)	0.249 (0.428)	0.019
35–39	0.272 [1] (0.305)	0.174 (0.716)	0.098
40–44	0.271 [1] (0.688)	0.262 (0.709)	0.009
45–49	0.320 [1] (0.832)	0.319 (0.834)	0.001
50–54	0.690 [1] (0.816)	0.871 (0.706)	−0.181
55–59	1.071 [1] (0.795)	0.986 (0.827)	0.085
60–64	1.745 [1] (0.786)	1.737 (0.789)	0.008
65–69	2.007 [1] (0.813)	2.173 (0.782)	−0.166
70–74	9.327 [1] (0.697)	8.134 (0.774)	1.193

N.B.: Degree of polynomial regression is indicated within [ ] and degree of fit ($R^2 = 1-$Error sum of square/Total sum of square) is indicated within ( )

Table 13.2 ASMR projections by age group for 2010, 2015, and 2020

	ASMR per 1,000		
Age interval (years)	2010	2015	2020
0–4	13.41	10.0	9.32
5–9	0.91	0.41	0.04
10–14	0.85	0.70	0.59
15–19	1.34	1.15	1.01
20–24	1.85	1.64	1.47
25–29	2.07	1.88	1.74
30–34	2.47	2.47	2.47
35–39	3.34	3.32	3.32
40–44	3.94	3.63	3.40
45–49	5.37	4.38	3.76
50–54	8.59	8.59	8.59
55–59	11.95	10.52	9.44
60–64	20.22	18.08	16.48
65–69	31.21	28.44	26.37
70–74	52.40	52.40	52.40

Table 13.3 The degrees of fit of polynomial regression models along with degrees of polynomials and the degrees of fit for exponential smoothing to estimates on age specific fertility rates in different years along with RMSEs and their differences

Age interval (years)	RMSE using polynomial regression [RMSE(reg.)]	RMSE using exponential smoothing [RMSE(exp)]	RMSE(reg.) − RMSE(exp)
15–19	0.613 [1] (0.976)	0.699 (0.969)	−0.085
20–24	2.378 [1] (0.843)	2.808 (0.780)	−0.431
25–29	1.605 [1] (0.971)	1.699 (0.967)	−0.094
30–34	2.305 [1] (0.951)	2.643 (0.936)	−0.338
35–39	1.219 [1] (0.979)	1.453 (0.970)	−0.234
40–44	0.454 [1] (0.986)	0.790 (0.957)	−0.335
45–49	0.202 [1] (0.962)	0.270 (0.933)	−0.068

N.B.: Degree of polynomial regression is indicated within [ ] and degree of fit ($R^2 = 1 -$ Error sum of square/Total sum of square) is indicated within ( )

Table 13.4 ASFR projections by age group for 2010, 2015, and 2020

	ASFR per 1,000		
ASFR	2010	2015	2020
15–19	35.5	30.6	26.3
20–24	202.5	194.8	187.1
25–29	154.6	142.9	131.2
30–34	72.0	57.0	42.0
35–39	28.1	16.7	5.2
40–44	10.3	4.7	1.0
45–49	4.3	1.7	1.0

the degrees of fit obtained by exponential smoothing to estimates on ASMFR in different years along with RMSE and their differences. It shows that polynomial regressions are better choices for projection of ASMFR. Table 13.6 below gives

Table 13.5 The degrees of fit of polynomial regression models along with degrees of polynomials and the degrees of fit for exponential smoothing to estimates on age specific marital fertility rates in different years along with RMSEs and their differences

Age interval (years)	RMSE using polynomial regression [RMSE(reg.)]	RMSE using exponential smoothing [RMSE(exp)]	RMSE(reg.) − RMSE(exp)
15–19	8.216 [3] (0.998)	5.648 (0.999)	2.568
20–24	3.072 [4] (0.910)	10.251 (0.870)	−7.1790
25–29	1.780 [2] (0.970)	2.516 (0.940)	−0.736
30–34	2.457 [2] (0.950)	2.983 (0.927)	−0.527
35–39	1.177 [2] (0.983)	1.529 (0.971)	−0.353
40–44	0.558 [2] (0.983)	0.915 (0.953)	−0.357
45–49	0.403 [2] (0.891)	0.287 (0.945)	0.116

N.B.: Degree of polynomial regression is indicated within [ ] and degree of fit ($R^2 = 1 -$ Error sum of square/Total sum of square) is indicated within ( )

Table 13.6 Age group wise ASMFR projections for 2010, 2015, and 2020

	ASMFR per 1,000		
Age interval (years)	2010	2015	2020
15–19	245.0	223.6	182.5
20–24	295.9	275.3	246.5
25–29	172.9	158.8	144.8
30–34	75.8	60.8	45.9
35–39	29.7	17.6	5.5
40–44	11.2	5.9	0.5
45–49	4.7	3.0	1.2

the ASMFR projections for 2010, 2015, and 2020 for each reproductive age group by using polynomial regressions. The projections for each age group again show a downward trend with varying rate of falling.

The age distributions of the female's population based on the World Bank estimates in 2010 and the Census estimates in 2011 are shown graphically in Fig. 13.1 along with age distribution of female population based on the World Bank estimates in 2025 and the Census estimates in 2026. These show major differences in the Census and the World Bank estimates inclusive of the major shifts of the age distribution towards the right in the case of the Census estimates. In the same figure, the age distributions of female population in 2015 and 2025, projected based on our current methodology and the ASFR estimates, show that the change in the age distribution will be only in the age interval of 0–10 years. The distributions match closely with the age distribution obtained as per the World Bank estimates of 2010 for the remaining ages. It indicates that the change occurs mainly in respect of gradual increase in chances of survival of the children aged below 10 years.

The age distributions of the female population based on the World Bank estimates in 2010 and the Census estimates in 2011 are shown graphically in Fig. 13.2 along with age distribution of female population based on the World Bank estimates in 2025 and the Census estimates in 2026. In the same figure, the age distributions of the female population in 2015 and 2025, projected based on our methodology and the ASMFR estimates, show much more clearly that major changes will occur only

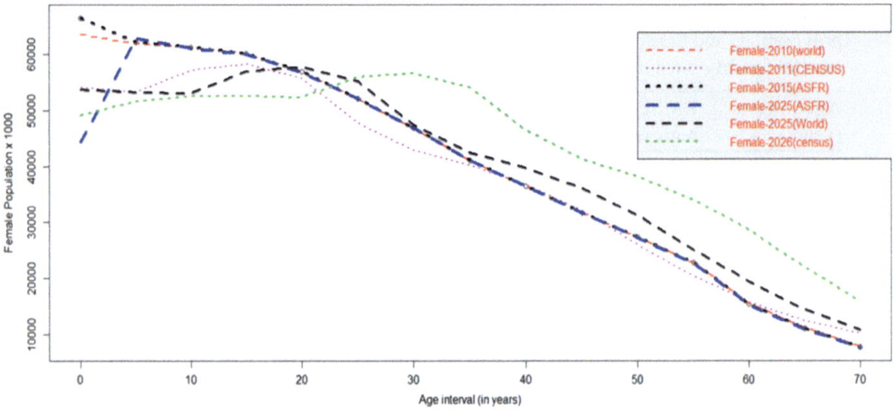

Fig. 13.1 Age distribution of female population in 2010 and 2025 (2011 and 2026 for the Census) based on the World Bank and Census estimates compared with the projected age distributions of females in 2015 and 2025 using ASFR

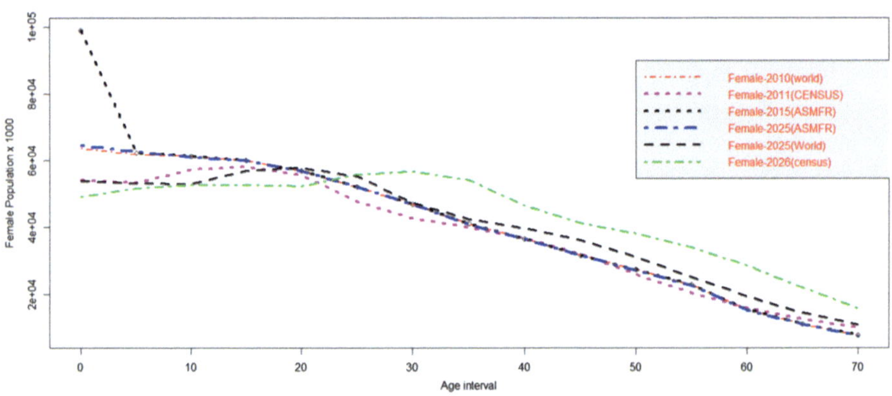

Fig. 13.2 Age distributions of female population in 2010 and 2025 (2011 and 2026 for the Census) based on the World Bank and the Census estimates compared with the projected age distributions of females in 2015 and 2025 using ASMFR

in the age interval 0–10 years. For all other ages, the distributions match closely with the age distribution, obtained based on 2010 estimates of the World Bank.

The age distributions of the total population, obtained as per the World Bank and the Census estimates, show major shift towards the age range of 20–45 years in the period 2010–2025 (2011–2026 for the Census estimates), indicating an aging of the population (as shown graphically in Fig. 13.3). Our methodology, however, does not establish such major shift except increase of population at ages below 10 years when total population is estimated for each age group from the age distributions of the female population by utilizing temporally changing sex ratio for corresponding age group.

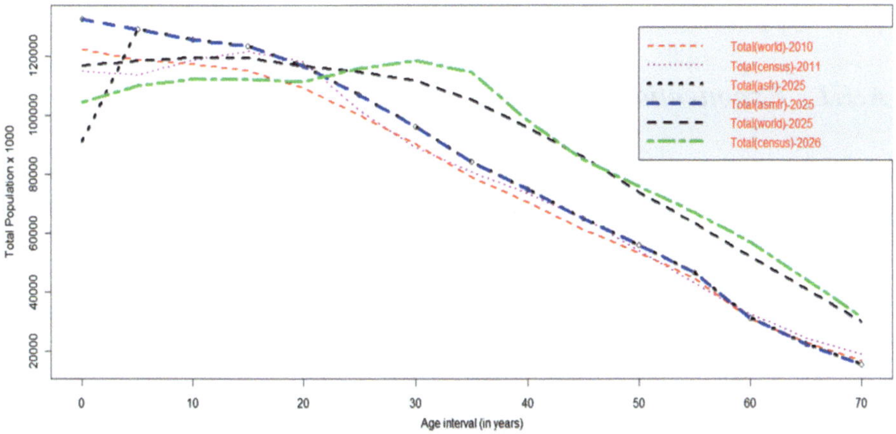

Fig. 13.3 Age distribution of total population in 2010 and 2025 (2011 and 2026 for the Census) based on the World Bank and the Census estimates compared with the projected age distribution of the total population in 2025 using ASFR and ASMFR

13.4 Conclusion

The projections, in general, indicate downward trend of ASMR, ASFR, and ASMFR. The study establishes that the projections of population made by the World Bank and the Census Directorate at higher ages are generally much on the higher side, indicating significant aging of the population. This might be due to prior assumptions on achieving targeted life expectancy and total fertility rate at the end of the projection period, following some functional path. Our projections are not based on such assumptions and it does not indicate such significant aging of the population; rather the age distribution is likely to follow the 2010 age distribution at ages above 10 years and the population at ages below 10 years are likely to increase significantly due to increase in their chances of survival and decreasing gender discrimination under ongoing reproductive and child health programmes. Our method incorporates temporal changes in the ASMR and ASFR/ASMFR, effected by changes in health and population policies and also health and family welfare programmes. It also incorporates by age group, temporal changes in the sex ratio for more precise projection of the population.

A.1 Appendix

A.1.1 Exponential Smoothing

It is one of the most popular types of automatic forecasting algorithms, which is used for the determination of an appropriate time series model, estimate of the parameters and compute the forecasts. Exponential smoothing methods were originally classified by Pegels' (1969) taxonomy. This was later extended by Gardner (1985), modified by Hyndman et al. (2002), and extended again by Taylor (2003), giving a total of 15 methods seen in Table A.1.

A.1.2 Point Forecasts for All Methods

We denote the observed time series by $y_1 \ldots y_n$. A forecast of y_{t+h} based on all of the data up to time t is denoted by $\hat{y}_{\mathrm{t+h|t}}$. To illustrate the method, we give the point forecasts and updating equations for method (A, A), the Holt–Winters' additive method:

$$\text{Level}: \quad l_t = \alpha\left(y_t - s_{t-m}\right) + (1-\alpha)\left(l_{t-1} + b_{t-1}\right)$$

$$\text{Growth}: \quad b_t = \dot{\beta}\left(l_t - l_{t-1}\right) + \left(1 - \dot{\beta}\right) b_{t-1}$$

$$\text{Seasonal}: \quad s_t = \gamma\left(y_t - l_{t-1} - b_{t-1}\right) + (1-\gamma)\, s_{t-m}$$

$$\text{Forecast}: \quad \hat{y}_{\mathrm{t+h|t}} = l_t + b_t h + s_{t-m+h_m^+}$$

where m is the length of seasonality (e.g., the number of months or quarters in a year), l_t represents the level of the series, b_t denotes the growth, s_t is

Table A.1 The fifteen exponential smoothing methods

	Seasonal component		
Trend component	N (none)	A (additive)	M (multiplicative)
N (none)	N, N	N, A	N, M
A (additive)	A, N	A, A	A, M
A_d (additive damped)	A_d, N	A_d, A	A_d, M
M (multiplicative)	M, N	M, A	M, M
M_d (multiplicative damped)	M_d, N	M_d, A	M_d, M

the seasonal component, $\hat{y}_{t+h|t}$ is the forecast for h periods ahead, and $h_m^+ = [(h-1) \mod m]+1$. To use this mentioned method, we need values for the initial states l_0, b_0 and $s_{1-m}, \ldots, s_0$, and for the smoothing parameters α, $\dot{\beta}$ and γ. All of these will be estimated from the observed data (Table A.2).

The triplet (E, T, S), which refers to the three components: Error, Trend, and Seasonality, is added to the models for discrimination among the additive and multiplicative errors. So the model ETS (A, A, N) has additive errors, additive trend and no seasonality—in other words, this is Holt's linear method with additive errors. Similarly, ETS (M, M_d, M) refers to a model with multiplicative errors, a damped multiplicative trend and multiplicative seasonality. The notation ETS (•, •, •) helps in remembering the order in which the components are specified.

A.1.3 Models for All Exponential Smoothing Methods

The general model involves a state vector $\boldsymbol{x}_t = (l_t, b_t, s_t, s_{t-1}, \ldots, s_{t-m+1})$ and state space equations of the form

$$y_t = w(\boldsymbol{x}_{t-1}) + r(\boldsymbol{x}_{t-1})\,\varepsilon_t \tag{13.17}$$

$$\boldsymbol{x}_t = f(\boldsymbol{x}_{t-1}) + g(\boldsymbol{x}_{t-1})\,\varepsilon_t \tag{13.18}$$

where $\{\varepsilon_t\}$ is a Gaussian white noise process with mean zero and variance σ^2 and $\mu_t = w(\boldsymbol{x}_{t-1})$. The model with additive errors has $r(\boldsymbol{x}_{t-1}) = 1$, so that $y_t = \mu_t + \varepsilon_t$. The model with multiplicative error has $r(\boldsymbol{x}_{t-1}) = \mu_t$, so that $y_t = \mu_t(1+\varepsilon_t)$. Thus, $\varepsilon_t = (y_t - \mu_t)/\mu_t$ is the relative error for the multiplicative model.

All of the methods in Table A.2 can be written in the form (13.17) and (13.18). The specific form for each model is given in Hyndman et al. (2008).

A.1.4 Estimation

In order to use these models for forecasting, we need to know the values of x_0 and the parameters α, β, γ, and φ. It is easy to compute the likelihood of the above innovations state space model, and so obtain maximum likelihood estimates. Ord et al. (1997) show that

$$L^*(\boldsymbol{\theta}, \boldsymbol{x}_0) = n \log\Big(\sum_{t=1}^{n} {\varepsilon_t}^2\Big) + 2\sum_{t=1}^{n} \log|r(\boldsymbol{x}_{t-1})|$$

Table A.2 Formula for recursive calculations and point forecasts

	Seasonal		
Trend	N	A	M
N	$l_t = \alpha y_t + (1-\alpha)\, l_{t-1}$ $\hat{y}_{t+h\|t} = l_t$	$l_t = \alpha\,(y_t - s_{t-m}) + (1-\alpha)\, l_{t-1}$ $s_t = \gamma\,(y_t - l_{t-1}) + (1-\gamma)\, s_{t-m}$ $\hat{y}_{t+h\|t} = l_t + s_{t-m+h_m^+}$	$l_t = \alpha\,(y_t / s_{t-m}) + (1-\alpha)\, l_{t-1}$ $s_t = \gamma\,(y_t / l_{t-1}) + (1-\gamma)\, s_{t-m}$ $\hat{y}_{t+h\|t} = l_t / s_{t-m+h_m^+}$
A	$l_t = \alpha y_t + (1-\alpha)\,(l_{t-1} + b_{t-1})$ $b_t = \dot{\beta}\,(l_t - l_{t-1}) + (1-\dot{\beta})\, b_{t-1}$ $\hat{y}_{t+h\|t} = l_t + h\, b_t$	$l_t = \alpha(y_t - s_{t-m}) + (1-\alpha)(l_{t-1} + b_{t-1})$ $b_t = \dot{\beta}(l_t - l_{t-1}) + (1-\dot{\beta})b_{t-1}$ $s_t = \gamma(y_t - l_{t-1} - b_{t-1}) + (1-\gamma)\, s_{t-m}$ $\hat{y}_{t+h\|t} = l_t + h\, b_t + s_{t-m+h_m^+}$	$l_t = \alpha(y_t / s_{t-m}) + (1-\alpha)(l_{t-1} + b_{t-1})$ $b_t = \dot{\beta}(l_t - l_{t-1}) + (1-\dot{\beta})b_{t-1}$ $s_t = \gamma(y_t / l_{t-1} - b_{t-1}) + (1-\gamma)\, s_{t-m}$ $\hat{y}_{t+h\|t} = (l_t + h\, b_t)\, s_{t-m+h_m^+}$
A_d	$l_t = \alpha y_t + (1-\alpha)\,(l_{t-1} + \varphi b_{t-1})$ $b_t = \dot{\beta}\,(l_t - l_{t-1}) + (1-\dot{\beta})\, \varphi b_{t-1}$ $\hat{y}_{t+h\|t} = l_t + \varphi_h\, b_t$	$l_t = \alpha(y_t - s_{t-m}) + (1-\alpha)(l_{t-1} + \varphi b_{t-1})$ $b_t = \dot{\beta}(l_t - l_{t-1}) + (1-\dot{\beta})\varphi b_{t-1}$ $s_t = \gamma(y_t - l_{t-1} - \varphi b_{t-1}) + (1-\gamma)\, s_{t-m}$ $\hat{y}_{t+h\|t} = l_t + \varphi_h\, b_t + s_{t-m+h_m^+}$	$l_t = \alpha(y_t / s_{t-m}) + (1-\alpha)(l_{t-1} + \varphi b_{t-1})$ $b_t = \dot{\beta}(l_t - l_{t-1}) + (1-\dot{\beta})\varphi b_{t-1}$ $s_t = \gamma(y_t / (l_{t-1} - \varphi b_{t-1})) + (1-\gamma)\, s_{t-m}$ $\hat{y}_{t+h\|t} = (l_t + \varphi_h\, b_t) s_{t-m+h_m^+}$
M	$l_t = \alpha y_t + (1-\alpha)\, l_{t-1}\, b_{t-1}$ $b_t = \dot{\beta}\,(l_t / l_{t-1}) + (1-\dot{\beta})\, b_{t-1}$ $\hat{y}_{t+h\|t} = l_t\, b_t^{\,h}$	$l_t = \alpha(y_t - s_{t-m}) + (1-\alpha) l_{t-1}\, b_{t-1}$ $b_t = \dot{\beta}(l_t / l_{t-1}) + (1-\dot{\beta})b_{t-1}$ $s_t = \gamma(y_t - l_{t-1}\, b_{t-1}) + (1-\gamma)\, s_{t-m}$ $\hat{y}_{t+h\|t} = l_t\, b_t^{\,h} + s_{t-m+h_m^+}$	$l_t = \alpha(y_t / s_{t-m}) + (1-\alpha) l_{t-1}\, b_{t-1}$ $b_t = \dot{\beta}(l_t / l_{t-1}) + (1-\dot{\beta})b_{t-1}$ $s_t = \gamma(y_t / (l_{t-1}\, b_{t-1})) + (1-\gamma)\, s_{t-m}$ $\hat{y}_{t+h\|t} = l_t\, b_t^{\,h}\, s_{t-m+h_m^+}$
M_d	$l_t = \alpha y_t + (1-\alpha)\, l_{t-1}\, b_{t-1}{}^{\varphi}$ $b_t = \dot{\beta}\,(l_t / l_{t-1}) + (1-\dot{\beta})\, b_{t-1}{}^{\varphi}$ $\hat{y}_{t+h\|t} = l_t\, b_t^{\,\varphi_h}$	$l_t = \alpha(y_t - s_{t-m}) + (1-\alpha) l_{t-1}\, b_{t-1}{}^{\varphi}$ $b_t = \dot{\beta}(l_t / l_{t-1}) + (1-\dot{\beta})b_{t-1}{}^{\varphi}$ $s_t = \gamma(y_t - l_{t-1}\, b_{t-1}{}^{\varphi}) + (1-\gamma)\, s_{t-m}$ $\hat{y}_{t+h\|t} = l_t\, b_t^{\,\varphi_h} + s_{t-m+h_m^+}$	$l_t = \alpha(y_t / s_{t-m}) + (1-\alpha) l_{t-1}\, b_{t-1}{}^{\varphi}$ $b_t = \dot{\beta}(l_t / l_{t-1}) + (1-\dot{\beta})b_{t-1}{}^{\varphi}$ $s_t = \gamma(y_t / (l_{t-1}\, b_{t-1}{}^{\varphi})) + (1-\gamma)\, s_{t-m}$ $\hat{y}_{t+h\|t} = l_t\, b_t^{\,\varphi_h} + s_{t-m+h_m^+}$

l_t denotes the series level at time t, b_t denotes the slope at time t, s_t denotes the seasonal component of the series at time t, and m denotes the number of seasons in a year; α, $\dot{\beta}$, γ and φ are constants, $\varphi_h = \varphi + \varphi^2 + \cdots + \varphi^h$ and $h_m^+ = [(h-1)\, mod\ m] + 1$

is equal to twice the negative logarithm of the likelihood function (with constant terms eliminated), conditional on the parameters $\theta = (\alpha, \beta, \gamma, \varphi)$ and the initial states $x_0 = (l_0, b_0, s_0, s_{-1}, \ldots, s_{-m+1})$, where n is the number of observations.

References

Brown, R. G., Meyer, R. F., & D'Esopo, D. A. (1961). The fundamental theorem of exponential smoothing. *Operations Research Society, 9*(5), 673–687.

Cheke, R. A. (1978). Theoretical rates of increase of gregarious and solitarious populations of the desert locust. *Oecologia, 35*(2), 161–171.

de Jong, P., & Tickle, L. (2006). Extending Lee–Carter mortality forecasting. *Mathematical Population Studies, 13*, 1–18.

Gardner, E. S., Jr. (1985). Exponential smoothing: The state of the art. *Journal of Forecasting, 4*, 1–28.

Gauthier, G., Panagiotis, B., Lebreton, J.-D., & Morgan, B. J. T. (2007). Population growth in snow geese: A modeling approach integrating demographic and survey information. *Ecology, 88*(6), 1420–1429.

Gayawan, E., Adebayo, S. B., Ipinyomi, R. A., & Oyejola, B. A. (2010). Modeling fertility curves in Africa. *Demographic Research, 22*(10), 211–236.

Gross, K., Morris, W. F., Wolosin, M. S., & Doak, D. F. (2006). Modeling vital rates improves estimation of population projection matrices. *Population Ecology, 48*, 79–89.

Hyndman, R. J., Koehler, A. B., Ord, J. K., & Snyder, R. D. (2008). *Forecasting with exponential smoothing: The state space approach*. Berlin: Springer. http://www.exponentialsmoothing.net/.

Hyndman, R. J., Koehler, A. B., Snyder, R. D., & Grose, S. (2002). A state space framework for automatic forecasting using exponential smoothing methods. *International Journal of Forecasting, 18*(3), 439–454.

Islam, R. (2009). Mathematical modeling of age specific marital fertility rates of Bangladesh. *Research Journal of Mathematics and Statictics, 1*(1), 19–22.

Islam, R. (2011). Modeling of age specific fertility rates of Jakarta in Indonesia: A polynomial model approach. *International Journal of Scientific & Engineering Research, 2*(11), 1–5.

Jensen, A. L. (1974). Leslie matrix models for fisheries studies. *Biometrics, 30*(3), 547–551.

Kajin, M., Almeida, P. J. A. L., Vieira, M. V., & Cerqueira, R. (2012). The state of the art of population projection models: From the Leslie matrix to evolutionary demography. *Oecologia Australis, 16*(1), 13–22.

Lee, R. D., & Carter, L. R. (1992). Modeling and forecasting U. S. mortality. *Journal of the American Statistical Association, 87*(419), 659–671.

Leslie, P. H. (1945). The use of matrices in certain population mathematics. *Biornetrika, 3*(3), 183–212.

Lewis, E. G. (1942). On the generation and growth of a population. *Sankhya, 6*, 93–96.

O'Neill, B. C., Balk, D., Brickman, M., & Ezra, M. (2001). A guide to global population projections. *Demographic Research, 4*, 203–288.

Ord, J. K., Koehler, A. B., & Snyder, R. D. (1997). Estimation and prediction for a class of dynamic nonlinear statistical models. *Journal of the American Statistical Association, 92*, 1621–1629.

Pegels, C. C. (1969). Exponential forecasting: Some new variations. *Management Science, 15*(5), 311–315.

Taylor, J. W. (2003). Exponential smoothing with a damped multiplicative trend. *International Journal of Forecasting, 1*, 715–725.

Thomas, J. R., & Clark, S. J. (2008). *More on the cohort-component model of population projection in the context of HIV/AIDS: A Leslie matrix representation and new estimation methods*. http://www.csss.washington.edu/Papers/wp88.pdf.

Trigg, D. W., & Leach, A. G. (1967). Exponential smoothing with an adaptive response rate. *Operational Research Society, 18*(1), 53–59.

Van Groenendael, J., de Kroon, H., & Caswell, H. (1988). Projection matrices in population biology. *TREE, 3*(10), 264–269.

Werner, P. A., & Caswell, H. (1977). Population growth rates and age versus stage-distribution models for teasel (Dipsacus Sylvestris Huds.). *Ecology, 58*(5), 1103–1111.

Yashin, A. I., Iachine, I. A., & Begun, A. S. (2000). Mortality modeling: A review. *Mathematical Population Studies, 8*(4), 305–332.

Zakria, M. (2009). *Stochastic models for population of Pakistan*. http://prr.hec.gov.pk/Thesis/681S.pdf.

Chapter 14
Growth Curve Analysis of Cumulative Automobile Defects

Ratan Dasgupta and Avinash Dharmadhikari

Abstract We study properties of cumulative defects for an industrial production through growth curve model in presence of nondecaying correlation structure within the number of successive defects accumulated over time. In automobile industry incidence per thousand vehicles (IPTV) is computed as number of failures observed in a month divided by sale quantity of a specified production batch for a specific lag period (from production to sale); the ratio is then multiplied by 1,000 to be called IPTV. For a batch of sold items it is seen that IPTV and the cumulative IPTV up to a time point may be approximated by Weibull distributions to a first degree. Such intriguing phenomena rules out independence of variables, as Weibull distribution is not closed under convolution. Postulating a simple model we show that such phenomena may be attributed to the presence of long range correlation in reported defects over consecutive time segments of sold objects from a production batch due to a common cause that regulates the growth pattern of cumulative IPTV. The observed correlation structure may be explained by the model.

MS subject classification: Primary : 62E17, secondary : 62P30.

14.1 Introduction and Some Empirical Observations

Weibull distribution has applications in reliability theory among others and may explain distribution of different industrial characteristics, see Engineering statis-

---

R. Dasgupta (✉)
Theoretical Statistics and Mathematics Unit, Indian Statistical Institute, Kolkata, India
e-mail: ratandasgupta@gmail.com

A. Dharmadhikari
Tata Motors, Pune, India
e-mail: avinash.d@tatamotors.com; avinash.dh@gmail.com

R. Dasgupta (ed.), *Advances in Growth Curve Models*, Springer Proceedings in Mathematics & Statistics 46, DOI 10.1007/978-1-4614-6862-2_14,

tics handbook (2008), Huang et al. (2012), Sagias and Karagiannidis (2005), and Weibull (1951). This may also be an appropriate model for number of defects in some cases, e.g., see http://arxiv.org/ftp/arxiv/papers/0905/0905.2288.pdf.

The number of defects across computer programs is seen to follow Weibull distribution when the programs are ranked by their sizes in Lines of Code metric.

In this paper we consider an automobile production process where the produced cars in different production batches are serviced within the warranty period for different types of reported defects viz., EE (Electronics and electrical), PT (Power train), UB (Upper body), LB (Lower body) defects. Aggregate defects under these categories may sometimes be represented by A, B, C, D or, 1, 2, 3, 4; under some sequencing.

A defective car sent for servicing is repaired and sent back to be exposed again in the traffic condition outside. The repaired car may not behave like a new car and may come back for further processing in case of subsequent problems, if any; especially under warranty period. A car may have more than one problem to be rectified. In Huang et al. (2012) event tree analysis method is used to determine the risk flow of automobile defects. The characteristics of the production process in factory and the intensity of exposure to road traffic may continue to affect subsequent life period of a vehicle.

Failure data observed was classified using lag month (difference between sale month and production month) and aggregate in terms of EE, PT, UB and LB. Further the sale quantity was used to calculate the IPTV i.e., incidents per thousand vehicles, which is computed as number of failures observed in that month divided by sale quantity of specified production batch for that lag, the ratio is then multiplied by 1,000 to get IPTV per thousand. This measure is similar to hazard rate. IPTV for different lag months were calculated separately. Next for various production months, $U(t)$ the monthly IPTV; i.e., IPTV reported in t = first month, second month, etc., over different "months in service" (MIS=Complaint reported Month−Sale Month) are considered; $X(t)$ the cumulative IPTV up to time t for various production months and over different MIS were also considered. The measure $X(t)$ of cumulative IPTV is similar to integrated hazard rate. The variable KMS, kilometres covered by a vehicle is also recorded. Data structure consist of chassis no, month of production, date of sale, complaint report date, aggregate type of defect, complaint description, kilometres at which failure occurred. Classification of the data thus looks like: (Chassis ID, Lag, Type of failed aggregate, MIS, KMS). For cleaning the data the production batches for which sale is less than 10% of the median for that lag are removed. Cars having frequent problems are analysed separate of this analysis. During the lag months, unsold cars are stored in a protective environment. We analyse the data to study the failure pattern of a vehicle type.

Weibull distribution $F(x) = 1 - \exp(-(x/\sigma)^{\alpha}), \alpha > 0, \sigma > 0; x > 0$ is a well-known model to explain item failure. This distribution is fitted for $U(t)$ and $X(t)$ separately with respect to lag, aggregate, MIS. Analysis is also done with respect to KMS, the kilometres covered by a vehicle.

It was observed that Weibull distribution provides a good fit for the IPTV $U(t)$, as well as the cumulative IPTV $X(t)$. Such a result rules out the possibility of $U(t)$ being independent, as Weibull distribution is not closed under convolution.

The correlation structure computed from observed failure data indicates that the observations arise from a long memory stochastic process with nondecaying correlation, see Table 14.1. The correlation between number of defects in two successive months may sometimes turn out to be negative e.g., when some remedial measures are taken based on reported defects in a previous month. Here maximum likelihood estimates of the Weibull parameters are considered and computed by Minitab software.

14.2 A Simple Model

As seen from the observed data correlation structure, there seem to be a common cause that affects the observations in general. For an individual automobile, let X_j be the j-th item (aggregate) characteristic of the assembled components in the job that lingers throughout the lifetime of the automobile. The assembled items in an automobile face exposure to traffic conditions outside, coupled with expertise of driver.

At the i-th instant of problem reporting of a specified kind by j-th item aggregate, assume that the magnitude of the characteristic of interest be

$$Y_{ij} = c_i X_j + e_{ij} \tag{14.1}$$

where c_i s are constants and e_{ij} may be considered error components. Then

$$\sum_i Y_{ij} = X_j \sum_i c_i + \sum_i e_{ij} = \delta X_j + e_{.j} \tag{14.2}$$

with $\delta = \sum_i c_i$ and $e_{.j} = \sum_i e_{ij}$. It therefore follows that if X_j i.e., Y_{ij} in (14.1) are approximately Weibull, then the cumulative sum $\sum_i Y_{ij}$ is also approximately Weibull, if the error components are negligible compared to the main part. One can compute the correlation between two values of Y as follows.

$$\rho(Y_{ij}, Y_{kj}) \approx \operatorname{sgn}(c_i c_k) \left\{ 1 - \frac{1}{2\sigma^2} \left(\frac{\sigma_{ij}^2}{c_i^2} + \frac{\sigma_{kj}^2}{c_k^2} \right) \right\} \tag{14.3}$$

to a first degree of approximation, where $\sigma^2 = Var(X_j), \sigma_{ij}^2 = Var(e_{ij})$.

The above simple model indicates that it is possible for correlated variables those are approximate Weibull to have sum as an approximate Weibull random variable.

Table 14.1 Correlation matrix of IPTV

		Monthly IPTV																	
		1	2	3	4	5	6	7	8	9	10	11	12	13	14	15	16	17	18
Monthly IPTV	1	*1.00*	0.20	0.28	*0.67*	0.36	*0.62*	*0.74*	*0.73*	0.25	*0.86*	*0.59*	−0.43	*0.83*	*0.57*	*0.43*	*0.73*	*0.40*	*0.70*
	2	0.20	*1.00*	0.30	−0.22	*0.45*	−0.45	0.36	−0.36	0.39	−0.08	−0.46	*0.63*	−0.20	−0.49	*0.58*	−0.31	*0.47*	0.11
	3	0.28	0.30	*1.00*	−0.26	*0.51*	−0.24	0.32	−0.10	*0.64*	−0.07	0.15	−0.11	−0.05	0.07	*0.64*	−0.24	0.01	−0.20
	4	*0.67*	−0.22	−0.26	*1.00*	0.26	*0.69*	0.37	*0.71*	0.14	*0.82*	*0.65*	−0.40	*0.74*	*0.74*	0.15	*0.83*	0.30	*0.56*
	5	0.36	*0.45*	*0.51*	0.26	*1.00*	−0.33	0.16	−0.21	*0.75*	0.02	−0.05	−0.05	−0.13	−0.08	*0.88*	−0.17	0.30	−0.25
	6	*0.62*	−0.45	−0.24	*0.69*	−0.33	*1.00*	*0.48*	*0.99*	−0.35	*0.82*	*0.81*	−0.47	*0.92*	*0.83*	−0.35	*0.93*	0.16	*0.77*
	7	*0.74*	0.36	0.32	0.37	0.16	*0.48*	*1.00*	*0.56*	0.04	*0.58*	*0.56*	−0.07	*0.64*	*0.44*	0.19	*0.53*	*0.79*	*0.80*
	8	*0.73*	−0.36	−0.10	*0.71*	−0.21	*0.99*	*0.56*	*1.00*	−0.22	*0.88*	*0.84*	−0.48	*0.96*	*0.84*	−0.19	*0.94*	0.19	*0.79*
	9	0.25	0.39	*0.64*	0.14	*0.75*	−0.35	0.04	−0.22	*1.00*	0.10	0.03	0.07	−0.07	0.07	*0.92*	−0.12	−0.10	−0.27
	10	*0.86*	−0.08	−0.07	*0.82*	0.02	*0.82*	*0.58*	*0.88*	0.10	*1.00*	*0.71*	−0.40	*0.97*	*0.76*	0.14	*0.95*	0.22	*0.81*
	11	*0.59*	−0.46	0.15	*0.65*	−0.05	*0.81*	*0.56*	*0.84*	0.03	*0.71*	*1.00*	−0.54	*0.79*	*0.89*	−0.03	*0.80*	0.21	*0.59*
	12	−0.43	*0.63*	−0.11	−0.40	−0.05	−0.47	−0.07	−0.48	0.07	−0.40	−0.54	*1.00*	−0.47	−0.51	0.04	−0.45	0.18	−0.10
	13	*0.83*	−0.20	−0.05	*0.74*	−0.13	*0.92*	*0.64*	*0.96*	−0.07	*0.97*	*0.79*	−0.47	*1.00*	*0.82*	−0.04	*0.97*	0.23	*0.85*
	14	*0.57*	−0.49	0.07	*0.74*	−0.08	*0.83*	*0.44*	*0.84*	0.07	*0.76*	*0.89*	−0.51	*0.82*	*1.00*	−0.07	*0.83*	0.07	*0.55*
	15	*0.43*	*0.58*	*0.64*	0.15	*0.88*	−0.35	0.19	−0.19	*0.92*	0.14	−0.03	0.04	−0.04	−0.07	*1.00*	−0.12	0.11	−0.16
	16	*0.73*	−0.31	−0.24	*0.83*	−0.17	*0.93*	*0.53*	*0.94*	−0.12	*0.95*	*0.80*	−0.45	*0.97*	*0.83*	−0.12	*1.00*	0.21	*0.83*
	17	*0.40*	*0.47*	0.01	0.30	0.30	0.16	*0.79*	0.19	−0.10	0.22	0.21	0.18	0.23	0.07	0.11	0.21	*1.00*	*0.56*
	18	*0.70*	0.11	−0.20	*0.56*	−0.25	*0.77*	*0.80*	*0.79*	−0.27	*0.81*	*0.59*	−0.10	*0.85*	*0.55*	−0.16	*0.83*	*0.56*	*1.00*

14.3 Model for Data Analysis

After cleaning the data, monthly IPTV are calculated for each aggregate of each production batch for each lag. Similarly, cumulative IPTVs are also calculated. Denote U_j to be the j MIS (month in service) IPTV of a particular production batch, X_i to be the cumulative IPTV upto i MIS of a particular production batch, then $X_i = \sum_{j=1}^{i} U_j$.

We mimic the simple model of Sect. 14.2 to write for the cumulative IPTV of i-th month, l-th lag, s-th subsystem (or aggregate) for p-th production batch in the following.

$$X_{ilsp} = c_{ils} X_p + e_{ilsp} \tag{14.4}$$

where $i = 1, 2, \cdots, 18$ represents MIS (month in service); $l = 1, 2, 3, 4$ represents "lag month" i.e., the lag period a produced car remains in factory before being sold; $s = 1, 2, 3, 4$ is the type of reported defect EE, PT, etc., and $p = 1, 2, \cdots, n$ is the production batch in month; e represents error components. A structure based on X_p similar to (14.4) when written for U_{ilsp}, the individual IPTV, leads to the proposed model (14.4), since cumulative IPTV leaves the model unchanged under summation; see Eq. (14.2). Thus the observed phenomena that IPTV and cumulative IPTV are all approximately Weibull rest on the fact that in the proposed model X_p, for a particular production batch p, is assumed to be a Weibull random variable, and this determining factor in the model is responsible for growth pattern in IPTV and cumulative IPTV.

Here c_{ils} are unknown constants and X_p is a Weibull random variable whose value has to be estimated from realised data by

$$\hat{X}_p = \overline{X}_p = \sum_{ils} X_{ilsp} / \sum_{ils} 1 \tag{14.5}$$

Thus, from (14.4) one may compute

$$\hat{c}_{ilsp} = \frac{X_{ilsp}}{\overline{X}_p} \tag{14.6}$$

For every production batch p, the coefficients $c_{ils} = c_{ilsp}$ may be separately estimated from data within that batch. One may then have a pooled estimate

$$\hat{c}_{ils} = \sum_{p=1}^{n} c_{ilsp} / n \tag{14.7}$$

Since e in (14.4) represents error components, one may also write

$$\hat{X}_p = X_{ilsp} / \hat{c}_{ils} \tag{14.8}$$

Growth model (14.4) has the following interpretation. For every production batch there is an (unknown) Weibull random variable X_p, such that IPTV/cumulative IPTV is essentially a fraction c of it, where c depend on other indices. Such a model identifies the associated weights that determine the magnitude of number of defects attributed factors represented by different indices. All the unknown quantities in the model are estimable as explained above.

14.4 Data Analysis

We observed that there are lot of variation in U_j and X_i for every aggregate A, B, C and D.

Next, we plot graphs for the cumulative IPTV for different lag for each aggregate.

An example of cumulative IPTV vs. Month in service is given in Fig. 14.1. It may be noticed that there are four curves each of which is for production batch of April 2010. The graph at the bottom and nearest to X axis indicates the pattern of IPTV occurrence of subsystem A of vehicle which was sold in the same month (Lag 1),

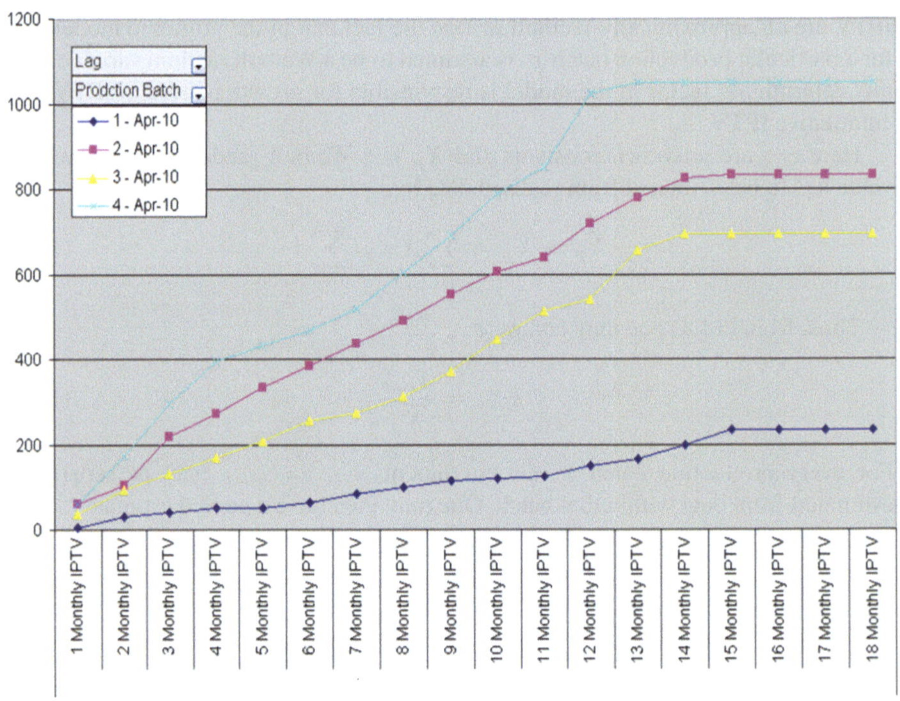

Fig. 14.1 Cumulative IPTV graph for subsystem A

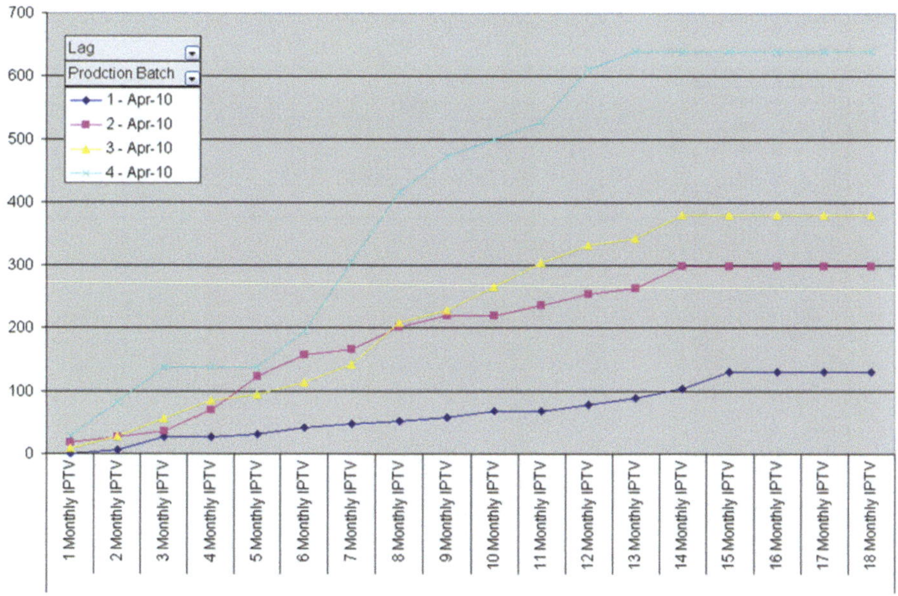

Fig. 14.2 Cumulative IPTV graph for subsystem B

whereas the topmost graph indicates the occurrence of IPTV of subsystem A with Lag 4.

Similar graphs for subsystem B, C, D are shown in Figs. 14.2–14.4. Lag effect can be observed in the IPTVs. Lag 1 IPTVs are observed to be less as compared to that of lag 2, lag 3 and lag 4 IPTV.

However, this pattern was not observed to be true for all subsystem and in all production batches.

It is observed in Figs. 14.5–14.8 that U_js and X_is follow approximately Weibull distribution, which is not expected under independence, as sum of independent Weibull variables is not Weibull. It is also observed that U_js are correlated, that is IPTV for j-th month depends on IPTV of $1, 2, 3, \cdots, (j-1)$ month.

Based on the data of 4–15 MIS for 13 production batches, the number of observations is $N = 13 \times 12 = 156$ and the estimated value of shape parameter for aggregate A is $\alpha = 3.498$, the scale parameter is $\sigma = 377.2$; the regression line fitted in Fig. 14.5 is $\log(-\log(1 - i/n)) = -16.6 + 2.79 \log(A)$, with $R^2 = 0.935$.

For B, the estimated parameters are $\alpha = 1.309, \sigma = 499.7$, the regression line fitted in Fig. 14.6 with $N = 156$ is $\log(-\log(1 - i/n)) = -10.8 + 1.75 \log(B)$, with $R^2 = 0.863$.

For C, the estimated parameters are $\alpha = 1.266, \sigma = 474.8$, the regression line fitted in Fig. 14.7 with $N = 156$ is $\log(-\log(1 - i/n)) = -9.51 + 1.55 \log(C)$, with $R^2 = 0.877$.

For D, the estimated parameters are $\alpha = 1.047, \sigma = 480.5$, the regression line fitted in Fig. 14.8 with $N = 156$ is $\log(-\log(1-i/n)) = -7.51+1.22 \log(D)$, with

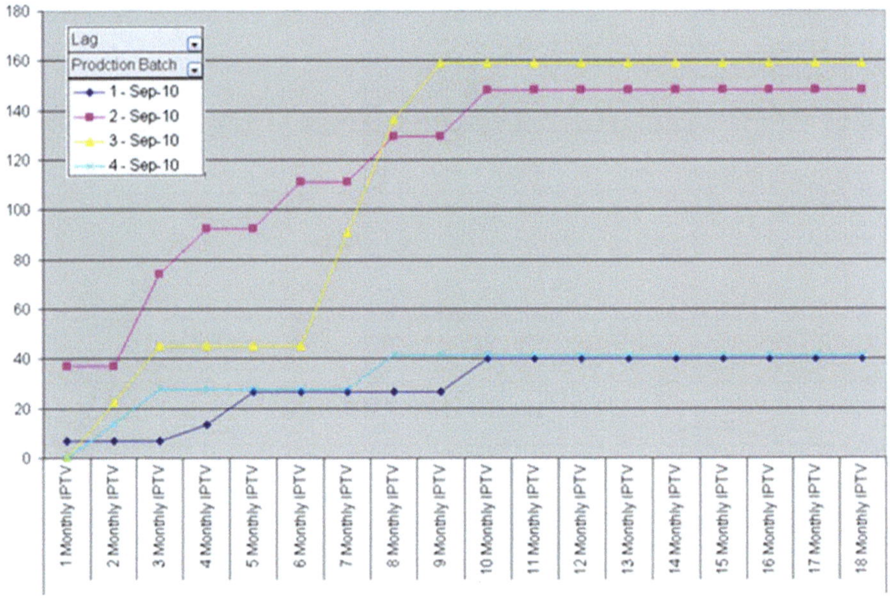

Fig. 14.3 Cumulative IPTV graph for subsystem *C*

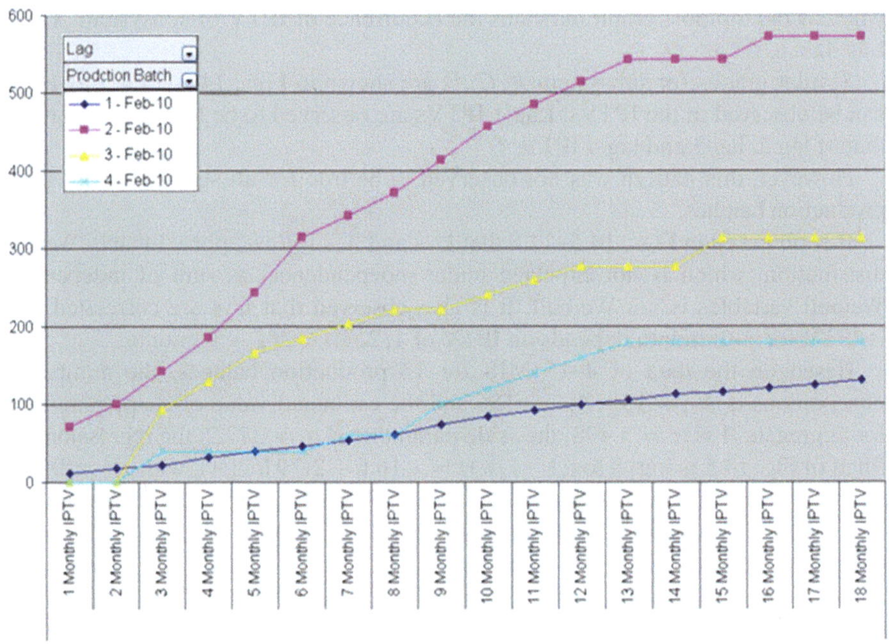

Fig. 14.4 Cumulative IPTV graph for subsystem *D*

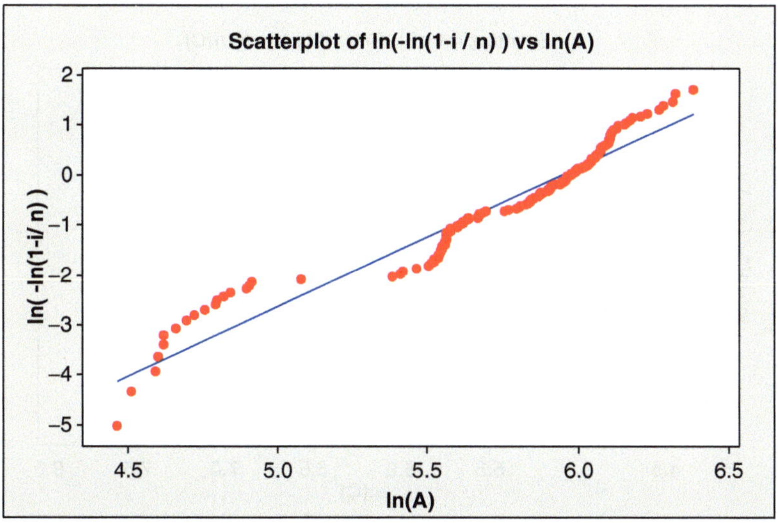

Fig. 14.5 Regression of ln(−ln(1−i/n)) on ln(A)

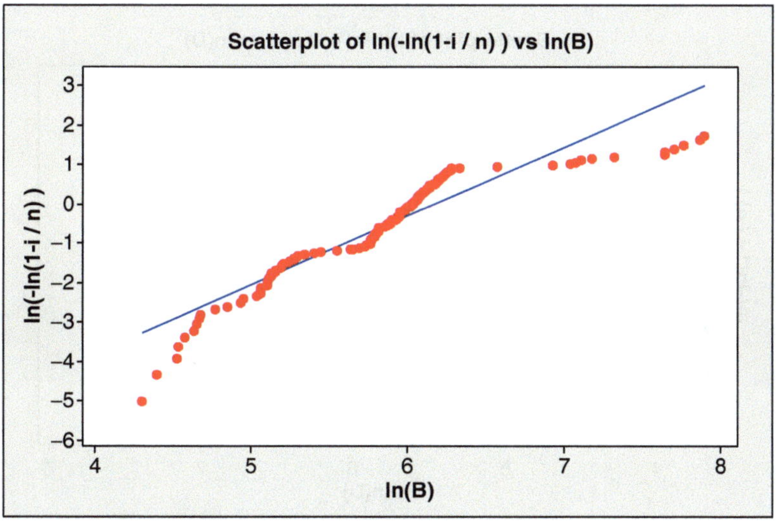

Fig. 14.6 Regression of ln(−ln(1−i/n)) on ln(B)

$R^2 = 0.885$. To a first degree of approximation the points are close to regression lines, except near the start in some figures. Just like Weibull plot by Minitab, the above regression analysis indicates the region where departure from model is prominent.

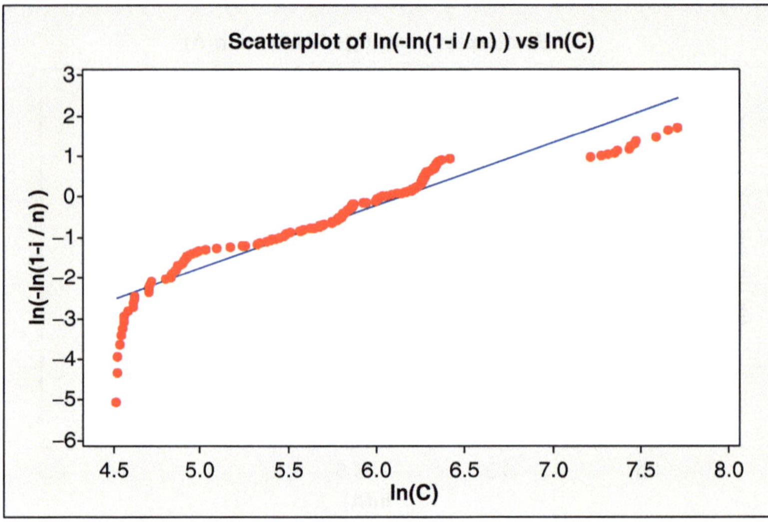

Fig. 14.7 Regression of ln(−ln(1−i/n)) on ln(C)

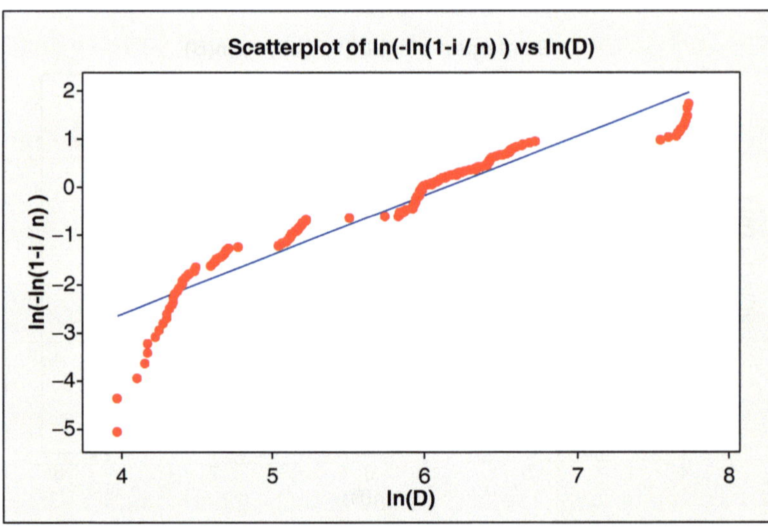

Fig. 14.8 Regression of ln(−ln(1−i/n)) on ln(D)

Next, for 13 production batches for A (12 batches for B–D, four entries in Table 14.2 are considered outliers and kept outside the analysis of graphical plots) averaged over different lags, we estimate X_p values for subsystems A–D, see

Table 14.2 Estimates of Xp and Xps for different production batches

		Xps			
Xp	Xp Estimates for	*A*	*B*	*C*	*D*
237.54	Production Month 1	395	308	118	180
260.77	Production Month 2	299	488	186	70
425.10	Production Month 3	500	280	328	592
361.14	Production Month 4	404	370	426	245
443.11	Production Month 5	477	511	323	461
*1344.16	Production Month 6	103	*1,454	*1,601	*2,219
124.44	Production Month 7	155	136	139	68
470.50	Production Month 8	436	526	427	494
396.63	Production Month 9	360	440	300	486
274.92	Production Month 10	443	166	321	170
254.91	Production Month 11	354	335	214	117
392.41	Production Month 12	119	337	546	567
393.05	Production Month 13	119	337	546	570

*Not considered while plotting, as these seem to be outliers

Table 14.3 Regression characteristics of log(−log(1 − i/n)) on log(aggregate)

Fitted line	Intercept	Slope	R^2	n	SSE	MSE
For aggregate $A, B, C\&D$ over different MIS, Lag & aggregates	−3.76	0.01	0.954	12	0.221	0.018
For A	−9.92	1.69	0.88	13	1.56	0.142
For B	−14.51	2.44	0.94	12	0.729	0.07
For C	−12.2	2.08	0.97	12	0.315	0.031
For D	−7.23	1.23	0.92	12	0.849	0.085

Eq. (14.4) specialised for different subsystems A–D with $X_p = X_{ps}$; a formula similar to (14.5) viz., the following formula (14.9) is used to compute X_{ps}

$$\hat{X}_{ps} = \overline{X}_{ps} = \sum_{il} X_{ilsp} / \sum_{il} 1 \tag{14.9}$$

The $\hat{X}_{ps}$ values are shown in Table 14.2. Characteristics of the regression lines fitted on $\{\log x, \log(-\log(1 - i/n))\}$ are shown in Table 14.3, the values of R^2 are high, indicating that the Weibull fit is satisfactory to a first approximation.

Plots for Weibull fit of $\hat{X}_{ps}$ by Minitab are shown in Figs. 14.9–14.12, estimated Weibull parameters are also shown in corresponding figures.

Next, we compute the X_p values for 13 production batches *averaged over different subsystems and different lags* by Eq. (14.5), and the corresponding Weibull fit with parameters are shown in Fig. 14.13.

The coefficients c are computed from Eq. (14.6), reflect the weight/importance of different indices to the reported defects in automobiles for taking corrective measures. These are shown in Table 14.4. With increase in MIS the c values

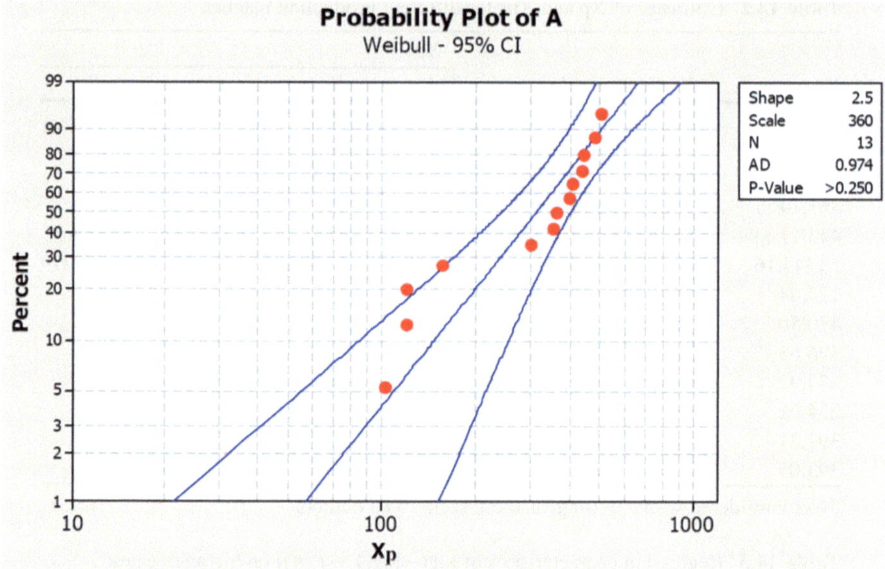

Fig. 14.9 Weibull plot of Xp for sub system *A*

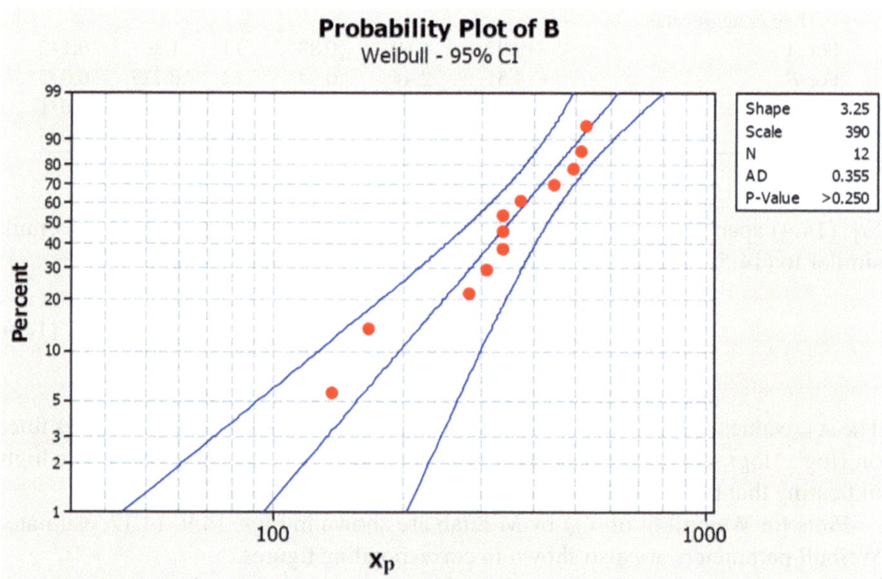

Fig. 14.10 Weibull plot of Xp for sub system *B*

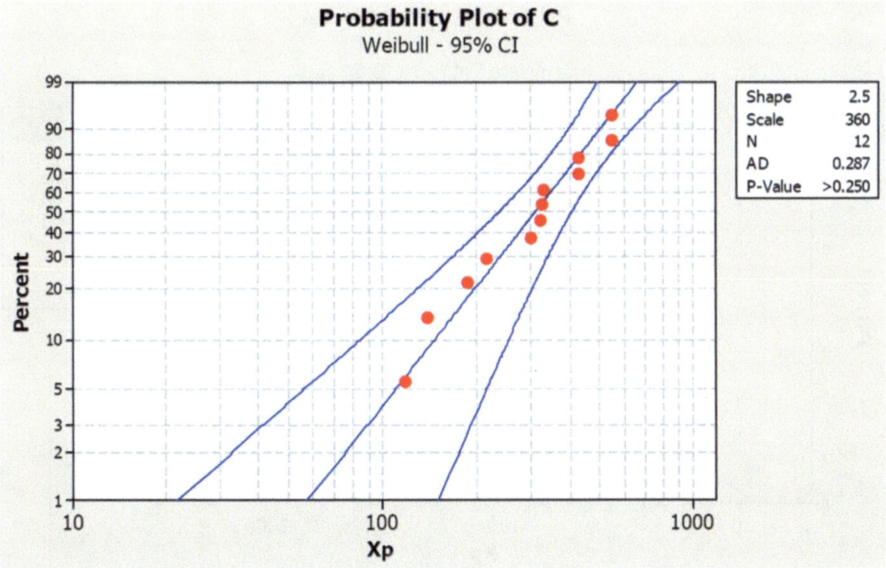

Fig. 14.11 Weibull plot of Xp for sub system *C*

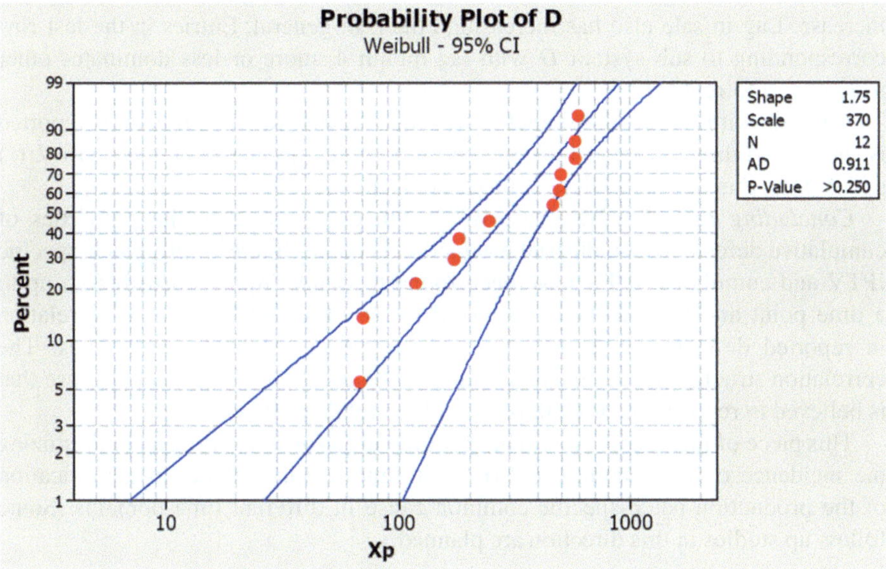

Fig. 14.12 Weibull plot of Xp for sub system *D*

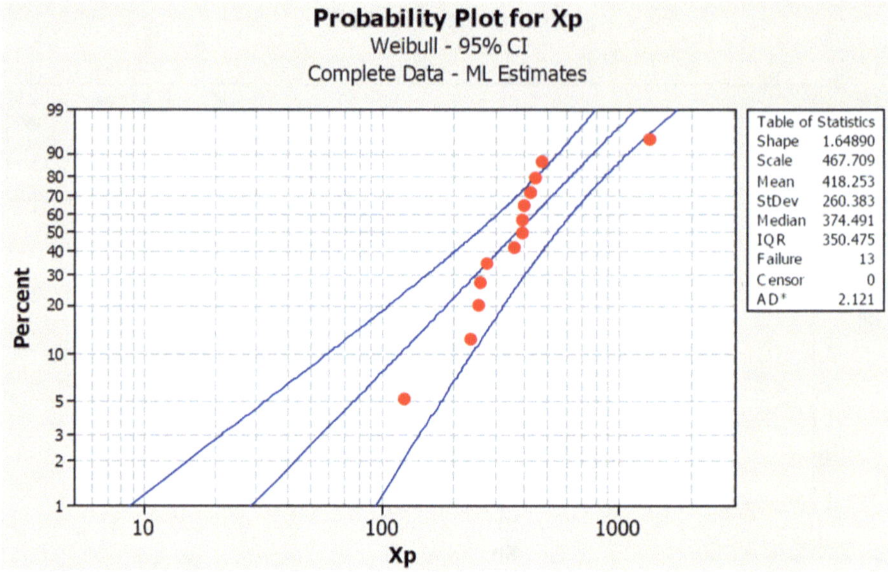

Fig. 14.13 Weibull plot for Xp calculated over different aggregates and different lags

increase. Lag in sale also has increasing effect, in general. Entries in the last row corresponding to sub system D with lag month 4, more or less dominates other rows over different MIS.

Analysis with respect to kilometers travelled (KMS) when complaint is reported is shown in Tables 14.5 and 14.6. Weibull fit to a first approximation hold for $U(t)$ and $X(t)$ in this case as well, like the earlier analysis.

Concluding remarks: Here, use of growth curve to study the properties of cumulative defects for a industrial production is made. Interestingly, batch-specific IPTV and cumulative IPTV have been found to follow Weibull distribution up to a time point and this has been attributed to the observed long range correlation in reported defects over consecutive time segments of sold automobiles. The correlation structure led to the assumption of the presence of a common cause that is believed to regulate the growth pattern of cumulative IPTV.

This piece of information could be gainfully used for controlling and minimizing the incidence of IPTV/cumulative IPTV in automobiles subject to identification of the production-batch-specific common cause in different time domains. Some follow up studies in this direction are planned.

Acknowledgements Ms. Aditi Savant and Ms. Upasana Roychoudhury of Tata Motors helped in data analysis. Thanks to a reviewer for constructive comments.

Table 14.4 Coefficients C_{ils} indicating weights of different indices towards automobile defects

		Month In Service (MIS)														
Lag	Subsystem	1	2	3	4	5	6	7	8	9	10	11	12	13	14	15
1	**A**	0.03	0.11	0.16	0.19	0.21	0.25	0.31	0.35	0.38	0.42	0.43	0.50	0.53	0.57	0.62
2	**A**	0.23	0.41	0.60	0.73	0.88	0.98	1.12	1.24	1.47	1.63	1.75	1.85	1.96	2.09	2.20
3	**A**	0.29	0.54	0.66	0.77	0.86	1.12	1.27	1.36	1.73	2.17	2.40	2.51	2.59	3.06	3.21
4	**A**	0.21	0.38	0.45	0.57	0.73	0.80	0.90	1.03	1.09	1.23	1.30	1.40	1.49	1.52	1.60
1	**B**	0.08	0.12	0.21	0.29	0.37	0.44	0.53	0.59	0.67	0.74	0.79	0.87	0.94	0.98	1.00
2	**B**	0.16	0.28	0.34	0.47	0.67	0.86	1.08	1.31	1.50	1.81	2.03	2.31	2.52	2.66	2.79
3	**B**	0.06	0.23	0.31	0.42	0.59	0.78	0.86	1.03	1.20	1.32	1.43	1.69	1.80	1.92	2.01
4	**B**	0.08	0.12	0.23	0.48	0.60	0.68	0.88	1.05	1.37	1.72	1.84	1.98	2.08	2.49	2.64
1	**C**	0.03	0.07	0.11	0.15	0.18	0.22	0.26	0.29	0.34	0.46	0.55	0.58	0.66	0.71	0.72
2	**C**	0.02	0.08	0.15	0.19	0.25	0.32	0.42	0.49	0.56	0.65	0.72	0.81	0.90	1.00	1.10
3	**C**	0.04	0.04	0.06	0.14	0.34	0.48	0.59	0.72	0.77	0.83	0.98	1.11	1.20	1.26	1.41
4	**C**	0.05	0.18	0.27	0.35	0.45	0.51	0.60	0.75	0.81	0.92	1.10	1.29	1.39	1.51	1.58
1	**D**	0.05	0.16	0.25	0.33	0.36	0.41	0.48	0.55	0.62	0.70	0.76	0.88	0.91	0.94	0.97
2	**D**	0.16	0.27	0.42	0.59	0.71	0.87	1.01	1.17	1.31	1.38	1.46	1.61	1.71	1.78	1.82
3	**D**	0.06	0.14	0.29	0.51	0.64	0.73	0.98	1.13	1.33	1.41	1.55	1.65	1.82	2.11	2.19
4	**D**	0.21	0.44	0.52	0.99	1.29	1.39	1.52	1.68	1.74	1.89	1.93	2.05	2.12	2.29	2.30

Table 14.5 Kilometer analysis for A when Weibull was fitted to $U(t)$

KMS	Shape	Scale	AD	P
less than 3,000	2.348	56.41	0.182	> 0.250
5,000–8,000	3.01	47.67	0.179	> 0.250
8,000–12,000	2.366	30.75	0.209	> 0.250
12,000–17,000	1.558	34.39	0.442	> 0.250
17,000–23,000	2.011	31.52	0.364	> 0.250
23,000–28,000	1.201	38.13	0.492	0.21
28,000–33,000	1.675	32.5	0.318	> 0.250
33,000–38,000	0.874	27.67	0.73	0.046
38,000–43,000	0.8043	21.92	1.107	< 0.010
43,000–48,000	1.723	25.9	0.437	> 0.250
48,000–53,000	1.628	29.96	0.171	> 0.250
53,000–58,000	2.661	6.414	0.375	> 0.250
58,000–63,000	1.131	12.97	0.339	> 0.250
63,000 Onwards	0.8121	12.55	0.736	0.042

Table 14.6 Kilometer analysis for A when Weibull was fitted to $X(t)$

KMS	Shape	Scale	AD	P
less than 3,000	2.348	56.41	0.182	> 0.250
5,000–8,000	2.74	95.12	0.219	> 0.250
8,000–12,000	2.581	113.3	0.571	0.139
12,000–17,000	2.637	146.5	0.556	0.155
17,000–23,000	2.526	173.7	0.495	0.212
23,000–28,000	2.179	198.5	0.593	0.119
28,000–33,000	2.021	222.5	0.422	> 0.250
33,000–38,000	1.774	234.4	0.577	0.134
38,000–43,000	2.196	218.3	0.366	> 0.250
43,000–48,000	2.105	232.2	0.407	> 0.250
48,000–53,000	2.006	240.6	0.447	> 0.250
53,000–58,000	2.009	241.7	0.427	> 0.250
58,000–63,000	1.995	242.5	0.429	> 0.250
63,000 Onwards	1.987	243.7	0.429	> 0.250

References

Engineering statistics handbook (2008). National Institute of Standards and Technology.

Huang, G.Z., Wang, Y., Chen, Y., & Wang, J. (2012). Risk assessment model of automobile defect based on weibull. *Advanced Materials Research, 383–390*, 7496–7502.

Sagias, N.C., & Karagiannidis, G.K. (2005). Gaussian class multivariate Weibull distributions: theory and applications in fading channels, Institute of Electrical and Electronics Engineers. *Transactions on Information Theory, 51*(10), 3608–3619.

Weibull, W. (1951). A statistical distribution function of wide applicability. *Journal of Applied Mechanics-Transactions of The Asme, 18*(3), 293–297.

Chapter 15
Growth and Nutritional Status of Pre-School Children: A Comparative Study of Jharkhand, Bihar and West Bengal

Susmita Bharati, Manoranjan Pal, and Premananda Bharati

Abstract This paper compares the growth and nutritional status of pre-school children of three states of India, namely, Jharkhand, Bihar and West Bengal using third National Family Health Survey (NFHS-3) data. The sample sizes of Jharkhand, Bihar and West Bengal are 951, 1,373 and 1,600, respectively. Data on socio-demographic background of the households such as sex composition, place of residence, religion, level of education of mothers, mother's age groups, and wealth index of the family are taken to see the differential effects of these variables on the child health status.

It has been found that the distributions of weight and height around the means remain remarkably stable over age in those three states. It has also been found that the rates of growth of mean weights and heights are far lower in Bihar and Jharkhand than in West Bengal and India. The low growth rates of the mean values during the first year for both weight and height translate to high rates of undernutrition and stunting. It is also seen that high rate of stunting and underweight in Jharkhand and Bihar starts from 9 months and onwards while in West Bengal and India it starts from 12 months and onwards. Percentage of undernourished children is the highest in Bihar followed by Jharkhand and West Bengal. Comparatively higher growth rate of nutritional status and the low intensity of under nutrition of children are found in the socio-economic groups of male gender, urban areas, other communities and of secondary and higher educated mothers. Another notable finding is seen that only in West Bengal, reduction of underweight is directly related to upward movement of literacy along with wealth index but in Jharkhand and Bihar, there is no impact of literacy on reducing underweight and only higher wealth index is responsible for reducing underweight and stunting.

S. Bharati (✉) • M. Pal • P. Bharati
Indian Statistical Institute, 203 BT Road, Kolkata 700108, India
e-mail: sbharati60@hotmail.com

R. Dasgupta (ed.), *Advances in Growth Curve Models*, Springer Proceedings in Mathematics & Statistics 46, DOI 10.1007/978-1-4614-6862-2_15,

15.1 Introduction

Undernutrition during childhood may affect the growth potential and risk of morbidity and mortality in future life of an adult. Undernourished children are more likely to grow into undernourished adult who face high risk of disease and death. South Asia has the highest prevalence of underweight and stunted children (Bamji 2003). India is home to the largest number of underweight and stunted children in the world. India is a vast country with 1/6th of the population living in 35 states and union territories. There are substantial variations in the economic, social, nutrition and health profile between states. More than half of the children suffer from under nutrition in the states of Madhya Pradesh, Orissa, Rajasthan, Uttar Pradesh, Bihar and Jharkhand. In Bihar, prevalence of both under weight and stunting among children is very high. In fact malnutrition rose by 4% from NFHS-2 to NFHS-3 (Bihar Road map 2007).

Late introduction of semisolid foods is a major problem with high risk of morbidity and under nutrition in Bihar and Jharkhand (Ramchandran 2007). In Bihar, more than 50% people are below poverty line, illiteracy among women was above 50% (Census 2001), and because of poor access to health and nutrition services, there is high under nutrition among children. In West Bengal poverty ratios declined substantially during 2000 but in Bihar it remained the same. One of the reasons is that a substantial portion of the people is still engaged in manual work for their livelihood and requires higher energy intake than is actually consumed. In this context, it is necessary to investigate the socio-economic condition such as education of parents, place of residence or economic conditions in order to understand the retardation of growth and nutrition.

Here the main objectives of the paper are (1) to study the growth and nutrition status of 0- to 59-month children on the basis of different age-groups in three states of India, namely, Jharkhand, Bihar and West Bengal and to see their comparative account with that of all India results and (2) to delineate the responsible socio-economic factors leading to growth and nutrition.

15.2 Materials and Methods

The data on growth and nutritional status of children was accessed from the third round National Family Health Survey (NFHS-3) of 2005–2006. The survey was co-ordinated by International Institute for Population Sciences (IIPS) in collaboration with the Ministry of Health and Family Welfare. Children of age 0–59 months are taken to form eight age-groups in our study. The sample sizes for the three states—Jharkhand, Bihar and West Bengal—are 951, 1,373 and 1,600, respectively and for India, it is 31,105. This survey collected data on weight and height of the children as well as computed "z" scores of under nutrition through weight for age and height for age indices.

Z-score value "-2" was used as a cut-off point for prevalence estimation (WHO 1995). Z-score is defined as the deviation of the value observed for an individual from the median of the reference population, divided by the standard deviation (SD) of the reference population i.e.

$$\text{Z-score} = \frac{(\text{observed value}) - (\text{median of the reference population})}{\text{SD of the reference population}}$$

The classifications of Z-score (followed by NCHS/WHO) are "below normal" (< -2), "normal" (≥ -2 & $\leq +2$) and "above average" ($> +2$). Places of residence are taken as "rural" and "urban". Caste and community wise four groups have been taken namely scheduled castes, scheduled tribes, other backward classes and "others". Mother's educational status is grouped into four categories such as illiterate (those who can neither read nor write), primary (literate up to class IV standard), secondary (class V to class X standard) and the fourth group is class XI and onwards (i.e. higher secondary, graduate or postgraduate, etc.). Age groups of mother during child birth are grouped into three categories such as 15–24 years (younger mother) and 25–34 years (middle-aged) mother and 35 years and above (older mother). Wealth index is a measure of the economic status of the household (Rutstein 1999). Though it is an indicator of the level of the wealth in the household, it is consistent with expenditure and income measure. It is based on 33 household assets and housing characteristics such as household electrification, type of windows, sources of drinking water, types of toilet facility, flooring, roofing, cooking fuel and house ownership, material of exterior walls, number of household members per sleeping room, ownership of a bank or post-office account, ownership of a mattress, a pressure cooker, a chair, a cot/bed, a table, an electric fan, a radio/transistor, a black & white television, a colour television, a sewing machine, a mobile telephone, and any other telephone, a computer, a refrigerator, a watch or clock, a bicycle, a motorcycle or scooter, an animal-drawn cart, a car, a water pump, a thresher and a tractor. Here each household asset was assigned a weight generated through principal component analysis and the resulting score was standardized in relation to a normal distribution and each household was assigned a score for each asset and the scores were summed for each household and individuals were ranked according to the score of the household and the scores were divided into five quintile groups starting from lower strata to higher strata like poorest, poorer, medium, richer and richest.

To see the relative and effective intervention, the risk of Z-score value for under nutrition was regressed on socio-economic variables using categorical logistic regression analysis. Dependent variables are taken as binary. Children with Z-scores below -2 are coded as "1" and with Z-scores -2 or higher are coded as "0". An estimated odd ratio of "1" indicates that the nature of dependent variable is not different from the reference category. If the estimated odd ratio is >1, the probability of becoming affected is more in this category compared to the reference category, and if it is <1, then it is just opposite to that of ">1" case.

15.3 Results

Table 15.1 and Fig. 15.1a, b describe the mean and SD of weight and height for each of eight sub-groups of age of 0- to 59-month children in three states of India, namely, Jharkhand, Bihar and West Bengal as well as total India. It is seen that all along the age-groups, there has been a positive trend for both the weight and height though the magnitude of changes was different in different age groups. West Bengal is seen to be in a better position at least in the higher age groups compared to Bihar and Jharkhand for the height and weight. Even the figures of West Bengal outperform the All India figures in the higher age-groups.

Table 15.2 and Fig. 15.2a, b give the percentage distribution of age-group wise different categories of under nutrition like underweight and stunting among 0- to 59-month children in the three states of India as well as total India. The percentages of stunted children are 46.2%, 49.9%, 37.8% and 40.7%, respectively in Jharkhand, Bihar, West Bengal and All India and the corresponding percentages of underweight are 53.3%, 54.0%, 32.9% and 35.7% respectively. It is evident that in case of stunting, percentage changes are positive from 9 to 35 months children for all the three states as well as India. After 35 months, the percentages go downwards. Same trend is found for underweight also. Percentages of underweight and stunted children are below all India level only for West Bengal whereas the percentages are quite high above the all India percentages for Jharkhand and Bihar.

Tables 15.3 and 15.4 describe the relationship of mean weight and height by socio-economic variables. It is seen that mean weight and height are significantly different among the categories irrespective of different socio-economic variables. For example, positive highest mean weight and height are seen among the children of the male gender, urban areas, other communities (i.e., which do not belong to scheduled tribes, scheduled castes or other backward classes), secondary and higher educated mothers. These results are statistically significant at 1% level of significance except height for gender differences in Jharkhand and for religion in all the three states.

Now we turn to the changes in the incidence of undernutrition by socio-economic variables (Tables 15.5 and 15.6). It is seen that children are invariably affected by stunting and underweight and show degrees of variation among the different categories of socio-economic variables. It is seen through the table that the children of rural areas, scheduled caste & scheduled tribe communities, illiterate and aged mothers and belonging to the poorest wealth index households are mostly affected. The results are statistically significant at 1% level of significance. Gender differences of the nutritional level were not found to be significant in any of the states.

Results of Table 15.7 demonstrate the effect of socio-economic variables on stunting and underweight. Underweight is significantly regulated by education and wealth index in West Bengal but in Bihar and Jharkhand, only highest wealth index is responsible for reducing stunting.

Table 15.1 Mean height and weight of 0- to 59-month children of Jharkhand, Bihar, WB and India

	Jharkhand					Bihar					West Bengal					India				
		Height		Weight			Height		Weight			Height		Weight			Height		Weight	
Age-groups	N	Mean	SD	Mean	SD	N	Mean	SD	Mean	SD	N	Mean	SD	Mean	SD	N	Mean	SD	Mean	SD
0–2	37	55.34	4.60	4.06	1.04	73	54.12	3.61	3.97	0.79	59	54.42	5.42	4.11	1.09	1,289	54.76	4.54	4.18	1.02
3–5	59	61.82	3.19	5.65	0.99	96	61.68	4.04	5.58	0.88	95	60.67	3.80	5.73	1.08	2,032	61.83	4.21	5.84	1.07
6–8	68	65.98	3.34	6.45	0.98	151	65.51	3.84	6.42	1.07	94	66.29	3.89	7.05	1.12	2,203	66.44	4.22	6.93	1.15
9–11	86	68.56	4.53	7.11	1.23	78	69.17	3.87	7.04	1.07	92	70.30	3.87	7.83	1.12	1,990	69.65	4.12	7.59	1.21
12–23	251	74.72	5.45	8.09	1.42	389	75.12	4.55	8.24	1.21	387	75.86	4.93	8.90	1.49	7,820	75.29	5.11	8.78	1.43
24–35	193	83.65	6.40	9.92	1.53	255	82.57	5.60	9.72	1.53	333	84.69	5.57	10.68	1.65	6,142	84.03	5.85	10.61	1.68
36–47	153	90.70	6.26	11.59	1.69	188	89.44	6.27	11.40	1.74	304	92.49	5.83	12.61	1.97	4,768	91.54	6.22	12.29	1.93
48–59	104	98.33	7.51	13.37	2.04	143	96.69	6.81	13.19	2.11	236	99.47	6.13	14.34	2.37	3,861	98.44	6.58	13.97	2.28

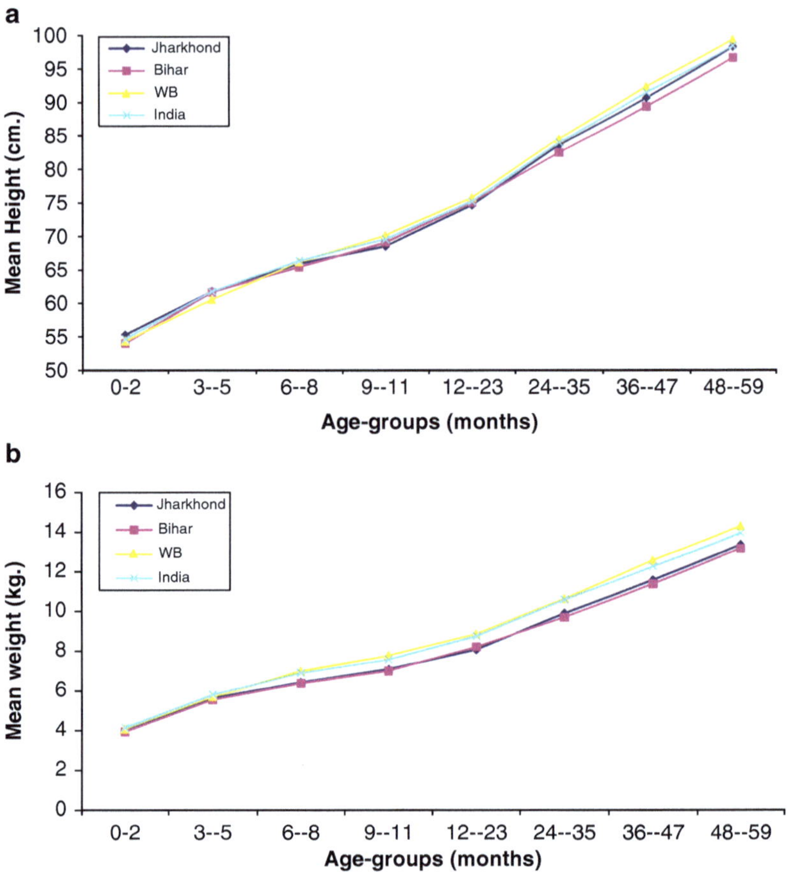

Fig. 15.1 (**a**) Mean height of 0–59 months children. (**b**) Mean weight of 0–59 months children

15.4 Discussion

This paper makes a comparative study of growth and nutrition of 0- to 59-month children in three states of India and also compares the deviation of growth and nutrition in the perspective of total India. It is seen that means of both weight and height of pre-school children in West Bengal are greater than all India figures whereas the corresponding average for Jharkhand, Bihar are much lower than all India figures. It is also seen that high rate of stunting and underweight in Jharkhand and Bihar starts from 9 months and onwards while in West Bengal and India it starts from 12 months and onwards. The percentage of undernourished children is highest in Bihar. It is followed by Jharkhand. West Bengal is in the lowest rung of the ladder. Magnitude of higher growth and low intensity of undernutrition in respect of socio-economic variables are seen among the children of the male gender, urban areas,

Table 15.2 Age-group wise percentage distribution of stunted and underweight children across three states and India

	Jharkhand			Bihar			West Bengal			India		
Age group (Months)	N	Stunted	Under-weight	N	Stunted	Under-weight	N	Stunted	Under-weight	N	Stunted	Under-weight
0–2	37	13.5	27.0	73	17.8	27.4	59	28.8	27.1	1, 289	19.2	23.7
3–5	59	13.6	32.2	96	19.8	38.5	95	24.2	27.4	2, 032	18.0	26.1
6–8	68	19.1	39.7	151	29.1	48.3	94	20.2	25.5	2, 203	22.7	28.0
9–11	86	37.2	48.8	78	37.2	48.7	92	29.3	20.7	1, 990	28.3	31.0
12–23	251	58.2	59.8	389	53.5	56.0	387	43.2	35.9	7, 820	47.1	36.9
24–35	193	59.1	62.7	255	65.1	63.9	333	44.4	38.7	6, 142	50.3	40.1
36–47	153	53.6	57.5	188	63.8	60.6	304	38.8	33.2	4, 768	46.4	39.7
48–59	104	37.5	48.1	143	60.1	55.2	236	36.0	30.9	3, 861	41.6	37.7
0–59	951	46.2	53.3	1, 373	49.9	54.0	1, 600	37.8	32.9	30, 105	40.7	35.7

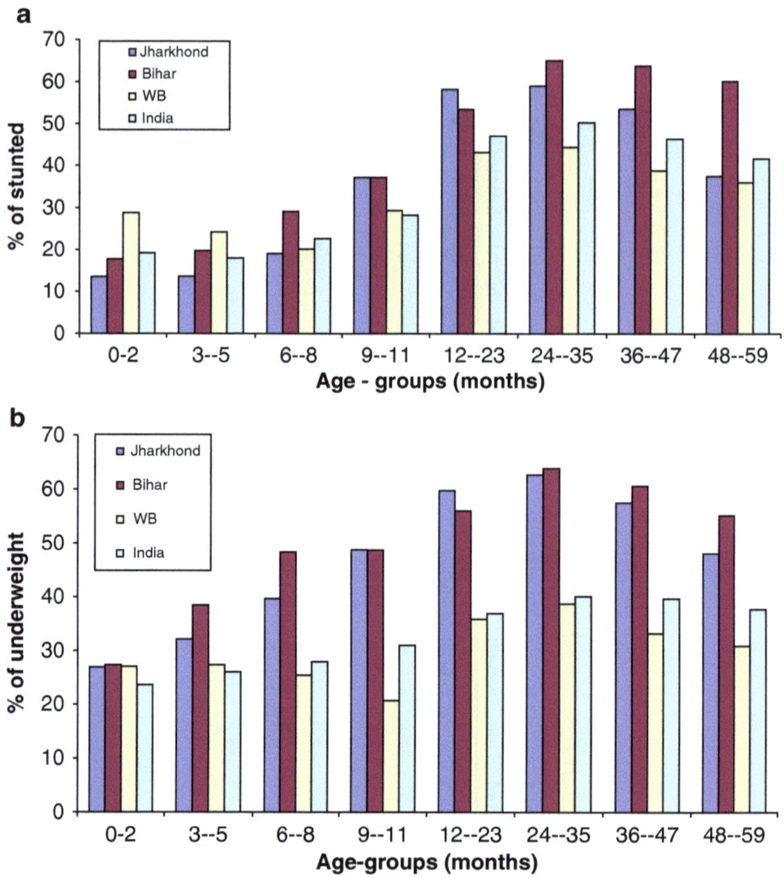

Fig. 15.2 (**a**) Percentage distribution of stunted children. (**b**) Percentage distribution of under-weight children

other communities and secondary and higher educated mothers. Another notable finding is that only in West Bengal, reduction of under-weight is directly related to upward movement of literacy but in Jharkhand and Bihar, there is no impact of literacy on reducing underweight. It is also seen that wealth index is inversely related to stunting and underweight in West Bengal, Jharkhand and Bihar implying that wealth index has a significant effect on reduction of both stunting and underweight. In Bihar and Jharkhand, under nutrition is very high which may be due to low per capita income and poor access to health that increase the morbidity. The reason for high underweight state may also be due to high illiteracy among women, causing low women status (Bihar Road map 2007). Besides this, the coverage of ICDS developmental programme for children in Bihar has ranked in the bottom ten (World Bank Report 2006).

Table 15.3 Relationship between (0–59) month children's mean height with different socio-economic variables in three states of India

	Jharkhand			Bihar			West Bengal		
Socio-economic variables	N	Height	ANOVA 'F' value	N	Height	ANOVA 'F' value	N	Height	ANOVA 'F' value
Sex of the children			5.3310.021df = 1			8.5340.004df = 1			7.5330.006df = 1
Male	466	79.93		742	78.17		826	82.66	
Female	485	78.00		631	76.19		774	80.81	
Place of residence			8.5340.004df = 1			8.5340.004df = 1			8.5340.004df = 1
Rural	665	78.23		949	76.60		965	80.07	
Urban	286	80.61		424	78.73		635	84.35	
Religion			2.1920.087df = 3			3.1780.023df = 3			2.9200.020df = 3
Scheduled caste	119	76.93		241	75.83		355	81.33	
Scheduled tribe	250	78.45		005	73.12		073	78.58	
Other backward class	427	79.02		847	77.11		048	80.88	
Others	152	80.84		278	79.05		898	83.07	
Women's education			4.9770.002df = 3			7.9320.000df = 3			12.6550.000df = 3
Illiterate	571	77.85		866	76.06		577	80.39	
Primary	105	78.35		139	78.16		336	80.99	
Secondary	237	81.46		317	79.89		597	82.37	
Higher	038	81.47		051	78.74		090	89.41	
Women's age group			26.8040.000df = 2			52.8040.000df = 2			69.2020.000df = 2
15–24	440	75.74		539	73.11		755	77.80	
25–34	429	81.57		679	79.73		747	84.97	
35–49	082	82.44		155	80.85		098	87.87	
Wealth index			7.5370.002df = 4			7.5370.002df = 4			7.5370.002df = 4
Poorest	470	77.19		372	75.53		394	78.78	
Poorer	135	78.79		403	75.98		337	78.95	
Middle	122	81.67		232	77.68		282	82.21	
Richer	119	78.96		210	79.34		337	83.55	
Richest	105	83.86		156	81.25		250	87.38	

Table 15.4 Relationship between (0–59) month children's mean weight with different socio-economic variables in three states of India

	Jharkhand			Bihar			West Bengal		
Socio-economic variables	N	Weight	ANOVA 'F' value	N	Weight	ANOVA 'F' value	N	Weight	ANOVA 'F' value
Sex of the children			15.0030.000df = 1			19.3310.000df = 1			15.9270.000df = 1
Male	466	9.46		742	9.08		826	10.55	
Female	485	8.74		631	8.42		774	9.91	
Place of residence			18.2490.000df = 1			11.3840.001df = 1			65.8190.000df = 1
Rural	665	8.83		949	8.61		965	9.72	
Urban	286	9.69		424	9.16		635	11.03	
Religion			4.7310.003df = 3			6.3510.000df = 3			5.3090.000df = 3
Scheduled caste	119	8.66		241	8.30		355	10.0	
Scheduled tribe	250	8.79		005	8.10		073	9.07	
Other backward class	427	9.12		847	8.74		048	10.38	
Others	152	9.78		278	9.35		898	10.62	
Women's education			11.7340.000df = 3			12.5470.000df = 3			24.8890.000df = 3
Illiterate	571	8.73		866	8.44		577	9.68	
Primary	105	8.91		139	9.02		336	10.08	
Secondary	237	9.84		317	9.48		597	10.54	
Higher	038	10.42		051	9.55		090	12.58	
Women's age group			25.4890.000df = 2			51.2290.000df = 2			62.7400.000df = 2
15–24	440	8.39		539	7.86		755	9.33	
25–34	429	9.68		679	9.33		747	10.99	
35–49	082	9.74		155	9.56		098	11.57	
Wealth index			14.1540.000df = 4			14.1540.000df = 4			39.5570.000df = 4
Poorest	470	8.58		372	8.27		394	9.31	
Poorer	135	9.02		403	8.50		337	9.37	
Middle	122	9.67		232	8.88		282	10.28	
Richer	119	9.16		210	9.29		337	10.86	
Richest	105	10.72		156	9.87		250	12.01	

Table 15.5 Percentage of stunted among 0- to 59-month children in relation to different socio-economic variables

	Jharkhand			Bihar			West Bengal		
Socio-economic variables	N	Stunted	ANOVA 'F' value	N	Stunted	ANOVA 'F' value	N	Stunted	ANOVA 'F' value
Sex of the children			1.0790.583df = 2			0.0450.978df = 2			0.4780.787df = 1
Male	466	47.9		742	49.9		826	37.7	
Female	485	44.5		631	49.9		774	37.9	
Place of residence			22.0630.000df = 2			12.8540.002df = 2			50.4010.000df = 2
Rural	665	51.1		949	53.0		965	44.7	
Urban	286	34.6		424	42.9		635	27.2	
Religion			19.4990.003df = 6			30.0200.000df = 6			18.8480.016df = 6
Scheduled caste	119	53.8		241	62.7		355	40.3	
Scheduled tribe	250	53.6		005	40.0		073	53.4	
Other backward class	427	44.5		847	49.7		048	29.2	
Others	152	33.6		278	39.6		898	33.9	
Women's education			36.8890.000df = 6			56.7540.000df = 6			83.2270.000df = 6
Illiterate	571	52.0		866	56.6		577	48.2	
Primary	105	43.8		139	46.0		336	43.8	
Secondary	237	39.2		317	38.2		597	28.0	
Higher	038	7.9		051	19.6		090	13.3	
Women's age group			0.8410.933df = 4			22.2300.000df = 4			4.7190.317df = 4
15–24	440	45.9		539	45.6		755	40.3	
25–34	429	46.4		679	49.5		747	34.9	
35–49	082	46.3		155	66.5		098	39.8	
Wealth index			57.6330.000df = 8			73.9990.000df = 8			161.9290.000df = 8
Poorest	470	54.0		372	56.7		394	54.3	
Poorer	135	48.1		403	54.8		337	46.6	
Middle	122	43.4		232	57.8		282	37.9	
Richer	119	43.7		210	40.0		337	28.2	
Richest	105	14.3		156	22.4		250	12.4	

Table 15.6 Percentage of under nutrition among 0- to 59-month children in relation to different socio-economic variables

Socio-economic variables	Jharkhand			Bihar			West Bengal		
	N	Under-weight	Chi-square	N	Under-weight	Chi-square	N	Under-weight	Chi-square
Sex of the children			0.1130.945df = 2			1.5710.456df = 2			0.9600.619df = 2
Male	466	53.9		742	52.6		826	31.8	
Female	485	52.8		631	55.8		774	34.1	
Place of residence			38.0650.000df = 2			19.1390.000df = 2			55.6720.000df = 2
Rural	665	59.8		949	57.3		965	40.0	
Urban	286	38.1		424	46.7		635	22.2	
Religion			37.7760.000df = 6			43.0510.000df = 6			42.1050.000df = 6
Scheduled caste	119	55.5		241	69.3		355	37.5	
Scheduled tribe	250	65.2		005	60.0		073	60.3	
Other backward class	427	52.5		847	54.0		048	20.8	
Others	152	34.9		278	40.6		898	29.1	
Women's education			75.7380.000df = 6			74.6210.000df = 6			94.2120.000df = 6
Illiterate	571	61.6		866	62.0		577	45.4	
Primary	105	56.2		139	46.8		336	33.9	
Secondary	237	38.4		317	40.4		597	24.5	
Higher	038	13.2		051	23.5		090	5.6	
Women's age group			4.3640.000df = 4			16.3780.000df = 4			4.7380.000df = 4
15–24	440	53.9		539	52.7		755	35.4	
25–34	429	51.5		679	52.0		747	30.3	
35–49	082	59.8		155	67.7		098	34.7	
Wealth index			80.0530.000df = 8			77.2770.000df = 8			141.8290.000df = 8
Poorest	470	63.2		372	66.4		394	47.5	
Poorer	135	57.0		403	57.8		337	43.9	
Middle	122	48.4		232	54.3		282	31.2	
Richer	119	47.1		210	43.3		337	24.0	
Richest	105	17.1		156	28.8		250	9.2	

Table 15.7 Logistic regression analysis of under nutrition on different socio-economic variables in among 0- to 5-month children across three states of India

Socio-economic variables	Jharkhand		Bihar		West Bengal	
	Stunted	Under-weight	Stunted	Under-weight	Stunted	Under-weight
Sex of the children						
Female®	1.00	1.00	1.00	1.00	1.00	1.00
Male	1.199	1.116	0.962	0.871	0.981	0.882
Place of residence						
Rural®	1.00	1.00	1.00	1.00	1.00	1.00
Urban	0.980	0.966	0.988	0.956	1.311	1.079
Religion						
Scheduled tribe®	1.00	1.00	1.00	1.00	1.00	1.00
Scheduled caste	1.084	0.664	1.845	0.0.931	0.750	0.486**
Other backward class	0.799	0.675*	1.189	0.592	0.681	0.295**
Others	0.788	0.509*	0.955	0.415	0.870	0.499**
Women's education						
Illiterate®	1.00	1.00	1.00	1.00	1.00	1.00
Primary	0.848	0.922	0.795	0.663*	0.953	0.639**
Secondary	0.981	0.609*	0.835	0.700*	0.735	0.705*
Higher	0.320	0.353	0.501	0.461*	0.948	0.364*
Women's age group						
15–24®	1.00	1.00	1.00	1.00	1.00	1.00
25–34	1.083	0.948	1.317*	1.064	0.810	0.818
35–49	0.959	1.173	2.167**	0.646**	0.924	0.871
Wealth index						
Poorest®	1.00	1.00	1.00	1.00	1.00	1.00
Poorer	0.846	0.948	1.059	0.824	0.692*	1.021
Middle	0.731	0.727	1.317	0.813	0.477**	0.593*
Richer	0.729	0.768	0.709	0.617**	0.306**	0.454**
Richest	0.204**	0.281**	0.355**	0.411**	0.098**	0.154**

*0.01–0.05 = 5% level; **<0.01 = 1% level

So, it is thus said that India is far from being a homogenous country in terms of malnutrition as there is wide inter-state variation. It can also be said from the growth pattern of Bihar and Jharkhand that the poor economic status is the sole factor towards reduction in the under nutrition. West Bengal is in much better condition than Bihar and Jharkhand in respect of mother's literacy, which has also much prominent role in reducing the under nutrition.

Thus it can be concluded that mother's education and income may be adjudged as the most effective factors to reduce underweight whereas income is seen to be the only effective factor to reduce the stunting, i.e. long-term undernutrition.

References

Bamji, M. S. (2003). Early nutrition and health – Indian perspective. *Current Science, 85*(8), 1137–1142.

Bihar Road Map for Development of Health Sector: A Report of the Special Task Force on Bihar Planning Commission, Government of India, New Delhi, August, 2007.

Census. (2001). http://gov,bih.nic.in/Profile/censusstates-03.htm.

International Institute for Population Sciences (IIPS) and ORC Macro. (2007). *National Family Health survey (NFHS-3), 2005–06*, Vol. 1. Mumbai: IIPS.

Ramchandran, P. (2007). *Nutrition transition in India 1947–2007*. New Delhi: Nutrition Foundation of India.

Rutstein, S. (1999). *Wealth versus expenditure: Comparison between the DHS wealth index and household expenditures in your departments of Guatemala*. Calverton: ORC Macro.

World Health Organization. (1995). Physical Status: The use and Interpretation of Anthropometry. WHO Technical Report Series No. 854. Geneva: WHO.

World Bank (2006). News and Broadcast: Urgent Action Needed to Overcome Persistent Malnutrition in India, says World Bank Report, New Delhi. http://web.worldbank.org/WBSITE/EXTERNAL/NEWS/0,,contentMDK:20917151~pagePK:64257043~piPK:437376~theSitePK:4607,00.html, Retrived 2011-10-16.

MIX
Papier aus verantwortungsvollen Quellen
Paper from responsible sources
FSC® C105338

If you have any concerns about our products,
you can contact us on
ProductSafety@springernature.com

In case Publisher is established outside the EU,
the EU authorized representative is:
Springer Nature Customer Service Center GmbH
Europaplatz 3, 69115 Heidelberg, Germany

Printed by Libri Plureos GmbH
in Hamburg, Germany